MACHINE ELEMENTS

LIFE AND DESIGN

Mechanical Engineering Series

Frank Kreith & Roop Mahajan - Series Editors

Published Titles

MACHINE ELEMENTS

LIFE AND DESIGN

BORIS M. KLEBANOV

DAVID M. BARLAM

FREDERIC E. NYSTROM

CRC Press
Taylor & Francis Group
Boca Raton London New York

CRC Press is an imprint of the
Taylor & Francis Group, an **informa** business

CRC Press
Taylor & Francis Group
6000 Broken Sound Parkway NW, Suite 300
Boca Raton, FL 33487-2742

First issued in paperback 2019

ISBN-13: 978-0-367-38864-5

Library of Congress Cataloging-in-Publication Data

Klebanov, Boris M.
    Machine elements : life and design / Boris M. Klebanov, David M. Barlam, Frederic E. Nystrom.
        p. cm. -- (Mechanical engineering series)
    Includes bibliographical references and index.

    1. Machine parts. 2. Machine design. I. Barlam, David. II. Nystrom, Frederic E. III. Title.

TJ243.K543 2007
621.8'2--dc22   2006051883

Visit the Taylor & Francis Web site at
http://www.taylorandfrancis.com

and the CRC Press Web site at
http://www.crcpress.com

Table of Contents

Preface

This book describes the behavior of some machine elements during action, based on our understanding accumulated over many decades of machine design. We have sought to describe the mechanisms of interaction between the motion participants in as much detail and depth as the scope of our knowledge and the volume of the book allow.

Our understanding is based in many respects on the work of others, and we have made reference to all authors and publications known to us. But the literature of mechanical engineering is vast, and we welcome notification by any author inadvertently omitted to enable us to amend this omission in the future.

Chapter 1 to Chapter 11 were written mainly by Boris M. Klebanov. Chapter 12 was written mainly by David M. Barlam, who also performed all the calculations using the finite element method (FEM) that appears in the book. Chapter 13 was written jointly by Boris M. Klebanov and David M. Barlam. Frederic E. Nystrom edited the entire work, including the text, tables, and illustrations.

This work is dedicated to our teachers.

Boris M. Klebanov
David M. Barlam
Frederic E. Nystrom

Authors

Dr. Boris Klebanov has spent all 48 years of his professional life in the design of diesel engines and drive units for marine and land applications, reduction gears, hydraulic devices, and mine clearing equipment. His Ph.D. thesis (1969) was on the strength calculation and design of gears. He is the author of many articles and coauthor of two books in the field of machinery.

Dr. Klebanov worked from 1959 to 1990 in St. Petersburg, Russia, as a designer and head of the gear department in a heavy engine industry, and then he worked until 2001 at Israel Aircraft Industry (IAI) as a principal mechanical engineer. Currently, he is a consultant engineer at Israel Aircraft Industry.

Dr. David Barlam is a leading stress engineer and a senior researcher at Israel Aircraft Industry (IAI), specializing in stress and vibration in machinery — the field in which he has accumulated 37 years of experience in the industry and seven years in academia. He is an adjunct professor at Ben-Gurion University. Dr. Barlam's current industrial experience, since 1991, includes dealing with diversified problems in aerospace and shipbuilding. Prior to that, he worked as a stress analyst and head of the strength department in heavy diesel engine industry in Leningrad (today's St. Petersburg). David Barlam received his doctoral degree (1983) in finite element analysis.

Dr. David Barlam is coauthor of the book *Nonlinear Problems in Machine Design* (CRC Press, 2000), and numerous papers on engineering science.

Frederic Nystrom has since 1997 held the position of senior project engineer at Twin Disc, Inc. (Racine, WI). He is responsible for management of both R&D projects and new concept development, focusing on marine propulsion machinery for both commercial and military applications. Prior to that, beginning in 1989, he worked as a senior engineer at Electric Boat Corp., Groton, CT (a division of General Dynamics).

While at Electric Boat he accumulated wide experience in the design of propulsion systems, product life cycle support, and manufacturing support for U.S. Navy surface ships and nuclear submarines. He currently holds U.S. Patent No. 6,390,866, "Hydraulic cylinder with anti-rotation mounting for piston rod," issued May 2002.

Introduction

We know nothing till intuition agrees.

Richard Bach, *Running from Safety*

Possibly, poetry is in the lack of distinct borders.

Joseph Brodsky, *Post Aetatem Nostram*

This book is mostly intended for beginners in mechanical engineering. Undoubtedly, experienced engineers may find a plentiful supply of useful material as well. However, we conceived of this work primarily with novices in mind. We remember all too well how we joined the engineering workforce upon graduating from college, not knowing where to begin. Admittedly, there is still much we don't know, as the processes in working machines are numerous and complex in nature. Nevertheless, we hope that thoughtful engineers will profit from our experience.

As one doctor singularly expressed, "What we know is an enormous mass of information, and what we don't know is ten times greater." We are skeptical about the tenfold estimate; presumably, it is much more. The problem, however, lies not only in the volume of knowledge but also in the fact that most of our knowledge is based on experience in the manipulation of experimental data, whereas many of the laws that govern physical processes are known only partly or not at all. Furthermore, natural, physical processes are statistical in nature, so that as a rule we can't be completely confident that our actions will bring the desired result. Despite this, what we do know allows us in most cases to solve fairly difficult technical problems.

If it is agreed upon that life is movement, then the being of machines can also be called life. To concentrate on the "physiology" of machines, we generally will not refer very much to the change in location of a mechanism's parts in relation to each other. Instead, we will mainly consider elastic and plastic deformations of parts under applied forces, changes in the structure of metals under the influence of stress (in the crystals and on their borders), temperature fluctuations, aggressive environments, and the effects of friction combined with aggressive surroundings, and so on. In all, the life of the machines proves to be very diverse and deserves attentive study.

Anyway, machines are in many respects similar to living creatures. Their birth is laborious. They get afflicted with childhood illnesses (the period of initial trials) and undergo a sort of adolescence (the break-in period); then they work for a long time, get old, and eventually pass away. Machines ache from rough handling; their bodies collect scratches and dents which deteriorate their health and weaken their capacity for work. They suffer from dirt, overheating, and thirst from a lack of lubrication. They also overexert themselves when given loads that are beyond their strength and will perish if nobody looks after their well being. They get tired in the same way from hard work and require check ups, preventative maintenance, and treatment just as people do. They also suffer and become unwell if they are not protected against moisture, heat or cold, soiling, and corrosion. It is no wonder that such terms from the world of the living as "aging," "fatigue," "inheritance," "survivability," and others have entered the technical lexicon. Just as some books focus on the physiology of animals' bodies and habits, this book is concerned with the life phenomena of machines and their parts.

We tried to avoid recommendations as "Do this, it's good" or "Don't do this, it's bad." As with biological life, it is not always possible to say definitely what is good and bad irrespectively of the machine. Sometimes the changes made to improve the design have contradictory results. In addition, many cheaper design solutions are good for less demanding conditions (for example, under relatively small loads, or if the expected service life is brief, or if a higher risk is allowed), but they prove to be unacceptable for the more serious applications. This is why our efforts are directed toward forming the beginning specialist's understanding of the subtleties of the life and work of the machine. Exposure to this material will help them to develop an instinctive impulse to think of those subtleties based upon their own experiences, i.e., to have "mechanical aptitude." This understanding makes the processes of design and calculation more effective, and the work of the designer more sensible, interesting, and creative.

"Ages ago," in 1948, a group of teenagers visited a small electric power station in a small town. This town, just as thousands of other towns and cities in Russia at that time, had been virtually destroyed during the war, leaving many families living in makeshift shelters. And so in the midst of this deprivation, the small power station, with a steam turbine and alternator of only 3000 kW, was a wonder of engineering for the poor children. Everything was fantastic in this shining machine room, but the elderly operator was even more wonderful. He told us:

> A machine is like a person: it likes cleanliness and good, fresh oil [in Russian "oil" and "butter" are expressed by the same word]; it likes when you look after it and take care of it, and is happiest when you don't overload it ...

He spoke with inspiration, this unforgettable man, and his hand stroked the shining casing of the turbine ...

Part I

Deformations and Displacements

Working mechanisms captivate the imagination. Nice-looking paint and bright chrome please the eye. Mechanical parts move back and forth along their paths, impressive with the accuracy of their purposeful, incessant movement. Everything works beautifully, looks well organized, and delights us all with the gift of engineering and the power of the human mind.

The mechanism works and works, all day, all month, all year … and then suddenly, it ceases working. Something went wrong with it, something broke or became jammed. Or it started to make a heavy noise and vibrations, forcing you to shut it off. Or it exploded and frightened you terribly, so that you started thinking of the stupidity of engineering and, generally, of the imperfection of the human mind. But it was working perfectly well! That means that something had happened to it while it was working! It means that, in fact, the life of the mechanism is much more complicated than is apparent. The captivating, purposeful movement of the parts has been accompanied by harmful processes (side effects), that didn't show any outward evidence until, with time, their accumulated result became apparent.

The physiology of machines is quite complicated. The parts of a mechanism are subjected to working loads and inertial loads. These loads cause the parts to deform elastically and sometimes plastically as well. This leads to changes in the structure of the metal and the accumulation of internal defects within it.

In the connections of parts, where there is sliding or even minute relative motion, the surface layers undergo structural changes and deterioration. Many micro-processes are involved in this macro-process, such as the shearing of microasperities, the plastic deformation of the surface layers, the impregnation of these layers with the components of the lubricant and the mating parts, the formation of particles of oxides and other chemical compounds, and the particles' movement from the contact zone. The friction also creates electricity that interacts with the contacting surfaces and lubricant.

At first, the processes described above may improve the work of the mechanism. In the areas of high stress concentration, the local plastic deformation leads to a more uniform load distribution and lowers the local stress peaks. In the friction zones the microasperities become smoothed out, and form the new structure of the surface layers that is more suited to the friction conditions than the initial one. But as the processes continue, the mechanism becomes less serviceable. It ages. The structure of metal deteriorates … the hinges wear out … the back is hurting … the knees …

It seems that we've moved to another realm! Alas, dear reader, we humans are mechanisms too, and as such, we feel the mechanical problems of aging all too well....

The mentioned above micro-processes in the parts and connections are of vital importance for the "health" of a mechanism and its ability to operate successfully during its service life, which is always limited. Let's focus our attention on these fine matters.

1 Deformations in Mechanisms and Load Distribution over the Mated Surfaces of Parts

A mechanism is a combination of rigid or resistant bodies so formed and connected that they move upon each other with definite relative motion.

Excellent definition! We would not be able to explain it better, so we took this definition from a well-known book.[1]

A mechanism usually begins with a mechanical diagram. The designer draws it on a computer screen or on a piece of paper, depending on where he was caught by a surge of inspiration — sometimes his ubiquitous boss provides him with an initial concept. One day, he draws up a diagram of a parallel link mechanism intended for lifting and lowering a weight (see Figure 1.1a). In this chart, everything looks perfect: two lines (1 and 2) symbolize the upper and lower links of the mechanism hinged to weight 3 and to frame 4. The frame looks respectable compared to lines 1 and 2; such a solid, massive rectangle! Electrical winch 5 turns the links and shifts the weight up and down. The designer was not a beginner. He noticed that in the upper position the links were near dead center, and he checked forces F_1 and F_2 in the links. These forces proved to be large, but no problems concerning the strength of the links and the adjoined elements were found. The designer even checked the stability of the links under compressive load; everything was OK!

Everything was really OK until this mechanism was designed in detail, manufactured, and tested. At the first lifting test, when the mechanism was close to its upper position, shown in Figure 1.1b, the weight suddenly fell down with a great crash and came to a standstill in the position shown in Figure 1.1c. Fortunately, the testers were experienced guys, and they were standing at some distance; therefore, they were not injured. They were only a bit scared and very surprised. The subsequent investigation revealed the following:

Because the weight of the mechanism was required to be as low as possible, frame 4 was welded from thin sheets of high-strength steel (see Figure 1.1b) and was quite pliable; however, its strength was checked and found satisfactory.

In the hinges, "good" clearances were made in order to make mounting of the axles of the hinges easier.

Lower link 2 was designed as two rods connected by cross-members (see view "A"), and upper link 1 was made of one rod and placed in the middle of the lower link, so that the rods of the links were in different planes.

Under load, forces F_1 and F_2 were applied to lugs 6 of frame 4, which bent as shown in Figure 1.1c. The distance between the lugs became increased, and this, combined with the increased clearances between the axles and the lug bores, enabled the mechanism to pop like a convex membrane or a pop-top cap. Thus, this product is not a mechanism in the strict sense, because its members don't "move upon each other with definite relative motion."

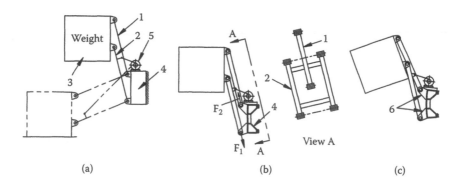

FIGURE 1.1 Lifting device.

From this accident, at least three conclusions should be drawn for the future.

1. The first is well known but worth repeating: Don't stand under the weight! In general, never mind how simple the mechanism, it is always better to stay a safe distance from it at first trials because you never know beforehand what its intentions and capabilities are.
2. The second conclusion: Link mechanisms, which have dead centers, need particularly cautious handling when used near these centers. Increased (for example, because of overload) elastic deformation of the links, enlarged (say, owing to wear) clearances in hinges, small deviations from the drawing dimensions, all become vitally important in these positions and may lead to unexpected and even perilous consequences either at the manufacturer's trials or later in service.
3. The third conclusion: The shape of the machine elements at work may differ significantly from those depicted in the drawing or built in the computer, even if they are manufactured in accordance with the drawing requirements. Therefore, it is expedient to perform a kinematic analysis taking into account the elastic deformations under load.

Let's consider one more occurrence: A designer drew a diagram of a block brake (Figure 1.2). In this brake, the rotation of drum 1 is retarded by blocks 2 with levers 3 and 4. The needed force is supplied by spring 5 through reverser 6. The brake is released by solenoid 7 connected to double-armed lever 8.

It is clear that the farther we want to get the blocks from the drum, the greater the stroke of solenoid 7 should be (or the less the ratio of lever 8 should be, but in this case the solenoid force must be greater). Increasing the stroke or the force of the solenoid leads to such a sizeable increase in its dimensions, weight, and cost that the designers usually make the distance between the blocks and drum (when released) very small, approximately 0.5–1 mm.

Our designer did just that. He had calculated levers 1 and 2 for bending strength only. When the brake was manufactured, it made a good impression on the workers; it was lightweight and smart. They tightened spring 5 by nut 9 to the needed length, and then pressed the release button. The solenoid clicked — the testers were certainly a little distance away, but not far — and they heard the click and saw lever 8 turn. But blocks 3 kept gripping the drum safely.

The post-test investigation revealed that the bending deformation of levers 3 and 4 was considerably greater than the designed displacement of the blocks. So when released, the deformation of the levers became less, but the blocks remained in touch with the drum, though the grip force decreased. As you see, in this case, making a kinematic analysis without taking into account deformations was erroneous.

When a designer uses a spring in a mechanism, he must take the length of the spring depending on its load. It is obvious. As far as other elements are concerned, there is some kind of inertia,

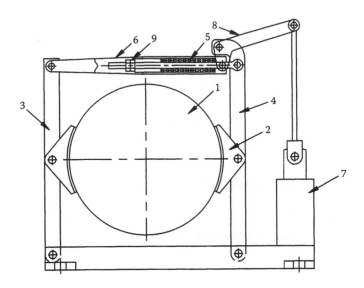

FIGURE 1.2 Block brake.

which possibly originates in calculations of beams for strength, where relatively small deformations are neglected. It should be noted that as the strength of materials increases, the deformations also increase, because the modulus of elasticity doesn't change, so the influence of the deformations grows.

In particular, frame 4, shown in Figure 1.1, was welded of thin sheets of steel of 900-MPa-yield strength, and under load, it was changing its form similar to a spring. It was actually visible!

In the kinematic analysis, even small deformations and clearances may cause important changes. Designers of mechanisms, which must have high kinematic accuracy, know that and take it into account. They design the machine's elements to be rigid, which often leads to increased weight. (For instance, levers 3 and 4 in Figure 1.2, after the trial had failed, were made much more massive, and this enabled the kinematics to be closer to the initial design.)

These two examples, which show how a lack of strain analysis leads to a mechanism's complete inability to work, relate rather to curious things, which are remembered by the participants with amusement. In practice, however, lots of examples may be found of how important it is to pay attention to relatively small deformations, even of microns. These deformations don't usually disable the mechanism, but they change the load distribution between mating parts as compared with the load distribution assumed in the strength calculations. This may result in the unsatisfactory functioning of the mechanism (increased noise, vibrations, overheating) or in premature failure. Such defects are often brought to light after a long period of time, when the mechanisms are being manufactured in quantity and their upgrade would require considerable expense.

Figure 1.3 depicts one end of a tie bar loaded with a variable axial force, F (the second end is similar). The tie bar consists of tube 1 and two lugs 2 welded to the ends of the tube. While in service, these tie bars have failed several times; the cracks were placed as shown: three cracks in 120° intervals. Investigation revealed that the cracks originated in plug welds 3. These welds are used for the preliminary attachment of the lugs to the tube before welding main seam 4. But the plug welds don't know that they are only needed to align the weld, and at work, they participate in load transmission between the lug and the tube. The plug welds' share of the load doesn't depend on their relative strength, but only on the ratio of compliances between the tube and shank 5 of the lug.

On the right of welds 3, the entire force F is transferred through tube 1. From the section where welds 3 are placed, part of the force is transferred through welds 3 directly to shank 5 of the lug,

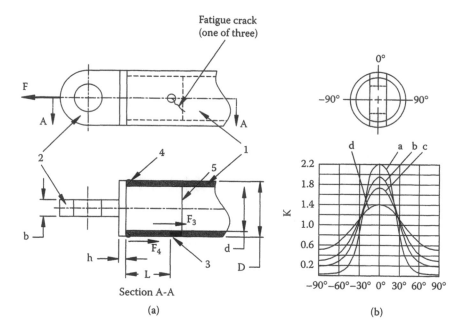

FIGURE 1.3 Tie bar (the left end is shown; the right end is identical).

and the rest of the force is transferred through tube 1 and weld 4 to the lug. The main thing we have to do to estimate the load distribution between the welds is to name the forces. Let's designate the forces transferred by welds 3 and 4 as F_3 and F_4 respectively. The rest is easy: just write the equations for deformations.

The elongation of shank 5 between welds 3 and 4 is

$$\delta_S = \frac{F_3 L}{E_S A_S}$$

The elongation of tube 1 in the same interval is

$$\delta_T = \frac{F_4 L}{E_T A_T}$$

In these equations, A_S and A_T are the areas of cross sections of the shank and the tube respectively, and E_S and E_T are the moduli of elasticity of materials.

Taking into consideration that $\delta_S = \delta_T$ and $E_T = E_S = E$ (because both the shank and the tube are made of steel), we find that

$$\frac{F_3}{A_S} = \frac{F_4}{A_T}$$

Because $F_3 + F_4 = F$,

$$F_3 = \frac{F}{1 + A_T / A_S} = \left(\frac{d}{D}\right)^2 F$$

In the case under consideration, the outer diameter of the tube $D = 80$ mm, and the shank diameter $d = 64$ mm, so

$$F_3 = \left(\frac{64}{80}\right)^2 F = 0.64F$$

This calculation is not exact because plug welds 3 don't connect the entire circumference of the shank but represent three local plug welds. Therefore, the actual compliance of the shank is greater and, consequently, the force F_3 should be smaller. The finite element method (FEM) gave $F_3 = 0.48\,F$. This more precise definition is not of principle in this case, because the conclusion remains unchanged: the plug welds should be avoided.

Now, after we have canceled welds 3, the load of weld 4 doubled, and we have to check its strength. Lug 2 is placed near the weld; thickness h of the shoulder looks fairly small, and it is clear that the load distribution must be sufficiently uneven. If we don't use FEM, we can calculate the mean tension stress by dividing the force F by the weld area, which is approximately equal to A_T. But this stress is undoubtedly less than the peak magnitude of it. We also can calculate the stress assuming that the entire force is transferred in the area of the lug width b. In our case, $b = 24$ mm, and the calculated stress will be 4.7 times the mean value. This is more than the real peak magnitude, but if the weld doesn't stand this stress, we need to know more exactly the real peak magnitude. In Figure 1.3b is represented the stress distribution in weld 4 found using the FEM model. Curves a, b, c, and d correspond to $h = 10$, 15, 20, and 30 mm, respectively. On the ordinate is plotted value $K = \frac{\sigma_{local}}{\sigma_{mean}}$. In the following pages, we will often face the problem of load distribution between parts and their elements.

Sometimes the interaction of two parts is influenced by many other parts connected with them, so the deformation analysis becomes multifarious. Figure 1.4 shows a draft of a gear. The load distribution along the teeth depends mainly on the parallelism of the shafts or, to put it more precisely, on the parallelism of the shafts' segments, which are adjoined to the gears. Figure 1.5 depicts factors that have an effect on the possible lack of parallelism:

The nominal position and forces applied to the gear and the pinion (Figure 1.5a)
The bending deformations of shafts (Figure 1.5b)

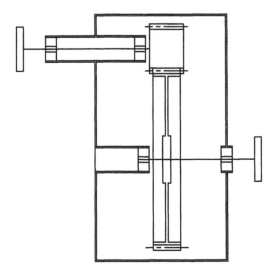

FIGURE 1.4 Gear sketch.

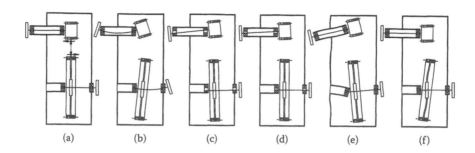

(a)   (b)   (c)   (d)   (e)   (f)

FIGURE 1.5 Displacements of shafts and gear rims caused by elastic deformations of parts and by bearings clearances.

The shafts' displacement caused by elastic deformation of bearings (Figure 1.5c)
The shafts' displacement caused by take-up of radial clearances in the bearings (Figure 1.5d)
The shafts' displacement caused by deformation of the housing (Figure 1.5e)
The gear wheel displacement caused by deformation of its body (Figure 1.5f)

Some factors which are hard to represent in the same manner should be added here, too:

Torsional deformation of the pinion, which may be considerable when the length of the pinion is greater than its diameter
Uneven radial deformations of the gear and the pinion caused by uneven heating and centrifugal forces (relevant mostly to high-speed gears)
Uneven rigidity of the teeth when the toothed rim is thin (see Chapter 8, Section 8.2)

Uneven load distribution along the teeth may also result from an unsuitable way of lubrication. The authors have observed deep pitting of the teeth profiles in the middle of the gear teeth, which took about 10% of the gear face width. It was placed exactly in the area where the lubricating oil was brought to the teeth by a narrow idler immersed into oil (called *rotaprint lubrication*; see Chapter 7, Section 7.10). To avoid such an effect, the width of the lubricating idler should be 70–80% of the gear to be lubricated.

Among the omitted factors are manufacturing errors and possible deformation of the housing while it is being attached to some foundation or substructure. These errors are very small, and the alignment of the teeth (bearing pattern) is finally checked by painting the teeth with dye and examining the pattern of dye transferred to mating teeth.

From Figure 1.5 we can see that the direction of displacements may change. This depends on the relative position of the bearings and gears, housing design, and direction of the applied forces. By means of proper design, the effects of some of the previously mentioned components of deformation might be mutually offset. This can be done by a reasonably chosen combination of gear element rigidities and direction of the axial force. But the main way is to decrease the deformations as far as possible.

From this point of view, the design shown in Figure 1.4 is extremely bad, because it is easily deformable. It is intentionally drawn to show the elements of deformations more clearly.

Analyses of deformation might be time consuming even if its most complicated components, such as the housing deformations, are made negligible by proper design. But the deformations of the majority of machine elements, such as shafts, bearings, gears, levers, etc., may be calculated by the so-called "engineering methods".

Sometimes the "engineering methods" are completely useless, and satisfactory results may be obtained by FEM only. Figure 1.6 shows a connection between piston 1 and connecting rod 2,

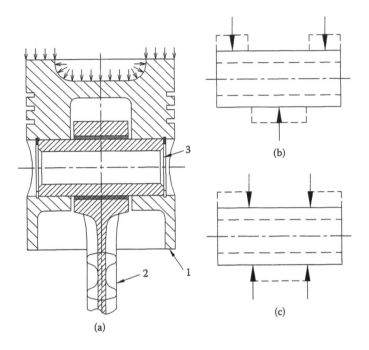

(b)

(c)

(a)

FIGURE 1.6 Loading of a piston pin.

provided by pin 3. The load distribution over the contacting surfaces of the pin depends on the elastic deformations of the parts and the hydrodynamic oil film parameters in the bearings. Trying to make strength calculation of the pin by engineering methods, we can consider two extremely simplified options of loading shown in Figure 1.6b and Figure 1.6c. The first option (Figure 1.6b) gives the stress almost twice as high as the second option (Figure 1.6c). But this pin is hollow, and, in addition to bending and shear stresses, it suffers from bending of its cross section (ovalization). These stresses can be easily calculated for a ring (two-dimensional problem), but the three-dimensional problem seems to be too hard for simplified analysis. What is important, the dependence between the stress and the load is nonlinear in this case. That means the stress increase in the pin is less than the increase of force F. The reliable determination of stresses can be achieved here only by FEM analysis and, finally, by measuring them on a working machine.

REFERENCE

1. Mabie, H.H. and Reinholtz, C.F., *Mechanisms and Dynamics of Machinery*, 4th ed., Virginia Polytechnic Institute and State University, VA.

2 Movements in Rigid Connections and Damage to the Joint Surfaces

The components of a mechanism must be somehow connected with each other; otherwise it is not a mechanism but only a group of separate machinery parts. The connections may be movable (sliding joints, such as hinges, telescopic joints, and so on), or immovable (rigid, or fixed, joints, such as bolted joints and others). In the latter case, the connected parts function as one part, which is made of two or more parts because of technological considerations or for assembly needs.

Complete immobility of a rigid connection is rarely achieved. In many cases, some microslip in the rigid joints under load is unavoidable, and this may cause damage to the joint surfaces and lead to the formation of fatigue cracks.

Among the many types of rigid connections, the two most commonly used are considered here: the *interference-fit connection* of a hub with a shaft and the *bolted connection* of two elements with a flat joint surface.

2.1 INTERFERENCE-FIT CONNECTIONS (IFCs)

2.1.1 IFCs LOADED WITH A TORQUE

Figure 2.1a depicts the simplest connection of a hub and a shaft, and in Figure 2.1b, these parts are shown separately. Both the hub and the shaft are loaded by a concentrated torque T and by distributed tangential friction forces, which balance the torque T.

The friction forces are distributed over the entire surface of the connection. In Figure 2.1, the arrows are located at the horizontal centerline, with each meant to represent a portion of the tangential friction force. The force at any location is understood to be uniform around the circumference, but varying in magnitude over the length of the joint.

Let's indicate by A the section where the shaft torque is maximal, and by B the free section of the shaft. Now let's assume that the tangential friction force is constant in the longitudinal direction. What will be the angles of torsion (the angle of rotation of section A relative to section B) for the shaft and for the hub? Usually, the range of D/d is from 1.5 to 1.6; hence, the torsional stiffness of the hub is about 4 to 5.5 times greater than that of the shaft. Therefore, lines 1 and 2 (Figure 2.1b), which depict the deformed (by twisting) generating lines, will be of dissimilar curvatures, because δ_1 is much larger than δ_2 (in inverse proportion to the stiffness). This means either there is local sliding in the connection loaded by a torque or the assumption of uniform load distribution in the longitudinal direction is not true. It is clear that when the loading torque is equal to the maximal total torque of the friction forces, the connection is completely spinning, and the tangential friction forces are equally distributed all around as shown in Figure 2.1b. If the load is decreased below the "breakaway" value, relative rotation of the two parts ceases, and they settle into an intermediate configuration (Figure 2.1c). But δ_1 is still unequal to δ_2.

It is reasonable to assume that there exists some torque magnitude that doesn't cause any sliding in the connection. To satisfy this condition, the tangential forces should be distributed in such a way that the torsion deformations of the shaft and the hub (in the area of IFC) are equal. Figure 2.1d qualitatively shows such a distribution, where most of the load is transferred in the area adjoined

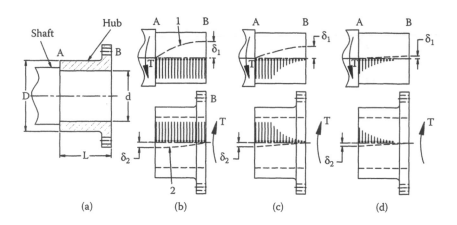

FIGURE 2.1 Torsional deformation in interference-fit connections (IFCs).

to section A. In this case, the shaft, which has a lesser moment of inertia, is twisted with a large torque only in the area near section A, and most of the shaft length is twisted with a relatively small torque. The hub, meanwhile, is twisted by a large torque through its entire length, which makes it possible to achieve the desirable equality $\delta_1 = \delta_2$.

Investigations made by analytical and experimental means[1,2] confirm the load distribution shown in Figure 2.1d. The factor of unevenness of load distribution along the IFC is given by

$$K_L = \frac{t_{max}}{t_{aver}} = \lambda L \frac{1+e^{-2\lambda L}}{1-e^{-2\lambda L}} \qquad (2.1)$$

where

$t_{aver} = T/L$ = average unit torque (N·mm/mm)

t_{max} = maximal unit torque (N·mm/mm)

$$\lambda = \sqrt{\frac{8G_H}{r^2 G_S}} \ \text{mm}^{-1}$$

where

G_H and G_S = shear modulus of the hub and the shaft materials (MPa)

$r = 0.5d$ = radius of the connection (mm)

L = length of the connection (mm)

If the hub and the shaft are made from identical materials, (i.e., $G_H = G_S$), then $\lambda = 2.83/r$. The K_L values for this case are shown in Figure 2.2. When $L/r = 2$, $\lambda L = 5.66$, and $K_L = 5.65$. That means that if (at $L = 2r$) the maximal torque transmitted at breakaway is 100%, only 17.7% can be transmitted without any local slippage. Therefore, in practice, we are usually forced to put up with some slip in the area adjoining section A. If the torque changes direction (torque reversal), the slip occurs repeatedly, coinciding with the frequency of torque reversals, and damages the joint surfaces.

2.1.2 IFCs LOADED WITH BENDING MOMENT

Deformations and variations of the surface pressure in IFC when the shaft is loaded with a *bending moment* are represented in Figure 2.3a and Figure 2.3b. When the shaft is bent, the pressure between

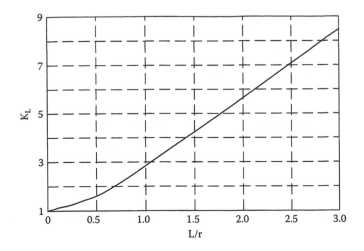

FIGURE 2.2 Nonuniformity of load distribution in interference-fit connections (IFCs).

the shaft and the hub decreases in the tensioned area (from above), with the shaft slipping out of the hub in this area. In the compressed area at the bottom, the contact pressure rises. The shaft in this area would like to slip into the hub, but the increased contact pressure (usually) prevents the slippage. When the connection rotates with respect to the vector of the bending moment, the tensioned area (that went out of the hub) moves around the circle. During one turn the entire shaft moves out of the hub by a tiny increment. But as the shaft continues to rotate, the micromovements accumulate to cause a macrosized shift in its axial position within the hub. This process is called self-pressing-out (without any axial force applied to the connection). The shaft shown in Figure 2.3b tries to move out of the hub in both directions with the same force, so the entire shaft remains in place. However, at the ends of the connection, the surface layers of the shaft are eventually stretched outward. This causes tension stresses in these areas, which bring down the fatigue strength of the shaft.

In a connection with an asymmetric load, in which the bending moment on one side is much greater than on the other (Figure 2.4), the shaft strives to go out of the hub toward the larger bending

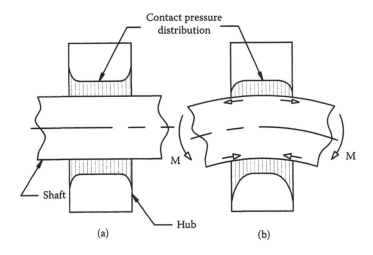

FIGURE 2.3 Slippage in an interference-fit connection (IFC).

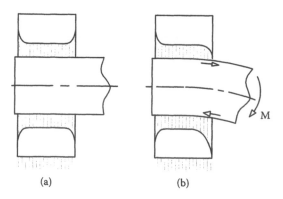

FIGURE 2.4 Self-dismantling of an interference-fit connection (IFC) under bending load.

moment. If the bending stress doesn't exceed a certain limit, the shaft is held in place by the friction forces in the connection, so the self-pressing-out is prevented. But if the bending moment is large enough, it is possible for the shaft to "self-press" itself completely out of the hub.

Such a failure mechanism can be demonstrated by pulling out a cork from a bottle; if the end of the cork protrudes from the neck of the bottle, you may easily get it out by just bending it from side to side.

In Chapter 5, Subsection 5.2.1, is shown how the bending load impairs the ability of the IFC to transmit torques and axial forces. Here, we are interested in the revelation of local slippage in a nominally rigid connection.

2.2 BOLTED CONNECTIONS (BCs)

In contrast to IFC, BC is a "spot" connection, similar to a spot weld. It doesn't necessarily apply pressure to the entire surface within the contact area. Figure 2.5a shows two parts contacting on a plane surface. If these parts had been pressed against each other by bolt force F_b, they would have become deformed as shown in Figure 2.5b. As applied to metal parts, the deformations in Figure 2.5b are greatly exaggerated. But if we take two rubber parts and press them together, the deformation will be as shown. What is important is that a bolt provides pressure between the contacting surfaces only inside some limited area around the bolt, but outside this area the surfaces are separated. If the connected parts are loaded in tension or in compression with force F_w, the dimensions of the compressed area will correspondingly decrease or increase as shown in Figure 2.5c. So an oscillating

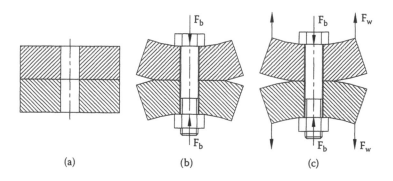

FIGURE 2.5 Bolted joint of two parts.

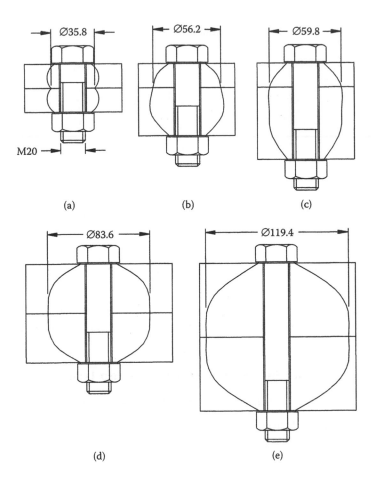

FIGURE 2.6 Size of compressed area depending on part thickness.

load will produce a cyclic movement of the border of the compressed area, and an inevitable microslip of the contacting surfaces will occur.

The size of the compressed area around the bolt depends mainly on the thickness of the connected parts. Figure 2.6 shows five examples of a bolted connection with different thickness of parts: 20/20 mm (Figure 2.6a), 20/40 mm (Figure 2.6b), 20/60 mm (Figure 2.6c), 40/40 mm (Figure 2.6d), and 60/60 mm (Figure 2.6e). (The volumes of the parts with compressive stresses are shaded.) These results, obtained by finite element method (FEM), show that the greater the thickness of the parts, the larger the diameter of the compressed area. Outside this area, the parts are separated.

The real parts contact through their surface asperities, so their surface layers have an increased compliance, as if a very thin pliable ("rubber") gasket is laid between the parts. That leads to an increased real compressed area as compared to that obtained using the FEM calculations.[3]

It is apparent that the local slip in the joint may be avoided only if the parts are pressed together strongly enough throughout the contacting surface and don't separate at any part under load. This may be achieved by reducing the contacting surface to the size of the compressed area around the bolt, by increasing the thickness of the parts in the bolted area, or by placing the bolts close to each other so that their compressed areas overlap. Often, all three methods are used.

There is an essential difference in the interaction of parts under load between connections with centrically applied load (Figure 2.7a) and those with eccentric load (Figure 2.7b). The latter case

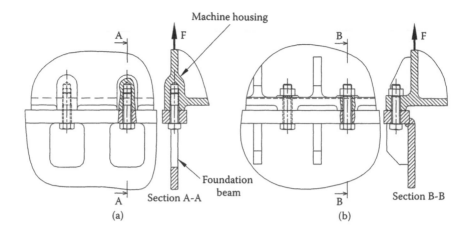

FIGURE 2.7 Concentric (a) and eccentric (b) loading of a bolted joint.

is much more problematic when the attainment of a safe connection and the strength of the bolts are concerned. But in practice, this case is a routine one. In Chapter 10, both of these variants are considered in more detail.

2.2.1 Forces in Tightened BC under Centrically Applied Load

Figure 2.8 shows parts 1 and 2, connected by a bolt and nut. The following notations are used here:

F_t = tightening force (preload) of the bolt (N)
F_b = tension force of the bolt at any load condition (N)
F_w = external (working) load (N)
F_c = contact force (between the contacting surfaces) (N)

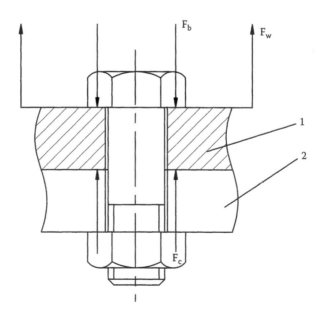

FIGURE 2.8 Forces in a tightened bolt connection.

Δ_b = elongation of bolt (mm)

Δ_f = negative deflection (contraction) of flanges (mm)

$$\lambda_b = \frac{\Delta_b}{F_b} \ (\text{mm}/\text{N}) = \text{compliance of the bolt}$$

$$\lambda_f = \frac{\Delta_f}{F_c} \ (\text{mm}/\text{N}) = \text{compliance of the connected parts (flanges)}$$

Well, the most boring section is behind us. Now we can write down the equation of equilibrium for part 1:

$$F_w + F_c = F_b \tag{2.2}$$

As long as there is no external load ($F_w = 0$), the only force in the connection is the bolt preload (tightening force), and in this situation, $F_c = F_b = F_t$. As the force F_w increases, the bolt force F_b increases as well. The bolt becomes longer then, relieving some of the compression on parts 1 and 2, and that in turn decreases the force F_c. Now let's rewrite Equation 2.2: when the force F_w increases by some amount δ_{Fw}, the force F_b increases by δ_{Fb}, and the force F_c decreases by δ_{Fc}. The new equation of equilibrium of part 1 is

$$F_w + \delta_{Fw} + F_c - \delta_{Fc} = F_b + \delta_{Fb} \tag{2.3}$$

Inserting Equation 2.2 into Equation 2.3, we obtain

$$\delta_{Fb} = \delta_{Fw} - \delta_{Fc} \tag{2.4}$$

From Equation 2.4, it is clear that the increase of the bolt force is less than the load increase due to the decrease of contact force in the joint. To numerically define the relations between F_b, F_w, and F_c, let's write Equation 2.4 in a more detailed form. Let's assume that because of the increase of the load by some amount, the bolt length is increased by a length Δ. In this case,

$$\delta_{Fb} = \frac{\Delta}{\lambda_b}; \quad \delta_{Fc} = \frac{\Delta}{\lambda_f}$$

Inserting these expressions into Equation 2.4, we obtain

$$\delta_{Fw} = \frac{\Delta}{\lambda_b} + \frac{\Delta}{\lambda_f} = \frac{\Delta}{\lambda_b}\left(1 + \frac{\lambda_b}{\lambda_f}\right) = \delta_{Fb} \frac{\lambda_f + \lambda_b}{\lambda_f}$$

And from there,

$$\delta_{Fb} = \delta_{Fw} \frac{\lambda_f}{\lambda_f + \lambda_b} = \delta_{Fw} \cdot \chi \tag{2.5}$$

with

$$\chi = \frac{\lambda_f}{\lambda_f + \lambda_b}$$

Here, χ is the coefficient of sensitivity of the bolt in the connection to the external load. On the basis of this equation, the bolt force in a loaded connection is

$$F_b = F_t + \delta_{Fb} = F_t + F_w \cdot \chi \qquad (2.6)$$

and the joint force is

$$F_c = F_t - F_w(1-\chi)$$

From this equation, we can determine the critical load F_{wcrit}, which initiates separation of the joint ($F_c = 0$):

$$F_{wcrit} = \frac{F_t}{1-\chi}$$

Hence, if the tightening force F_t is less than $F_w(1-\chi)$, the connected parts will separate completely. This is absolutely inadmissible; therefore, the tension force is determined from the equation

$$F_t = kF_w(1-\chi)$$

where k is a safety factor. It is recommended to take $1.2 \leq k \leq 1.6$ for static loads and $1.6 \leq k \leq 2.5$ for variable loads. (These values are preliminary.)

From Equation 2.5, we can see that by increasing the compliance of the bolt (λ_b) relative to that of the connected parts (λ_f), the dependence of the bolt force on the working force may be considerably decreased. This is very important for cyclically loaded connections, because the fatigue limit of the threaded parts is very low (usually within 50 to 80 MPa in terms of average stress in the bolt shank), and the strength of the bolts is often a real problem. But the compressive force in the joint must not be forgotten as well.

2.2.2 FORCES IN TIGHTENED BC UNDER AN ECCENTRICALLY APPLIED LOAD

Figure 2.9a shows a scheme of a BC consisting of two flanges. When the working load F_w is applied, the flanges become deformed. Rigid flanges rotate around the end of the connection (Figure 2.9b), and as soon as they separate completely (except on the line of rotation), the bolt force may be attained from the following equation of equilibrium of a lever:

$$F_b = F_w \frac{a+b}{b}$$

To decrease the bolt force, the distance a should be made as small as possible, taking into consideration the clearance needed for a wrench. Dimension b should be larger than a, but not too large, because the flange is flexible in bending (Figure 2.9c), and the excessive length a may be useless.

Note: If the load is concentric, the maximum bolt force (when the joint surfaces are separated and $F_c = 0$) will be $F_b = F_w$ (see Equation 2.2). Thus, the eccentrically applied load always results in greater bolt force.

The bending of the flange shown in Figure 2.9c brings about the separation of the flanges under lesser load, including part of the compressed area around the bolt, and the flanges should be thick

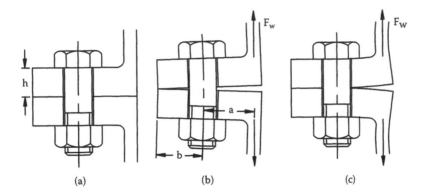

FIGURE 2.9 Deformations in an eccentrically loaded bolted connection (BC).

enough to achieve satisfactory strength of this connection. Usually $1.5d \leq h \leq 2.5d$, where d is the bolt diameter, but sometimes the flanges in the area of the bolts are made much thicker.

More detailed calculation of BCs is given in Chapter 10.

2.3 DAMAGE TO THE MATING SURFACES IN THE SLIP AREA

Working conditions of the joint surfaces in a rigid connection are very specific. They are characterized by the following parameters:

High pressure between contacting surfaces (tens or even hundreds of MPa)
Small amplitude of relative movements in the contact area (usually microns or fractions of microns)
Lack of access of lubricants to the area of contact

Under such conditions, the cyclic microslip brings about specific damage to the joint surfaces; this is called *fretting*. It begins with a smearing, which is a result of mechanical wear. Later, the smallest wear fragments oxidize, and then the surfaces look rusty (reddish brown color for ferrous alloys and black for aluminum alloys). The wear fragments are trapped at the locations where they were created, and because their volume increases owing to oxidization, the pressure in the connection also increases. Because the joint surfaces become roughened, the friction coefficient increases as well. Under these circumstances, the oscillating slip may result in a considerable increase in heat generation at the contact points, followed by local seizure (microwelding) of the contacting surfaces, and finally resulting in tear-off of the welds.

In contrast to macroslip, which results in wear and possibly seizure, the cyclic microslip (with very small amplitude) brings about the cyclic deformation (in bending and shear) of micro-asperities and the formation of microcracks in the surface layer (Figure 2.10). The microcracks that run up to macrosize mostly rise again to the surface (curves 1). The hatched particles separate from the surface, typically leaving pits on it. Some of the cracks may propagate deep into the part under the influence of the working stresses of bending or shear (curves 2 in Figure 2.10). If they continue to grow and spread, they may eventually bring about the formation of fatigue cracks, which can become large enough to completely fracture the part. Fretting may decrease the fatigue strength by a factor of 3 to 5.

The present state-of-the-art can't reliably predict the origination and rapidity of the development of fretting. But as the strength of the materials increases, elastic deformations under load grow correspondingly (as previously described), and cases of fretting-motivated failures are becoming much more widespread.

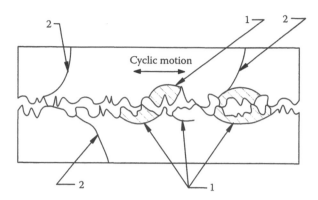

FIGURE 2.10 Strength-damaging effect of fretting.

Structural and technological means to increase the durability of parts in areas of rigid connections are aimed at achieving the following goals:

Increase of compressive stresses in the joint to prevent slip or at least diminish its amplitude.

Formation of a thin intermediate layer between joint surfaces (for example, by means of copper or lead coating) in order to avoid direct contact between two hard metals.

Inducing residual compressive stresses in surface layers of parts (by means of shot peening or burnishing — see Chapter 3); this treatment doesn't prevent fretting, but cracks originating in the surface layer don't propagate deeply into the part, so the strength-harming effect of fretting damage may be neutralized partly or completely.

Surface hardening by means of nitriding, carburizing, or induction hardening; these treatments not only produce high compressive stresses in the hardened layer (because of greater specific volume of the hardened metal), but also increase by far the mechanical strength of this layer and diminish the probability of crack initiation.

In movable joints (for example, in spline connections, or in gear-type couplings) working under load, relative immobility may result when high contact pressure and very small amplitude of oscillating slip are simultaneously present. These conditions are similar to those in rigid joints, and they also may result in fretting. For these joints, in addition to the methods described previously, friction in the connection may be diminished by making the lubricant flow through the connection or by coating the surfaces with molybdenum disulfide (MoS_2). These measures, however, can't be recommended for rigid connections.

REFERENCES

1. Müller, H.W., Drehmoment-Uebertragung in Pressverbindungen, *Konstruktion*, 14, H.2 u.3, 1962 (in German).
2. Klebanov, B.M., Load transmission by press-fit connections between shafts and massive discs, *Mechanics of Deformable Systems in Agricultural Engineering*, collected articles, Rostov-na-Donu, 1974 (in Russian).
3. Marshall, M.B., Dwyer, L.R., and Joyce, R.S., Characterization of contact pressure distribution in bolted joints, *Strain*, 42, 2006, Blackwell Publishing Ltd.

3 Deformations and Stress Patterns in Machine Components

A school teacher took two pieces of lead, cleaned their surfaces with a knife, and pressed them strongly against each other. The two pieces stuck together, and the teacher asked us to separate them. We were little boys, and this was a hard job for us. In that way, the teacher had clearly demonstrated to us the property of molecular adhesion. Now we have grown up, and we know that a metal is a complicated mass of crystals, grains, molecules, and atoms combined together by the forces of molecular and atomic bonding.

3.1 STRUCTURE AND STRENGTH OF METALS

Metals at a normal temperature are crystalline solids. The atoms in the crystals are located periodically and form a structure that represents a multitude of identical elements. If lines are drawn through the centers of atoms, these lines form a spatial lattice called *crystal lattice*. Metals and alloys used in engineering mostly have a cubic lattice. Figure 3.1a shows the elementary cube of the iron structure at a normal temperature. It is a body-centered cube (bcc) with atoms placed in the corners and in the center. The length of the cube side $a = 2.86$ Å, that is, $2.86 \cdot 10^{-7}$ mm. (The unit Å is the Angström, a linear dimensional unit of convenient size for atomic-scale measurements.)

If the metal is alloyed, not all the places in the cube are filled with the parent metal, but some of the places are taken by the alloying metal atoms. But this doesn't change the following considerations.

Figure 3.1b shows a small part of a perfect crystal lattice. Application of load (Figure 3.1c) deforms the lattice, but the forces of atomic bonding resist the possible sliding of the atomic layers. When the external load ceases, the atomic forces restore the initial balanced condition of the lattice. It is clear that the external force can be big enough, so as to overcome the atomic bonding and to displace the atomic layers by one step of the lattice. But this assumes that the entire layer of atoms would be moved at once. Calculations based on the strength of the atomic bonding show that the external force needed for this to take place would be greater than the real strength of the material by a factor of 100 and more. Therefore, something must explain the disparity between theory and reality, and it is as follows.

Numerous investigations have disclosed that the reason for the degraded strength of materials (as compared to its theoretical value) is the imperfection of the crystal structure. Several kinds of imperfection are shown in Figure 3.2:

Vacancies (Figure 3.2a), where one atom is absent.
Interstitials (Figure 3.2b), where one atom has squeezed itself into the lattice.
Edge dislocations (Figure 3.2c), where an incomplete row of atoms is squeezed into the
  lattice (or, if you prefer to be more pessimistic, part of a row is missing).
Screw dislocations. (The last consist of a row of atoms turned relative to the parent lattice by a
  small angle; they have a 3-D form and therefore are more difficult for 2-D graphic presentation.)

Of the reasons given previously, the greatest reduction of strength is due to dislocations. Figure 3.3a shows an unloaded part of lattice with edge dislocation 1. Near the end of the dislocation the crystal lattice is distorted, and some of the atomic bonding forces (for example, between atom 2 (fully shaded) and atoms 3 and 4 (both are half-shaded) are weakened. When the load is applied

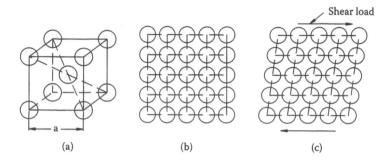

FIGURE 3.1 Crystal cell and crystal lattice.

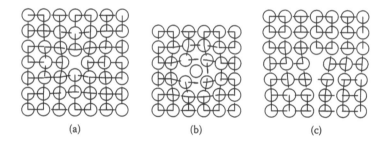

FIGURE 3.2 Crystal lattice imperfections.

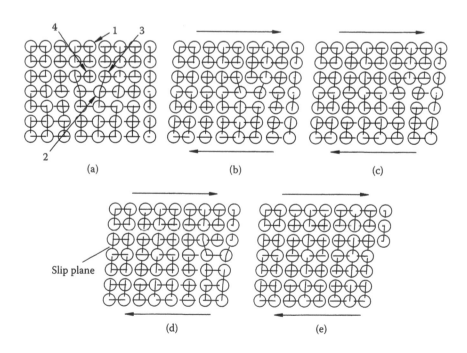

FIGURE 3.3 Dislocation and its movement under load.

(Figure 3.3b), most of it is taken by the elastically displaced atomic rows, but, in addition, atom 2, which has been already moved from atom 3 toward atom 4, jumps from 3 to 4, so that the dislocation moves to the right as shown in Figure 3.3b. This jerking (stepwise) motion continues under load until the dislocation goes out of the crystal (Figure 3.3c to Figure 3.3e). Thus, when a dislocation exists, the force needed to displace the atomic rows by one atom spacing is quite small. The plane of displacement is called *sliding plane* or *slip plane*.

The process of crystallization begins when the liquid metal is chilled somewhat below its melting temperature. In the liquid metal appear lots of nuclei of crystallization, and the neighboring atoms attach to them. The crystals grow until they meet the neighboring crystals. In the end of this process, which lasts several seconds, the metal passes into solid state in the form of a granular structure (see Figure 3.4). Because, in the nuclei centers, the orientation of the initial cubes was random, the same orientation remains in the grains of the solidified metal.

Figure 3.4 may give the impression that the quantity of the lattice cells in one grain is not so big. This impression is false. Let's calculate. The dimensions of the grains in steel, for example, range from about 0.015 mm (extra-fine-grained steel) to 0.220 mm (extra-coarse-grained). If we take the finest grain, the quantity of the lattice cells in one row is given by

$$\frac{0.015}{2.86 \cdot 10^{-7}} = 5.24 \cdot 10^4$$

It is clear that the presentation of grains in Figure 3.4 exaggerates the size of the lattice cubes for illustration.

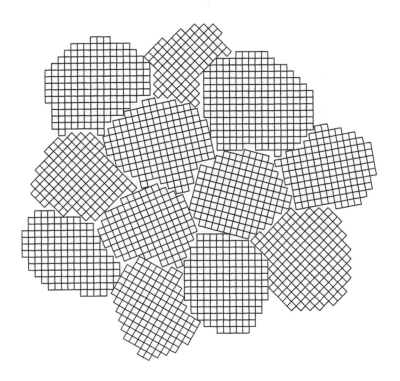

FIGURE 3.4 Crystals and grains in the metal structure.

The quantity of lattice cubes in 1 mm³ is

$$\frac{1}{(2.86\cdot10^{-7})^3} = 4.27\cdot10^{19}$$

Among this astronomical amount, there may be millions of dislocations and other defects initiated by alien atoms, free electrons, grain boundaries, internal stresses, etc. When an external load is applied, it induces stresses in the macrovolume of a part. These stresses have a certain pattern of direction and magnitude depending on the shape of the part, the kind of load, and the place of load application. In the microvolumes in which the direction of the stresses coincides with the direction of the slip planes of a grain, this grain may become plastically deformed by a very small amount, perhaps the magnitude of one atomic spacing (depending on the stress magnitude), owing to the movement of dislocations. Such "unlucky" crystals are always present, so metals with ideal elasticity don't exist; after even a small load is applied, some structural changes occur immediately. This kind of submicrodeformation is related to such phenomena as energy loss in materials in the course of elastic deformation (so-called internal friction) and a time delay in deformation as compared to the applied load.

When the stress is big enough, the dislocations multiply, and at the yield stress, they multiply very intensively, causing plastic deformation of the metal. Lots of dislocations interfere with each other, and their motion is hindered. This results in strengthening of the metal due to plastic deformation.

Dislocations are generated under smaller stress as well. At a cyclic load, the dislocations are constantly generated owing to movement on the grain boundaries and because of other processes in the crystal lattice, including their self-multiplication. The moving dislocations pile up on the grain borders and may finally form a microcrack.

The description of the mechanism of plastic deformation and fatigue given previously is very much simplified. The dislocations can move not only straight in direction (horizontally), as shown in Figure 3.3, but they also can move around a closed polygon in the sliding plane and even vertically (changing the sliding plane). The dislocations can unite within the crystal and divide it into parts forming subgrains. The atomic-scale processes occurring in the metal under load are multifarious in nature and are not yet studied enough to be used in strength calculations. But this description shows that the metal is similar to living matter that feels the load applied and reacts to it by structural changes.

Thus, perfection of elasticity depends on the accuracy of measurements. But in practice, when we deal with macrostresses and macrodeformations, it is generally agreed (and it is accurate for most practical applications) that the metals are perfectly elastic until the stress reaches the yield point; then it begins to get plastically deformed in macrovolumes, and the part changes its shape.

3.2 DEFORMATIONS IN THE ELASTIC RANGE

There is a linear dependence between stress and deformation called *Hooke's law*. It means that if we know the stress magnitude and the modulus of elasticity of the material, we can calculate the elastic deformation of the stressed piece. On the other hand, when the deformation is known, the stresses can be easily determined. This dependence is the basis for most stress-measuring instruments; they measure deformations and translate them into stresses.

But there is one more field to which Hooke's law can be applied: if we know (or are able to make a true assumption about) the distribution of the deformations over the cross section of a part, we can calculate the maximal stress. To illustrate this idea, let's consider the bending of a straight beam. The so-called "engineering approach" to strength calculation of beams is based on the hypothesis of plane cross sections (PCS). The idea of this hypothesis is that the cross sections of the beam, which are plane before bending (Figure 3.5a), remain plane after bending (Figure 3.5b). With bending, the neighboring cross sections turn relative to each other by a certain angle φ. The

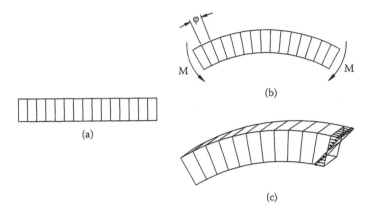

FIGURE 3.5 Bending deformation of a straight beam.

magnitude of this angle depends on the magnitude of the bending moment, the rigidity of the beam and, obviously, on the distance between the sections. Because the cross sections remain plane, the deformation of the beam between them is a linear function of the height. Across the width of the beam, the deformation doesn't change (Figure 3.5c).

Let's examine the word *hypothesis*. A hypothesis is a kind of assumption usually developed as follows: If we desire to work out the relation between appearances to explain or predict phenomena or processes and don't have sufficient data to establish an exact dependence, we may begin by making some simplifying assumptions. Then, to check the validity of the hypothesis, these assumptions and the dependence based on them are checked experimentally. If the correspondence between the experimental results and this invented dependence is not satisfactory, more appropriate assumptions should be made.

In our case, the hypothesis of PCS has been proved to be very exact and productive as applied to beams of a uniform cross section. But there are some exceptions to this rule (as with all rules). For example, in the thin-walled beams shown in Figure 3.6, the stresses across the width are not uniform; hence, the plane cross sections become distorted after application of a load. Use of formulas based on the hypothesis of plane cross sections for such beams (with wide and thin flanges) may lead to a considerable error (the real stresses will be bigger than the calculated ones). But let's get back to the usual case when this hypothesis works with high accuracy.

Figure 3.7a shows a rod with a rectangular cross section loaded by bending moment M. The cross section of the rod is of variable height: h on the right and H on the left. A fillet of radius r makes the transition from the smaller height to the bigger one.

First, let's define the stresses in the right part of the rod, where the cross section is of constant height h. For this part, the hypothesis of plane cross sections is valid. We consider a small element

FIGURE 3.6 Deformation of thin-walled beams with wide thin shelves or flanges.

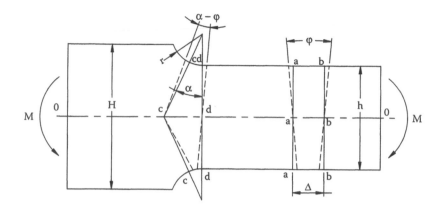

FIGURE 3.7A Plane sections and broken sections.

of the rod bounded by plane sections a-a-a and b-b-b (Figure 3.7a). Dashed lines show the shape of this element after bending. The boundary sections remain plane but they turn relative to each other by angle φ. With bending, the layers of the rod above the zero line (the neutral axis) become longer and the layers below the zero line become shorter, both by a variable amount δ (see Figure 3.7b).

Because the deformations are antisymmetric about the zero line, it is enough to consider only half of the element as shown in Figure 3.7b. (In this picture, one side, a-a, has been left unturned for convenience, and the other side, b-b, is rotated by the angle φ, which is assumed to be very small.)

The stress in the deformed layer according to Hooke's law

$$\sigma = E\varepsilon \tag{3.1}$$

where
  E = modulus of elasticity of the material (MPa)
  ε = relative deformation (in this case, it is elongation) determined by the equation

$$\varepsilon = \frac{\delta}{\Delta}$$

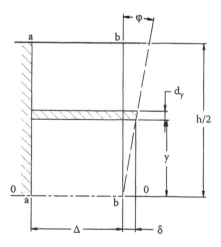

FIGURE 3.7B Strain distribution in a straight bar.

As is clear from the picture, the magnitude of δ depends on the ordinate y:

$$\delta = y\varphi$$

Therefore, the stress in each layer is proportional to the distance y of the layer from the zero line:

$$\sigma = E\varphi \frac{y}{\Delta}$$

As the values E, φ, and Δ are constant in this problem, all of them may be substituted by one coefficient, say, $K = E\varphi/\Delta$:

$$\sigma = Ky \tag{3.2}$$

Now we are ready to calculate the stresses in this part of the rod. Elementary moment produced by a slice dy about the zero line

$$dM = \sigma by \cdot dy = Kby^2 dy$$

Here b = thickness (width) of the rod (not shown in the picture).

The sum of all the elementary moments, taken above and beneath the zero line, should be equal to the loading moment M:

$$M = 2 \int_0^{h/2} dM = K \frac{bh^3}{12}$$

From here,

$$K = \frac{12M}{bh^3}$$

The maximal stress (at $y = h/2$) from Equation 3.2:

$$\sigma_{max} = Ky = \frac{12M}{bh^3}\frac{h}{2} = M \frac{6}{bh^2}; \tag{3.3}$$

Fortunately, we didn't make any mistakes and have got successfully the well-known equation for a rod of a constant rectangular cross section. Now, let's try to deal with the fillet area. In this area, the hypothesis of PCS doesn't work. For this case, the hypothesis of broken-line cross sections (BCS) suggested by Verhovsky[1] works better. (The BCS hypothesis is based on experiments with rubber parts exposed to bending. The "cross sections" are just drawn on the side surface of the rubber bar, so that their shape under bending can be easily observed.) According to this hypothesis, the cross sections, which continue to be plane, are directed at a perpendicular to the curvilinear surface (section c-c-c, Figure 3.7a).

We consider a small element of the rod bounded by cross sections c-c-c and d-d-d (Figure 3.7a). This element is adjoined to the beginning of the fillet and characterized by an angle α that is supposed to be very small. Also, here the dashed lines show the shape of this element after bending.

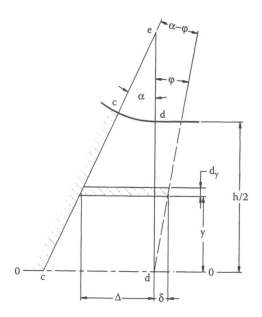

FIGURE 3.7C Strain distribution in a filleted bar.

In the same way as the previous case, we consider only half of the small element (c-c-d-d, Figure 3.7c). One side of the element (c-c) is left unturned, and the other side, d-d, is rotated because of bending by an angle φ. Notice that in this case, the elongation δ is also proportional to the ordinate y:

$$\delta = y\varphi$$

In the previous case, the length of the deformed layers Δ was constant, whereas in this case, the length is variable and determined as follows:

$$\Delta = (r + h/2 - y)\alpha$$

This result is easily obtainable, if we take into consideration that the length of the elementary layer depends linearly on its distance from point e. The layer with vertical coordinate y is at a distance of $(r + h/2 - y)$ from point e. (Don't forget: angle α is very small.)

That is, in the element c-c-d-d not only do the deformations grow as the layers get farther from the zero line, but also the length of the layers decreases, so the increase of the relative deformation $(\varepsilon = \delta/\Delta)$ with the increasing ordinate y is not linear but more intensive.

The stress in the layer at a distance y from the zero line is

$$\sigma^* = E\varepsilon = E\frac{\delta}{\Delta} = \frac{Ey\varphi}{(r + h/2 - y)\alpha}$$

Because in this problem, values E, α, and φ are constant, they can be substituted by one coefficient, say $L = E\varphi/\alpha$. Then the maximal stress (at $y = h/2$) is given by

$$\sigma^*_{max} = L\frac{h}{2r} \tag{3.4}$$

The elementary moment produced by a slice dy about the zero line is given by

$$dM = \sigma by \cdot dy = L\frac{y^2 b}{r + h/2 - y}dy$$

The sum of all the elementary moments should be equal to the loading moment M:

$$M = 2\int_0^{h/2} dM = 2\int_0^{h/2} Lb\frac{y^2}{r+h/2-y}dy = 2Lb\left[-\frac{1}{2}hr - \frac{3}{8}h^2 - (r+h/2)^2 \ln\frac{r}{r+h/2}\right]$$

After the L value is derived from this equation and inserted in Equation 3.4, the following formula for the maximal stress is obtained:

$$\sigma^*_{max} = \frac{Mh}{4rb}\frac{1}{\Phi} \tag{3.5}$$

where

$$\Phi = -\frac{1}{2}hr - \frac{3}{8}h^2 - (r+h/2)^2 \ln\frac{r}{r+h/2} \tag{3.6}$$

Let's do comparative stress calculations for the straight part of the rod by Equation 3.3 and for the beginning of the fillet by Equation 3.5 and Equation 3.6. For this calculation, we can take $b = h = 1$, because, for the comparison, it is not important if it is 1 mm or in. (or whichever unit of length you want).

For the straight section, the maximal stress obtained from Equation 3.3 is

$$\sigma_{max} = 6M$$

At the beginning of the fillet, the stress depends on the fillet radius. For example, if $r = 0.2h$, $\Phi = 0.1389$, and the stress

$$\sigma^*_{max} = \frac{M}{4 \cdot 0.2} \cdot \frac{1}{0.1389} = 9M$$

The stress concentration factor $K_t = 9/6 = 1.5$. If the fillet radius $r = 0.1h$, $\Phi = 0.22$, the maximal stress

$$\sigma^*_{max} = \frac{M}{4 \cdot 0.1} \cdot \frac{1}{0.22} = 11.36M$$

and the stress concentration factor $K_t = 11.36/6 = 1.89$.

Here, shown clearly (and even visually), is the mechanism of the local stress raise called *stress concentration*. It is caused by a change in the shape of the machine component that modifies the deformation pattern. The place where the shape changes is called a *stress raiser* (or *stress riser*). The sharper the change of the shape (in our case, the less the fillet radius), the higher the local stress.

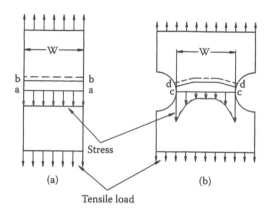

FIGURE 3.8 Stress concentration in a notched strip in tension.

The BCS hypothesis is good for a visual presentation of the stress raiser mechanism. But it gives a correct result only when the bigger section is sufficiently larger than the smaller one, and the stress raiser is not too sharp. Specifically, in the case considered previously, according to R. E. Peterson,[3] $K_t = 1.34-1.53$ for $r = 0.2h$, and $K_t = 1.48-1.93$ for $r = 0.1h$ depending on the ratio H/h. The BCS hypothesis doesn't take into account this factor, and in our calculation we got $K_t = 1.5$ and $K_t = 1.89$, respectively, i.e., close to the biggest values given by Peterson.

In most cases, the magnitude of the local stress can't be determined by an analytical procedure. Experimental methods and FEM are usually used for this purpose.

Another application of the BCS hypothesis is with a flat strip of rectangular cross section, in tension. If the strip is of constant cross section (Figure 3.8a), then obviously the hypothesis of PCS is valid. Under load, any small element of the strip bounded by sections a-a and b-b elongates evenly across the width of the strip (see the dashed line).

Figure 3.8b shows a strip with semicircular notches that has the same width w (in the narrow section). The cross sections, which keep their shape when loaded, look different (sections c-c and d-d). In the area of the notches, the length of the layers is significantly shorter than in the middle of the strip, and when section d-d is displaced under load (see dashed line), the relative elongation in this area will be greater. The stress distribution across the width of the strip will change correspondingly. Because in both cases the sum of all the stresses must be equal to the external load, the maximal stress in the notched specimen will be greater than that in the straight strip.

Let's discuss more fully the meaning of the stress concentration factor (SCF), because it is not always clear. The maximal stress magnitude in the area of the stress raiser is determined using theory of elasticity techniques, by FEM, or experimentally. To calculate the SCF value, the maximal stress is divided by some "nominal" stress determined by methods of the strength of materials (so-called engineering methods). For the rod shown in Figure 3.7a, the nominal stress is determined from Equation 3.3. For the strip presented in Figure 3.8b, the nominal stress is obviously the ratio of the stretching force to the minimal cross-sectional area (through the bottoms of the notches). But if, for example, you are going to use SCF diagrams for a splined shaft, or a threaded shaft, or a shaft with a keyway, you have to figure out which cross section had been taken by the author when calculating the "nominal" stress: was it a circular cross section corresponding to the inner diameter of the teeth or thread, the real cross section, or something else? To calculate strength of the machine part, you should know the maximal stress. For that you calculate the nominal stress and then multiply it by the SCF. So until you know which cross section should be taken for the "nominal" calculation, you can't use these diagrams.

The SCF obtained from such calculations is called *theoretical SCF* (K_t), and it enables (in principle) the designer to determine the maximal stress. At this point, the mathematics comes to an end, and some uncertainty begins. The fact is that in the trials for strength, the load capacity of the machine members reduces not by K_t times but less, by K_e times. The K_e value is called *effective SCF*, and that is what we really need to know but can't. The interrelation between K_t and K_e is complicated and multifactored. Much depends on the loading condition. For example, under a static load, ductile materials (such are all the structural steels) are not sensitive to stress raisers. When the local stress comes up to the yield point, this brings about a local plastic deformation that doesn't harm the strength of the material but redistributes the stresses in such a way that the less-stressed volumes became more stressed in the elastic range. As the static load increases, the plastically deformed volume increases until it spreads across the whole section. This is the limiting state under static load whether there is a stress raiser or not, and thus here, $K_e = 1$.

Under a cyclic load, the stress raisers reduce the load capacity of the machine parts, but less than the K_t value predicts, depending on the feature called *sensitivity to stress concentration*. In practice, the K_e value is often derived from the following equation:

$$K_e = 1 + q(K_t - 1)$$

where q is a coefficient of sensitivity to stress concentration.

That coefficient depends on a number of factors:

The kind of material and existence in its structure of internal stress raisers (such as discontinuities in the form of foreign inclusions, defects on the grain boundaries, and dislocations). For instance, gray cast iron is hardly sensitive to sharp stress raisers under cyclic load, because its structure is pierced with flakes of graphite, which are sharp stress raisers incorporated into the material. And yet, the gray cast iron is sensitive to the stress concentration under the static load, because it is brittle, and stresses can't be effectively redistributed by local plastic deformations.

The ductility of the material: The less the ratio of the yield stress to the ultimate stress of a steel, the less sensitive it is to the stress concentration. That means that high-strength steels are more sensitive to SCF.

The sharpness of the stress raiser: The sharper it is, the smaller the high-stressed volume and the stronger the effect of the stress release owing to local plastic deformations (for example, the sensitivity to the stress raiser is less).

The size of a part: Larger parts have greater (at the same sharpness of the stress raiser) high-stressed volume, yielding a less than helpful effect from local plastic deformation.

A host of other factors, such as quantity of load cycles and ratio of the mean stress to its amplitude.

Because most of these factors are difficult to take into account, the recommendations for choosing the q value are very approximate. But it is known that high-strength structural steels are very sensitive to stress concentration under cyclic load, and the recommended q value for them is about 0.90–0.95.

Some authors recommend (on the basis of experiments) coefficients K_e for a certain type of stress raiser (for example, threaded or splined shafts) related to certain steels. But in very important applications, in which totally reliable information about the permissible load is needed, the actual parts should be tested for strength and durability under real loading conditions.

3.3 ELASTOPLASTIC DEFORMATION (EPD) OF PARTS

As in the case of elastic deformation, to know the stresses in the part, we have to know the strain pattern and the relationship between the strain and the stress. It was found that for EPD of beams with uniform cross section, the PCS hypothesis is also valid to sufficient accuracy. But the strain–stress relationship beyond the yield point is much more complicated, and it varies for different metals. This relationship is usually derived from a tensile test of a standard specimen.

Figure 3.9a shows the initial part of the stress–strain diagram typical of low-carbon steels (it is also valid for some alloy steels). This diagram consists of three portions. Portion 1 represents the elastic deformation where the relation between strain and stress is linear and meets Hooke's law. Portion 2 represents a very specific feature of these steels: as the stress slightly exceeds the yield point, the specimen elongates plastically by approximately 1.5–2%, practically without any increase in the applied load.

The plastic elongation ε_p is much greater than the elastic elongation ε_e. The latter is easy to calculate from Equation 3.1. If the yield point for low-carbon steel is, for example, about 200 MPa, and the modulus of elasticity $E = 2 \cdot 10^5$ MPa, then the elastic elongation is

$$\varepsilon_e = \frac{\sigma}{E} = \frac{200}{2 \cdot 10^5} = 0.001 = 0.1\%$$

Thus, the plastic elongation ε_p is 10-fold or 20-fold the elastic elongation ε_e.

It is interesting to note that plastic deformation of a certain magnitude brings about an increase in the strength of the steel: yield stress S_{ys} instead of S_y.

Figure 3.9b represents the stress–strain diagram typical of most alloy steels. This kind of diagram is too complicated for our simplified analyses. Therefore, we put it aside for now and consider what could happen to a beam that is bent beyond the yield stress starting from Figure 3.9a.

Figure 3.10a shows the boundary load in the elastic range, when the maximum stress is equal to the yield point S_y. Bending moment M may be calculated from an equilibrium equation:

$$M = 2 \int_0^{h/2} \sigma \cdot b \cdot y \cdot dy$$

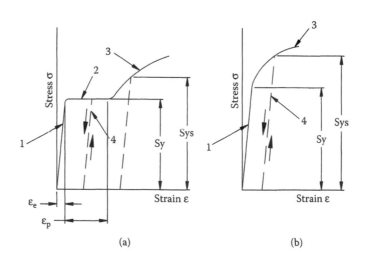

(a)   (b)

FIGURE 3.9 Stress–strain diagrams.

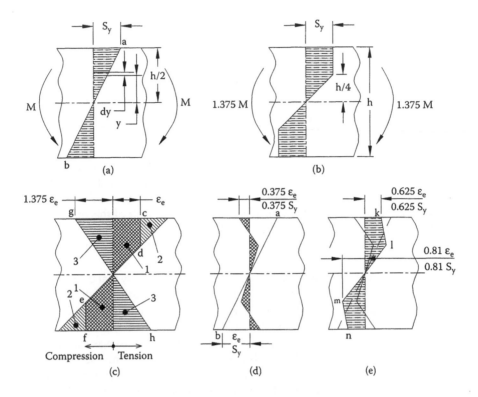

FIGURE 3.10 Deformation of a beam in elastic and elastoplastic ranges.

where

    b = width of the beam (not shown in the figure)

    σ = stress at a distance y from the zero line

$$\sigma = S_y \frac{2y}{h}$$

From this

$$M = \frac{1}{6} S_y b \cdot h^2$$

What will happen if the load is increased? As said earlier, the hypothesis of plane cross sections is valid in this case as well. That means that the angle φ between the two neighboring cross sections (Figure 3.7a and Figure 3.7b) will grow, and the layers will be additionally elongated by an amount that is proportional to the distance y of the layer from the zero line. But the deformation of the layers most distant from the zero line will grow without increase in stress beyond the yield point. Figure 3.10b shows the situation in which deformation is twice as much as in Figure 3.10a. In this case, the yielding area begins at $y = h/4$ and extends to $y = h/2$. The equilibrium equation for this case is

$$M^* = 2 \int_0^{h/4} S_y \frac{4y}{h} b \cdot y \cdot dy + 2 \int_{h/4}^{h/2} S_y b \cdot y \cdot dy = \frac{11}{48} S_y b \cdot h^2$$

As we can see, bending moment $M*$ is greater than M:

$$\frac{M^*}{M} = \frac{11 \cdot 6}{48} = 1.375$$

Hence, if such a yield that covers half of the cross-sectional area is permissible, the bending moment may be increased by 37.5%.

There may be some doubt about the plastic deformation in this case: is it still relatively small or is the metal deformed too much and already beginning to break? As we saw earlier, the maximum elastic deformation ε_e is about 0.1%. In our case, this layer is placed at a distance of $h/4$ from the zero line. The maximal deformation, elastic and plastic (at a distance of $h/2$), in accordance with the plane cross sections hypothesis, is twice as much, i.e., 0.2%. Half of this amount (0.1%) is the elastic deformation that remains in the material, and the other half is the plastic deformation. This magnitude of plastic deformation is very small and cannot yet impair the mechanical properties of the metal.

A comparison of Figure 3.10a and Figure 3.10b demonstrates clearly that the plastic deformation of layers more distant from the zero line brings about increased loading of the layers at a shorter distance from the zero line. So, the summary moment of the internal forces about the neutral axis of the cross section is obviously greater in the case of EPD.

It should be noticed, however, that the gain in load capacity (37.5%) was calculated for a beam with rectangular cross section (Figure 3.11b). If the beam is of a round cross section (Figure 3.11c), the increase in load capacity should be greater, because the layers become wider as they are closer to the zero line. For an I-beam (Figure 3.11d), though, EPD is less effective because the cross section is thin in the middle portion, and additional stressing does not add much.

Plastically deformed metal behaves in the same manner as the initial one: as it is unloaded, it follows Hooke's law with the same modulus of elasticity (line 4 in Figure 3.9a, arrow downwards). When loaded again, it follows the same line (arrow upwards) until it comes to the yield stress and continues plastic deformation as if the process of unloading and repeated loading has not been performed. This feature, plus the validity of the PCS hypothesis for EPD, allows us to predict the residual stress pattern in deformed machine parts. For example, after the beam was loaded in the elastoplastic range according to Figure 3.10b and then unloaded, the plastically deformed layers didn't allow the elastically deformed layers (in the middle of the section) to return to their initial state. This counteraction results in a balanced system of residual stresses in the unloaded beam. The distribution of these stresses may be found by superposition of deformations. Figure 3.10c shows the distribution of deformation over the height of the cross section when loaded by a positively directed bending moment of $1.375M$ (Figure 3.10b). Areas 1 (lines c-d-e-f) show the elastic portion of deformation, and areas 2 represent the plastic portion. The last doesn't take part in subsequent

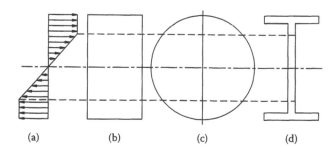

(a)   (b)   (c)   (d)

FIGURE 3.11 The effectiveness of elastoplastic deformation.

considerations, because it is not a "stored" deformation that is related to stress and can be released, but just a new condition of the material.

Note: The distribution of elastic deformation in Figure 3.10c is exactly the same as the stress distribution in Figure 3.10b. It is so as long as the metal behaves in accordance with lines 1 and 2 in Figure 3.9a.

Let's continue: Unloading the beam is the same as loading it with an additional bending moment $1.375M$ in the opposite direction. For the sake of simplicity, presuppose that the deformation caused by this opposite moment will be in the elastic range (a bit later we will see that this assumption is correct). Taking into account that the bending moment M brings about maximum stress S_y (see Figure 3.10a) and correspondingly brings about maximum deformation ε_e, then areas 3 (line g-h) should represent the distribution of deformation when the opposite (negatively directed) moment $1.375M$ is applied. The algebraic sum of the positive and the negative bending moments is an unloaded condition. The algebraic sum of deformations represented by areas 1 and 3 results in the residual elastic deformation (residual stresses) shown in Figure 3.10d.

Take note: In layers more distant from the zero line, the deformations have changed their algebraic sign. For instance, in the loaded state (Figure 3.10b), the upper layers are tensioned, and in the unloaded state (Figure 3.10d) these layers are compressed. Imagine now that the unloaded beam with the residual stresses is loaded gradually by a positively directed bending moment. In that case, as the moment increases, the compressive stresses in the layers most distant from the zero line will at first decrease up to zero. Then they will change their sign (revert to tensile stresses) and increase from zero to a positive value. If the final magnitude of the bending moment is M (as in Figure 3.10a), then the final stress pattern may be obtained by superposition of the residual stresses (Figure 3.10d) and the stresses caused by bending moment M (line a-b taken from Figure 3.10a). This results in broken line k-l-m-n, (Figure 3.10e).

As we see, in the considered case, the residual stresses subtract from the working stresses, so a considerable decrease of working stresses is achieved. It is important to emphasize that this effect may only be obtained if the load direction when performing EPD and when working is the same. If the working load is directed opposite to EPD, the effect will be negative (i.e., detrimental).

EPD may change the shape of the part; therefore, it is usually used for torsionally loaded round bars, which don't change their shape, and for springs (leaf and spiral as well), which are not required to have an exact shape. These parts are loaded so as to create plastic deformation in the outer layers of the cross section (Figure 3.12a). In the unloaded part remain the residual stresses shown in Figure 3.12b. In the surface layer, the sign of the stresses is opposite to that in the loaded state. Under load, these stresses subtract from the working stresses (Figure 3.12c), and then the load capacity of the part is increased.

The internal layers of the cross section are now more stressed than the surface layer, but the material inside is stronger than on the surface, because it doesn't have microscopic stress raisers and doesn't contact with the aggressive components of the environment (oxygen of the air, active

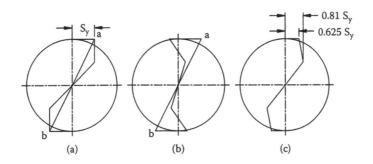

FIGURE 3.12 Torsional deformation of a round beam in elastoplastic range.

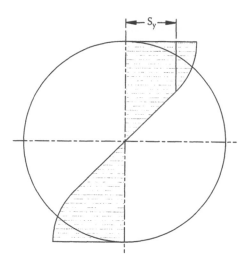

FIGURE 3.13 Torsional deformation of a round beam that yields according to Figure 3.9b.

additives to the lubricants, and others). So the surface layers remain the potential trouble spot, but they become less stressed.

Not all materials have a "plateau" (line 2 in Figure 3.9a) in their strain–stress diagrams. Figure 3.9b shows the strain–stress diagram typical for most of the alloyed steels. It doesn't change the principle of EPD, but the stress distribution is different (Figure 3.13). The residual stress pattern should be determined relying on the real strain–stress diagram of the material used.

It is important to note that the plastic deformation is not a momentary process; it takes time to change the structure of the material. Therefore, when carrying out EPD, the part is usually held in a deformed state for some time, or the deformation is repeated several times.

3.4 SURFACE PLASTIC DEFORMATION (SPD)

As we have seen previously, useful EPD may be done only if the working load is of constant direction relative to the part. At that, the direction of the load vector relative to other objects, such as the housing, the ground, and the like, should not be considered. For example, a radial force applied to a shaft from gear teeth is of constant direction relative to the housing, but the shaft is rotating, so the vector of the bending moment rotates with reference to the shaft. As this takes place, each point of the shaft is loaded alternately in tension and in compression, and the EPD by bending in any direction can't help but make the shaft crooked. In this case, it is useful to induce residual compressive stresses in the surface layer of the part — throughout or in the high-stressed portion of it.

Just a minute! A careful reader might say. If there is a residual compressive stress in the surface layer and the working stress changes cyclically from tension to compression, the stresses will alternately subtract and add. So the maximum tension stress will be less but the maximum compressive stress will be greater than without any residual stresses.

Yes, that is what is going to happen. But, what should be taken into account here is that the strength of steels is identical in compression and in tension, but only under a constant load. Under cyclic load, when materials fail by fatigue, it is not so.

Investigators made fatigue tests of T-shaped beams (Figure 3.14) bent by pulsating moment M. The direction of the bending moment was chosen so as to have the compressive stresses greater than the tension stresses. The needed relation between these stresses was achieved by controlling the width of flange W. (The wider the flange, the lower is the position of the neutral axis and the

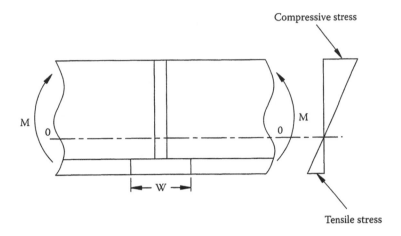

FIGURE 3.14 Bending stresses in a T-beam.

less the tension stress as compared to the compressive stress.) The fatigue cracks that appeared were mostly on the tensioned side. On the compressed side, cracks were obtained only when the compressive stress was about twice the magnitude of the tensile stress. Consequently, the fatigue strength of steels may be increased by decreasing the tensile stresses, even at the cost of an increase of the compressive stresses. But if the total compressive stress exceeds the yield point of the material, the residual stress decreases. The mechanism of residual stress loss is discussed in the following text.

The two methods most commonly used for SPD are burnishing (with balls or rollers) and shot peening. When burnishing, the surface layer is rolled out like dough on a table. To let the dough expand freely, the cook powders the table with flour. But in our case, the "table" (the layers underneath the plastically deformed layer) is bonded with the "dough" and resists its expansion. The interaction between them creates some balanced system of internal stresses: compressive in the surface layer and tension underneath it (Figure 3.15). In addition to the compressive residual stress, the plastically deformed layer is somewhat stronger (see Figure 3.9). These two factors enable increase in the load capacity of the parts.

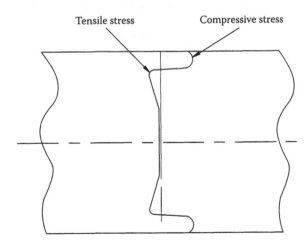

FIGURE 3.15 Residual stress distribution after shot peening or cold rolling.

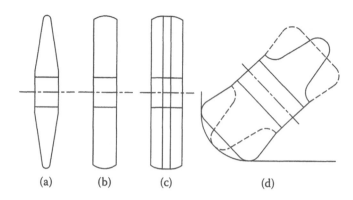

(a)   (b)   (c)   (d)

FIGURE 3.16 Shapes of typical rollers for burnishing shafts.

The burnishing roller should be pressed against the surface by a force that depends on the strength of the material, the diameters of the part and the roller, and the desired thickness of the deformed layer. The larger the diameter of the part, the thicker (as a rule) the plastically deformed layer should be (millimeters or even centimeters) to achieve a considerable increase in strength.

The shape of the roller also depends on the shape of the burnished surface. Usually, the working surface of the roller is shaped to a circular form (Figure 3.16a and Figure 3.16b), sometimes with a cylindrical portion (Figure 3.16c). To burnish a fillet, the radius of the roller profile should be less than the fillet radius. Figure 3.16d shows a special profile roller for burnishing fillets[2]: when the tool rotates with the shaft, the changing profile of the roller surface causes the contact points to alternately sweep from the center of the fillet to the sides, and back to the center. In this way, the contact spots of the fillet with the roller cover the entire surface of the fillet.

The same goal (creation of plastically deformed layer with residual compressive stress) is achieved by shot peening. A stream of hard spherical shots strikes the surface of the part at high velocity. Each shot dents the surface and produces a small plastic deformation. Thousands of shots create a plastically deformed layer. Its thickness doesn't exceed 0.4 – 0.5 mm. After shot peening, the surface has a specific "pock-marked" appearance. The geometry of the shot-peened surfaces changes slightly, requiring those surfaces that have to be exact and smooth (such as shaft necks for bearings) to have subsequent fine machining, usually grinding. This process may be harmful because heating is inevitable while grinding and may relax the residual compressive stress and even change it to tension stress. The result of the machining can't be controlled by nondestructive methods, and this reduces the reliability of this process.

Shot peening is successfully used for parts of complicated shape that can't be burnished and don't have to have an exact surface, such as springs, engine connecting rods, etc.

Burnishing allows creating a much thicker plastically deformed layer than shot peening, and, in case subsequent grinding is needed, the local overheating is much less dangerous. Besides, the surface after burnishing is very smooth, so that for some surfaces (for example, threads) SPD via burnishing may eliminate the need for further processing. And what is more, small and midsized threads are usually formed by rolling, without any cutting tools.

Finally, let's discuss the issue of the possible loss of residual stress.

Figure 3.17a shows a very simplified stress distribution in the initial state: the part is unloaded, and in the shot-peened layer, there is a residual compressive stress σ_r that equals approximately $0.5S_y$ (S_y is the yield point). Under this layer, there are tension stresses of relatively small magnitude. Dashed line 1 represents the stress pattern caused by bending the part with moment M. The sum of the two patterns shown in Figure 3.17b represents the stress distribution in the loaded part. In the lower portion of the part, the compressive residual stress σ_r subtracts from the tension bending stress σ_b, and this results in a sharp decrease of the stress in this layer. In the upper portion

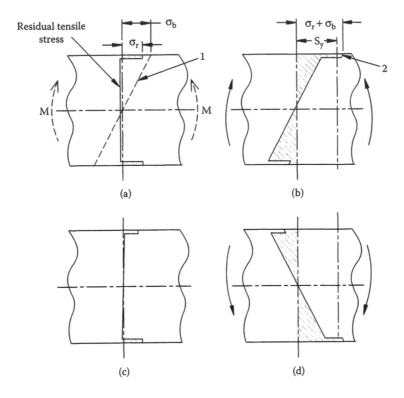

FIGURE 3.17 Loss of residual stress due to reversed bending.

of the part, the compressive residual stress σ_r and the compressive bending stress σ_b are additive, and their sum exceeds S_y. But in the material, the yield stress can't be exceeded at this magnitude of strain, and part of this sum (marked 2) is cut off. After the part is unloaded, the residual stress in the upper portion becomes less by exactly the cut off piece 2 (Figure 3.17c). In the lower portion, the residual stress remains the same until the part is bent in the opposite direction (Figure 3.17d). Then, the residual stress will be lost in the lower portion as well.

If the sum $\sigma_r + \sigma_b$ doesn't exceed the yield stress, the residual compressive stress in the plastically deformed layer remains unchanged.

REFERENCES

1. Verhovsky, A.V., Broken cross-section hypothesis and its application to calculation of bars of complicated shape, *Transactions of Tomsk Polytechnic Institute on the Name of S. M. Kirov*, Vol. 61/1, Tomsk, 1947 (in Russian).
2. Almen, J.O. and Black, P.H., *Residual Stresses and Fatigue in Metals*, McGraw-Hill, New York, San Francisco, Toronto, London.
3. Peterson, R. E., Stress Concentration Factors, John Wiley & Sons, New York, London, Sydney, Toronto, 1974.

Part II

Elements and Units of Machines

When designing a new machine, we usually start by looking for a patent and other information about what has been done in this field. It is a rare case when we need to make something completely new. Generally, we can often find a machine in service that closely meets our needs as evidenced by its function (and sometimes also by parameters such as capacity, output etc.). Such a machine is called a *prototype*.

The only documentation available for our prototype is often just a data sheet or some servicing instructions, and it may be difficult to obtain. So it definitely would be advantageous to go to the prototype and study it in service. In that way we begin to obtain more of an idea of the similar machine that we are to create.

What next? Let us suppose that we like the prototype, and it is highly desirable (in order to save time and money) to follow its design as closely as possible. We have only some visual information, and we will have to fill in the missing information using our own abilities. In other words, to understand, interpret and replicate the design of the prototype in every detail depends completely on our knowledge, our experience, and the intuition (called "mechanical aptitude") that we have developed from the experience.

Unfortunately, even if we know something, we may forget about it; also we can miss it when considering the numerous factors and options of the design. If it isn't at the forefront of our mind, there is little chance to think of it at the right time. Therefore, there is no substitute for studying typical design features of units and details beforehand, for future use. If practiced frequently, this should bring the intimate knowledge that becomes part of an instinctive feeling for the right solution. This is what Part II is all about.

4 Shafts

4.1 SELECTING THE BASIC SHAFT SIZE

A shaft is a straight, mostly round bar, which is designed to have other rotating parts of a mechanism mounted on it. To function as such, the shaft may include splines and keyways to transmit torque, journals for rolling or sliding bearings, cylindrical and conical sections, threads, shoulders, and so on. Some elements that are commonly made as separate items (e.g., pinions, cams, connecting flanges) are often made instead as integral parts of the shaft.

Sometimes a shaft is not a part of a mechanism but serves as a connection between mechanisms or machines placed at a distance from each other. Such a shaft, called *intermediate*, must be provided with elements transmitting torque on its ends; these are, as a rule, flanges, or splines, or necks for couplings.

Although the shape of a shaft depends on the kind and features of the elements adjoined to or mounted on it, the size (or, rather, the diameter) of the shaft is mainly dictated by strength considerations. Not only the strength of the shaft's body but also the strength of the connections transmitting torque and radial forces, necessary diameters of bearing journals, etc., should also be considered. All these elements should be designed as component parts of a unit. Then, the elements of the shaft should be proportioned so as to avoid large, and/or abrupt differences (changes) in diameter. This may be achieved by selecting a suitable material and heat treatment, changing relations between diameter and length of connections and journal bearings, varying types and dimensions of rolling bearings, changing types of connections (e.g., spline, key, or flange), and by other means. All in all, the process of designing a shaft is an integral part of designing the whole mechanism.

In this iterative process, it is important to make an initial appraisal of the shaft diameter as limited by its strength. The diameter may vary over a wide range depending on environmental conditions, materials used, heat treatment, surface treatment, thoroughness of smoothing out the stress concentrators, and other factors. Tentatively, for the beginning of design, the diameter of a shaft loaded with a torque and bending moments, and with stress raisers such as shoulders, key ways and the like can be estimated from:

$$d \approx (5 \ to \ 6) \sqrt[3]{\frac{T}{S_u}} \tag{4.1}$$

where
   T = transmitted torque (N·mm)
   S_u = tensile strength of the shaft material (MPa)

The diameter of a shaft that is loaded with torque only and doesn't have considerable stress raisers may be as small as one half of the above:

$$d \approx (2.5 \ to \ 3) \sqrt[3]{\frac{T}{S_u}} \tag{4.2}$$

EXAMPLE 4.1

A shaft transmits a torque of 15,000 N·m. In addition, it is loaded by bending moments generated by gears and couplings. The tensile strength of the shaft material S_u = 930 MPa. The shaft has stress raisers such as press-fit connections, keyways, and shoulders. Limited by fatigue strength, the diameter of the shaft should be approximately equal to

$$d \approx (5\ to\ 6) \sqrt[3]{\frac{15000 \cdot 10^3}{930}} = 125\ to\ 150\ mm$$

It should be clear that this approximate estimation can't replace the accepted strength calculation, and the shaft must finally be checked for strength. In some cases, the equations cited earlier are inapplicable, for example, when the torque is about zero (see Figure 6.3 and Figure 6.24). The diameter of each axle in these figures is determined by the inner diameter of the planet wheel bearing and by the bending strength of the axle, which is loaded by the gear teeth forces.

Although decreasing the dimensions of shafts is essential in the effort to limit weight, designers usually don't use the strongest material for the first design of a new mechanism, and there are three reasons for that. First, such materials are noticeably more expensive and more difficult in machining, and that raises the price. Second, unanticipated loads may be present, which result in breakdowns, forcing a redesign with an increase in shaft strength as the objective. Third, sometimes the customer wants to use the mechanism for greater loads and asks the manufacturer to increase its load rating. In these occurrences, the increase of the shaft diameter is usually undesirable, because it results in changing the mating parts, often including the housing. This increases by far the upgrade expenses. If alternatives are available, such as using some stronger material, application of surface hardening (carburizing or nitriding) and cold working, they will save a great deal of expense and effort in these situations.

Sometimes there is no alternative but to increase the shaft diameter to satisfy rigidity or stability requirements, mount larger bearings, or increase the load capacity of connections between the shaft and the attached parts.

EXAMPLE 4.2

Figure 4.20 depicts the main shaft of a subway tunnel escalator. Forces F_{cu} and F_{cl}, estimated in dozens of tons, load each of the two hauling sprockets. The diameters of the bearing necks of the shaft have been defined by the bearing's inner diameter. The diameter of the shaft between the sprockets, as calculated in compliance with the strength requirements, shall be of about 250 mm. (See dashed lines in Figure 4.20.) But in this case, the elastic bending of the shaft will result in inadmissible tooth misalignment in the reduction gear. To avoid this defect, the shaft diameter has been increased to 450 mm. Calculation of the resilience of this shaft is cited in Section 4.4.

Shafts, which are increased in diameter for the sake of rigidity, can be made of less strong (and consequently cheaper) materials. But even in these cases, the use of soft steels with hardness of HB 140–160 is not a good choice. While assembling tight or press fits, the mounting surfaces may be damaged (i.e., scored or crushed), and the shaft may be bent. An accidental hit with a hard object may leave a remarkable dent on such a shaft. In service, a short-term overload may cause turning of the inner race of a rolling bearing relative to the shaft neck. The soft material of the shaft will be worn away a bit, the bearing fit will become weaker, and the bearing unit will be unable to work (see Chapter 6, Section 6.2).

If the soft shaft is made as an integral part with a flange, the latter may be damaged while tightening bolts (see Chapter 10, Section 10.1). For these and other reasons as well, it is desirable that the shaft be sufficiently hard.

EXAMPLE **4.3**

An intermediate steel shaft of 2000 mm length should transmit a constant torque of 2000 Nm without considerable bending loads. Rotational speed is 1000 r/min. The tensile strength of the shaft material S_u = 930 MPa. From Equation 4.2, the shaft diameter should be

$$d \approx (2.5 \; to \; 3) \sqrt[3]{\frac{2000 \cdot 10^3}{930}} = 32 \; to \; 39 \; mm$$

Taking the shaft diameter value equal to 35 mm, let's define the natural frequency of its bending vibrations by the following equation[1]:

$$f_n = \frac{\pi n^2}{2L^2} \sqrt{\frac{EIg}{A\gamma}} \; Hz \qquad (4.3)$$

where
    n = number of the mode shape (n = 1, 2, 3, and so on)
    L = shaft length between bearings (mm)
    E = modulus of elasticity (MPa)
    g = 9.81·10³ mm/sec² = acceleration of gravity
    A = shaft cross-sectional area (mm²)
    γ = specific weight (density) of the shaft material (N/mm³)
    I = moment of inertia of the cross section (mm⁴)

For a solid round shaft made of steel,

$$E = 2.06 \cdot 10^5 \; MPa; \quad I = \frac{\pi d^4}{64} \; mm^4$$

$$A = \frac{\pi d^2}{4} \; mm^2; \quad \gamma = 7.7 \cdot 10^{-5} \; N/mm^3$$

Substitution of this data in Equation 4.3 gives us the fundamental frequency that is valid for all solid steel shafts at n = 1:

$$f_1 = 2.01 \cdot 10^6 \frac{d}{L^2} \; Hz \qquad (4.4)$$

In our case, d = 35 mm, L = 2000 mm,

$$f_1 = 2.01 \cdot 10^6 \frac{35}{2000^2} = 17.6 \; Hz$$

The rotational speed is 1000/60 = 16.7 revolutions per second, which is very close to the fundamental frequency f_1. This may cause resonance vibrations, and the shaft stiffness should be increased so as to make the f_1 value about 1.4 to 1.5 times more than the rotational speed. Inasmuch as the f_1 value of a solid shaft is in direct proportion to the shaft diameter (see Equation 4.4), the diameter should be multiplied by 1.4 to 1.5, i.e., to 50 mm roundly. At that diameter, the weight of the shaft doubles, from 15 kg to 30 kg. If the weight is limited, the shaft should be made hollow, for example, diameters outside (d_o) and inside (d_i) may be chosen as d_o/d_i = 40/30 mm or 45/38 mm. (These diameters are calculated so as to keep the same section modulus as for the solid shaft of 35 mm in diameter.) Substituting the data of these sections in Equation 4.3 gives us the following values of fundamental frequency:

For section 40/30 mm, f_1 = 25.1 Hz
For section 45/38 mm, f_1 = 29.6 Hz

Thus, the fundamental frequencies of the hollow shafts are greater than the rotational speed by a factor of 1.55 and 1.77. An additional benefit is that the weight is less (than that of the solid shaft of 35 mm diameter) by a factor of 1.7 and 2.0, respectively.

Striving for further decrease in weight may lead to a thin-walled shaft. For instance, at diameters of d_o/d_i = 75/73 mm (wall thickness is of 1 mm), the weight of the hollow shaft is only 25% of that of the solid one. But the ratio of the wall thickness t to the mean radius r of the wall [$r = (d_o + d_i)/4$] is now r / t = 37. Because this ratio is bigger than 10, the shaft should be checked for buckling. In our case, the shaft is loaded with a torque only, so the critical (for stability in torsion) shear stress may be obtained from the following equation[2]:

$$\tau_{cr} = 0.25E\left(\frac{t}{r}\right)^{1.5}$$

As mentioned, r = 37 mm, t = 1 mm, so the critical stress is

$$\tau_{cr} = 0.25 \cdot 2.06 \cdot 10^5 \left(\frac{1}{37}\right)^{1.5} = 229 \ MPa$$

The actual value of the shear stress

$$\tau_{act} = \frac{T}{2\pi r^2 t} = \frac{2000 \cdot 10^3}{2\pi \cdot 37^2 \cdot 1} = 233 \ MPa$$

As we see, the actual shear stress is bigger than the critical one. That means that the hollow shaft with diameters 75/73 mm may buckle under load, and its dimensions should be changed so as to keep the actual shear stress not bigger than about half of its critical value.

Desirable changes in weight, stiffness, and the natural frequencies of a shaft may also be attained by changing the material properties, for example, by using titanium or aluminum alloys instead of steel. The permissibility of this replacement should be considered with respect to all elements of the shaft, such as splines, bearing necks, and others.

4.2 ELEMENTS OF SHAFT DESIGN

Figure 4.1 shows a shaft with other parts mounted on it: a gear wheel, two bearings, a sleeve, a nut, and a half-coupling. This shaft has several sections of differing diameters. The stepped change of the shaft diameter from 150 mm to 140 mm provides shoulders, which represent location surfaces necessary for the proper installation of the gear wheel and roller bearing. The change in diameter from 140 mm to 135 mm is required because of the difference in fit tolerances for the roller bearing and the half-coupling. The latter should have a press-fit connection with the shaft, whereas the roller bearing needs a much lighter fit. If the nominal diameters of these two connections were the same,

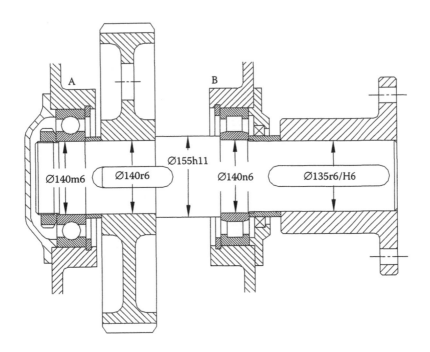

FIGURE 4.1 Gear output shaft.

the bearing would have been pushed (during assembly) through and beyond the half-coupling seat, and the increased interference while passing through this place would cause assembly problems.

In Figure 4.1, the diameter of the half-coupling seat is decreased to the nearest standard value (135 mm). But if a strength problem is encountered when using a round dimension and standard fit, the half-coupling seat can be made of the same diameter with the same tolerances as the bearing seat. The needed interference for the half-coupling fit should be achieved in that case by selecting an appropriate tolerance for the opening (bore) of the half-coupling.

Places where the section of the shaft is changing are also the places of stress concentration (see details in Chapter 3), and the sharper (more abrupt) the change in cross section, the stronger the effect of stress concentration. A smoother change of shaft sections is achieved by setting sufficiently large fillet radii.

Although the influence of fillet radii on stress concentration factors is well known, the authors in their practice have often seen drawings of shafts in which the fillet radii were dimensioned too small or even were not specified at all. When a designer is dimensioning a shoulder for a rolling bearing (Figure 4.2a), he usually takes the fillet radius from the bearing's catalog prescriptions.

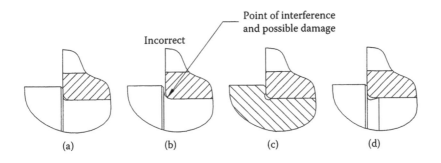

FIGURE 4.2 Various shaft shoulders and fillets.

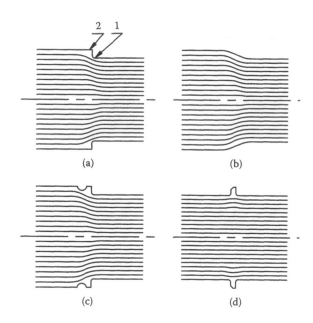

FIGURE 4.3 Lines of forces in shouldered shafts.

The designer keeps in mind that if the fillet radius is bigger than the chamfer on the inner race of the bearing (Figure 4.2b), the bearing will be mounted incorrectly, possibly with a skewness about the shaft axis. (Such a condition could lead to more than just misalignment or vibration, for the bearing may actually cut into the fillet, and cause premature shaft failure.) As far as the stress raisers are concerned, there is sometimes a lack of understanding of the mechanism of stress concentration, which is obviously more difficult for comprehension than the conjugation between a chamfer and a fillet. As a matter of fact, the shaft section becomes bigger, but the stress, for some reason, doesn't diminish but increases! In Chapter 3, the mechanism of the local stress increase is clearly demonstrated, using the broken section hypothesis, which describes the nature of the strain distribution in places where the part shape is changing. A convenient way of visualizing the stress concentration is the line of forces method (Figure 4.3a). It is assumed that the part is tensioned, and the field of forces in the body of the part is depicted as electric current lines or as fluid flow streamlines. Where the section of the part is constant, the current lines are evenly distributed over the cross section. Where the cross section is changed, the current lines, which aim at straightening, approach reentering corner 1, and the density of the lines near this corner increases. At the same time, they move from salient corner 2. Therefore, the stress increases in zone 1, and zone 2 proves to be less loaded.

To lessen the stress concentration, the shape of the part should be as smooth as possible. Figure 4.3b depicts a shaft with a gradual change from smaller to greater diameter. In this case, the stress concentration is noticeably less, but such an option is unacceptable if there is a need for a shoulder. If this is the case, there are two possible ways to better the existing conditions: machine an annular groove near the shoulder (Figure 4.3c) or lessen the length of the shoulder (Figure 4.3d). By these means the maximal stress (in corner 1) may be decreased by 5–10%. The most effective way is to increase the fillet radius as far as possible in the specific design. To achieve this, a removable shoulder which enables a considerable increase in the fillet radius may be used (Figure 4.4b), and an even more favorable form is the ellipse-shaped fillet shown in Figure 4.4c.

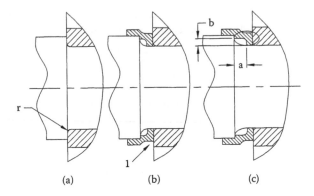

FIGURE 4.4 Removable shaft shoulders.

Example 4.4

Figure 4.5 presents the relation between the stress concentration factor K_t (in bending) and the relative value of the fillet radius (r/d) for $D/d = 1.2$.[3] Let's assume that $d = 100$ mm and $D = 120$ mm. For ball bearing 6220, the fillet radius r shown in Figure 4.4a should not exceed 2 mm. Because in that case $r/d = 0.02$, it follows from Figure 4.5 that $K_t = 2.63$. If the fillet radius is

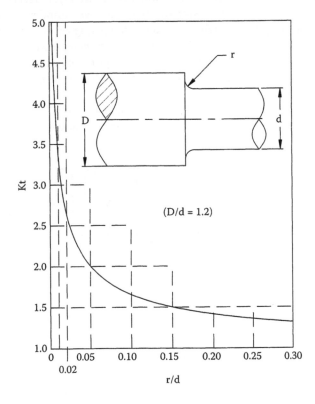

FIGURE 4.5 Stress concentration factor for a shaft in bending with a shoulder and fillet.

increased (using a removable shoulder, Figure 4.4b) to 8 mm, then $r/d = 0.08$, and $K_t = 1.75$. If an elliptical fillet is used (Figure 4.4c), the maximum fillet radius will be

$$r_{max} = \frac{a^2}{b}$$

where a and b are the major and the minor semiaxes of the ellipse (see Figure 4.4c). Keeping the fillet dimension unchanged in radial direction ($b = 8$ mm) and increasing it in axial direction (say, $a = 15$ mm), the max fillet radius may be obtained as big as

$$r_{max} = \frac{15^2}{8} = 28.1 \, mm$$

In this case, $r_{max}/d = 0.28$, and $K_t = 1.32$. That means that the strength of the shaft with elliptical fillet is twice as much as that with the fillet $r = 2$ mm.

When designing fillets, technological subtleties should be considered as well. For example, if the fillet is designed as shown in Figure 4.2a, the grinding of the neck and shoulder must be performed with a grinding wheel having a radius-shaped edge. When the wheel can't be shaped appropriately, the shape of the fillet should be changed to enable grinding of the neck and shoulder (Figure 4.2c) or of the neck only (Figure 4.2d). The design shown in Figure 4.2c is especially advisable for surface hardened shafts (carburized or nitrided), for it enables finishing (grinding) of the exact surfaces without touching the fillet.

Grinding is harmful for the strength of the hardened surface layer, because it takes off the part of the layer with largest residual compressive stresses. Moreover, grinding is accompanied by heat generation, which may bring about high-temperature tempering of the hardened layer. This may result in a changing of the sign of the residual stresses (from compressive to tension), and even in cracking of the surface layer. Therefore, it is advisable to grind only surfaces that must have exact geometry, and not to touch the others. In addition, surface-hardened parts should be checked for burns and cracks after grinding.

In general, shape modifications should be as smooth as possible. Edges of radial holes should be rounded off and polished (Figure 4.6a). If needed, the section with holes can be reinforced as

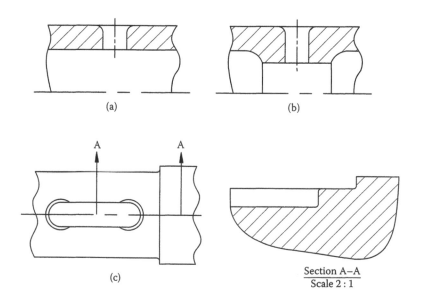

FIGURE 4.6 Strengthening of stress concentration areas.

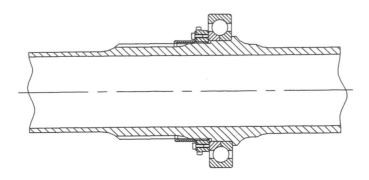

FIGURE 4.7 Hollow shaft with a ball bearing, locknut, and spline.

shown in Figure 4.6b. Rounding off the interior corners and sharp edges on the ends of a keyway (Figure 4.6c) noticeably decreases the stress concentration (on large keyways where the added cost may be justified, an elliptical form called *spooning* may be employed). When the safety factor is low, it is undesirable to combine two or more stress raisers in one place. For example, the end of a keyway should be moved away from the shoulder of the shaft (Figure 4.6c).

In Figure 4.1, the gear wheel hub is too short; to increase the working length of the key, the keyway is extended in both directions. It is supposed that the strength of the shaft is satisfactory. As an alternative solution, the length of the hub may be increased. If the strength of the shaft is crucial, it is better to shorten the keyway and not bring its end up to the shaft shoulder.

Sharp stress raisers, such as threads and annular grooves for snap rings, should be avoided on heavily loaded sections of shafts. (They may be used on the ends, where the stresses are almost zero; see Figure 4.1.) If the application of sharp stress raisers in the middle of the shaft is unavoidable, the shaft should be locally reinforced to compensate for the loss of strength caused by the stress raisers. An example of such a shaft taken from aircraft engineering is shown in Figure 4.7. Examples of reinforcement of shaft sections having stress raisers in the form of press-fit connections and pins are also shown in Figure 4.8c, Figure 4.8d, and Figure 4.9b.

It is clear that heavily loaded shafts made of high-strength materials must in particular be thoroughly designed with respect to making the stress raisers smoother. It is important to keep in mind, however, that fatigue failure is of a statistical nature. This means that the probability of failure is never zero, even when the safety factor is more than sufficient, and a sharp stress raiser increases this probability. Don't have doubts if you can make it smoother! It is cheaper to do that when designing than later after the shaft has been broken in service.

A special emphasis should be placed on the decrease in fatigue strength caused by press fits. This kind of stress raiser is not connected with any change in the shaft's shape, but it may reduce the fatigue strength even to 25–30% of that of the plain shaft. This effect is caused both by additional stresses where the hub and shaft surfaces make contact (see also Chapter 2, Section 2.1) and by fretting damage to the shaft surface layer. The formation of fretting is associated with the relative motion of the contacting surfaces under the influence of bending moments and torque. A mechanism in which this motion appears has been described in Chapter 2. The following design and technological means are commonly employed to increase the durability of shafts in press-fit connections:

1. Decreased contact pressure peaks on the connection ends is attained by forming thin lips 1 on the hub (compare Figure 4.8a and Figure 4.8b). The area of maximum pressure has moved from the connection end inwards, where the relative motion of the parts becomes lesser. The recommended thickness of the lips $t = (0.04 - 0.05)D$, and their length $L = 2t$.[4] With these dimensions, the fatigue limit of a shaft can be increased by 20–30%.

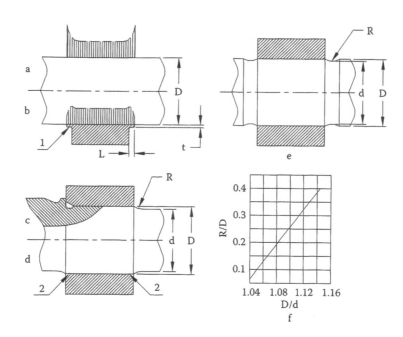

FIGURE 4.8 Some methods of decreasing the detrimental effect of a press fit on the fatigue strength of a shaft.

2. Decreased operating stresses in the area where the contact pressure and the relative movement of the surfaces are at a maximum may be attained by forming a salient angle in this area. This effect can be obtained by increasing the press-fit diameter a little (Figure 4.8c and Figure 4.8d) or by diminishing the shaft diameter near the press-fit connection ends by grooving (Figure 4.8e). In the last case, there is a risk of failure through the groove, but the stress concentration factor for the groove can be much less, and it can be estimated quite precisely. Furthermore, the groove can be cold-worked

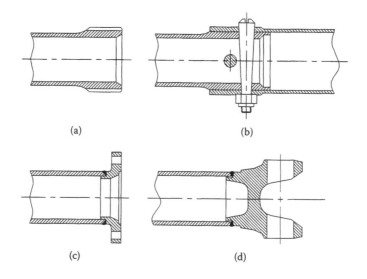

FIGURE 4.9 Design of hollow shafts.

(roll-burnished) to exclude the possibility of fatigue cracking through it. The recommended relation of the diameters is $D/d = 1.05$–1.10 and by this means the fatigue limit of the shaft in the press-fit connection may be increased by 25–50%.

The larger the quantity D/d and the smaller the fillet radius r, the greater is the positive effect of decreasing the operating stresses in area 2 (Figure 4.8c and Figure 4.8d) and likewise, the greater the negative effect of increasing stresses in the fillet. Experiments have shown[4] that the optimal result is achieved when the fillet radius is chosen according to the diagram in Figure 4.8f.

3. Producing residual compressive stresses in the surface layer of the shaft using shot peening, roll burnishing, or laser treatment. By these means, the fatigue limit of a shaft in a press-fit connection may be doubled. It is to be noted that these methods don't prevent formation of fretting and microcracks on the shaft surface inside the connection, but the residual compressive stresses block the spreading of the cracks into the shaft's body.
4. Surface hardening using induction hardening, carburization, carbonitriding, or nitriding. These surface treatments not only induce high compressive stresses into the surface layer but also increase by far its mechanical properties, reducing the formation of fretting and microcracks. The fatigue strength of the shaft in the area of the press-fit connection may be increased three- to fourfold, approaching the value appropriate to a single shaft (without press-fitted parts).

The aforementioned methods can't always be used. Surface hardening may be impossible because of large shaft dimensions, or if the shaft material is not suitable for this kind of treatment. Local increase in diameter may be unacceptable for some reasons as well. What is almost always accessible and useful is cold working (surface plastic deformation) — on condition that the shaft is not dedicated for a high-temperature application. But the influence of this treatment on the surface geometry and roughness should be considered. Final grinding, if needed, should be very fine to keep most of the residual compressive stresses, which are initially (before grinding) about 50% of the yield stress of the material. (See Chapter 3, Section 3.4.)

Shot peening creates a hardened layer not more than 0.4–0.5 mm thick. It is effective enough for relatively thin sections without final grinding. For larger parts, roll burnishing is preferable, because the thickness of the hardened layer may be as deep as needed. Roll burnishing in combination with decreasing the contact pressure peaks on the connection ends (see preceding text) may completely neutralize the harmful influence of the press-fit connection on shaft strength.

The only absolutely reliable way to prevent fretting in connections is elimination of connections. For example, if the gear wheel is made as an integral part of the shaft (Figure 8.3a), the problem of fretting will not occur. Furthermore, the weight of the parts will be considerably reduced. Such a part may be more expensive, but sometimes these expenses are considered as an acceptable payment for reduced weight and increased reliability.

4.3 HOLLOW SHAFTS

This design is widespread in lightweight machines, see examples in Figures 4.7 and Figure 4.9, as well as Figure 5.19 (Chapter 5), Figure 7.31 (Chapter 7), and Figure 8.3a-b (Chapter 8). Hollow shafts are made mostly from tubular stock (Figure 4.9a and Figure 4.9b), or just from a tube with welded ends intended for connections; for example, a flange (Figure 4.9c) or a fork (yoke) of a universal joint (Figure 4.9d). Welded shafts are much cheaper than forged, but their strength is lower for the following reasons: First, weldable steels have low to medium strength; high-strength steels have poor weldability. Second, the strength of the weld usually doesn't exceed 75–80% of the original metal strength.

In addition, quality control of the weld can't always be satisfactory because access to the inner side of the weld may be difficult or impossible. In lots of applications, in which the risk of failure

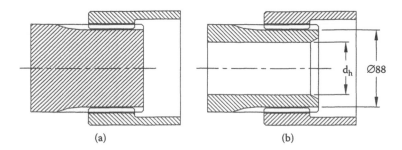

FIGURE 4.10 Splined shafts.

doesn't endanger people and can't result in heavy material losses, it is enough to occasionally check one piece by cutting. If high reliability is of great importance, the part should be so designed as to enable full quality control, even if it makes the part more complicated and expensive. When the strength and reliability are expected to be superior, the shaft should be made from a forged or rolled blank and controlled carefully for surface and internal defects.

One should also take into consideration that stress concentration factors are larger in hollow shafts than in solid shafts. For example, for the solid shaft taken from Example 4.4, Figure 4.4b (r/d = 0.08), the stress concentration factor in torsion is $K_t = 1.40$. If this shaft has a central hole of ⌀80 mm, then $K_t = 1.69$.[3] For a splined solid shaft in torsion (Figure 4.10a), $K_t = 1.79$. For this shaft with a central hole of $d_h = 58$ mm or $d_h = 68$ mm, $K_t = 1.87$ and $K_t = 1.89$, respectively. (These results were obtained using finite element method [FEM].)

4.4 SELECTION OF A LOADING LAYOUT FOR STRENGTH ANALYSIS

The first stage of a strength analysis is to work out a suitable loading layout, i.e., to define supporting conditions and places where the load is applied to the shaft (for example, by constructing a "free-body diagram"). Chapter 1 (on the example of a piston pin, Figure 1.6) showed that stress and deflections of the shaft might depend utterly on the deformations of the mating parts. In this special case, it would be too optimistic to schematize this complicated three-dimensional structure in the form of a simply supported beam and to expect a correct result. Sufficient accuracy may be obtained here only by using finite element analysis or by directly measuring the stresses.

Fortunately, the calculated stresses don't always depend so totally on the selected loading layout, but the difference may be considerable. Let's take, for example, the shaft depicted in Figure 4.11. It is loaded by gear force F, which develops torsion and bending of the shaft. Usually, the force F is considered to be applied in the middle of the gear face width (Figure 4.11a). Let's assume that the bending moment in this case is 100%. If we suppose that force F is transmitted to the shaft at the ends of the gear hub (Figure 4.11b) equally, the bending moment will be only 75%. If we place the whole force F on one end of the gear hub (Figure 4.11c), the bending moment will be 110%, but this case is completely unrealistic.

As we see, if force F is placed in the middle of the gear face width, the error is relatively small in the case being considered. But there are other less conservative cases, for example, if there is a loose fit, such as in Figure 4.12a. Here, force F should be applied to the shaft on the ends of the pinion. For the design in Figure 4.12b with the press fit, the shaft should be considered as fixed in the middle. Is it not peculiar that in both depicted cases, the maximal bending moment in the shaft is the same, a condition that may not be readily apparent at first glance? But the deformations of the shaft are different; in the case of press fit, the shaft is more rigid.

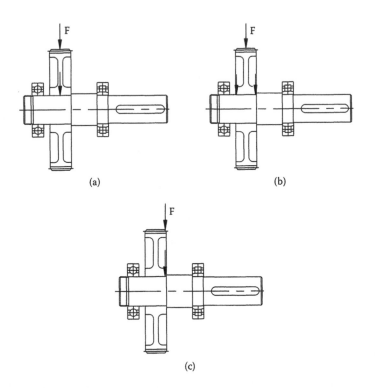

FIGURE 4.11 Shaft loading layout.

Now, to discuss the supporting conditions, the shaft in Figure 4.1 should be defined as a beam with one overhanging end, supported at two points. One (A) is a locating support (fixing the beam in the axial direction), and the other (B), a floating one. Both the ball bearing (A) and the roller bearing (B) should be considered spherical hinges. That means that the shaft in these supports may be skewed with respect to the supports, in the vertical and the horizontal planes.

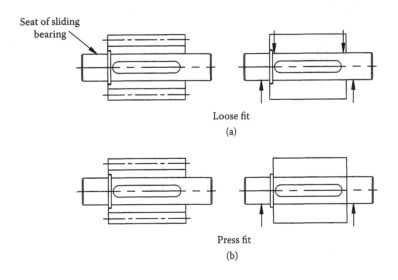

FIGURE 4.12 Gears of a gear pump.

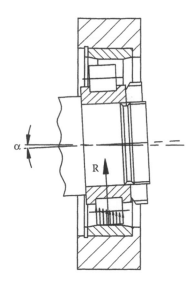

FIGURE 4.13 Tilt in bearings.

Some people think that a roller bearing (such as bearing B in Figure 4.1) hinders the shaft from skewing (inclining) relative to the housing. This assumption is erroneous. Figure 4.13 shows a roller bearing with the inner race skewed relative to the outer race by some angle α. It is clear from the picture that this skew is unimpeded, because in the unloaded part of the bearing (from above) there is a gap between the rollers and the races. True, the skewed position of the race results in an uneven load distribution along the rollers, and consequently, the bearing reaction R is displaced from the middle of the bearing towards its more loaded face. But the allowable skewness angle α is pretty small (about 0.001 rad), and the unevenness in load distribution as well as the reaction R displacement are negligible. Therefore, the reaction of a rolling bearing is usually placed in the middle of it.

In sliding bearings, the picture is somewhat different. These bearings should be considered as hinged supports too, because they don't hinder the shaft from skewing. Regarding the location of reaction R, it depends essentially on shaft rigidity. The elastic bending deformation of the shaft brings about a load concentration in a sliding bearing. One can find recommendations to place the R vector at a distance of $0.25L$ to $0.30L$ from the bearing's inner end (see Figure 4.14). This assumption decreases bending stresses in the shaft. But if the shaft must not be underdesigned, it is more conservative to place the vector R in the middle of the bearing's length.

When a support includes two bearings taking a radial load (Figure 4.15a), the load distribution between them should be considered, because it has an influence on the strength of the bearings to be selected, as well as on the bending stresses in the shaft. This problem becomes complicated owing to several parameters, which have some scattering region and can't be uniquely defined:

- Initial bearing clearances (in the radial direction)
- Amount of clearance or interference (i.e., the fit) in the connections of the bearing races with the shaft and housing
- Service temperatures of the shaft, the housing, and the bearings
- Alignment (coaxiality) of shaft journals
- Alignment (coaxiality) of the bearing seats in the housing

To diminish this uncertainty, the bearings to be installed together should be selected with respect to their actual diameters and clearances, or a double-row bearing should be used.

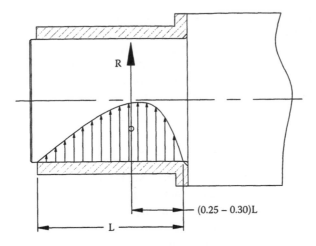

FIGURE 4.14 Position of the reaction force R in a slide bearing.

If the bearings and their fits are chosen in such a way that all the clearances in the bearings are taken up under small load, the load distribution between the bearings may be found by known means. To do this, the compliance of all the elements that take the load (shaft, housing, and bearings) must be taken into account. This analysis could be quite complicated and time consuming.

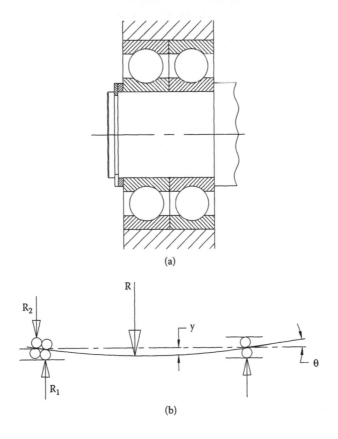

FIGURE 4.15 Load distribution between two bearings.

In view of these problems, the option of two radial bearings installed together has found practical use only in cramped places, where there is no place for one bearing of sufficient load capacity. Nevertheless, if this option is used, the shaft deflections should be checked to prevent jamming of the bearings, as shown in Figure 4.15b. Here, one of these two bearings (outer) is loaded by force R_2 in the opposite direction, so as the load R_1 of the other bearing is larger than it would be if there was only one bearing. In that way the second bearing may paradoxically decrease the total load capacity of the support.

4.5 ANALYSIS OF SHAFT DEFORMATIONS

Do you like calculations? If so, we envy you. The point is that most engineers like to draw all kinds of machines and to enjoy their pictures. Some of them avoid even the simplest calculation for strength and deformation, relying on their intuition and hoping for the best. What self-confidence!

Well, deformation analysis is not a luxury. Sometimes it takes quite some time, and we often tend to use our intuition and decide whether this work is a must or not. Occasionally, it is really obvious that deformations are negligible. For example, if a fly alights on a table, the latter will be bent, but we don't take this deflection into consideration when placing a plateful of soup on it! Our experience says that the real danger for the soup comes when the fly flies off the table. With this is mind, let's consider a shaft of 2 m length and of about 0.5 m in diameter. The first impression is that no power on earth can deform this iron log! No power on earth? Let's calculate the bending deformations of such a shaft.

Analysis of shaft deformations is time consuming. It is better to do this work using computer programs, particularly when the diameter of the shaft is variable. For our readers, who might not have such a program, the authors have decided to give formulas for calculation of deflection y and slope θ (shown in Figure 4.15b). These formulas are accompanied by graphs (Figure 4.16 to Figure 4.19) for estimation of slopes, which are usually important to know because they affect the alignment of mating parts and load distribution in their contact. Using these graphs, we are able to estimate very quickly the order of the slope magnitude and decide whether it is worthwhile to analyze the slopes precisely or not. In addition, the simplified estimation may serve as a reference point for the precise calculation and possibly prevent a mistake. For Figure 4.16 and Figure 4.17, the slopes are to be determined from the equation

$$\theta = K \frac{FL^2}{EI} \tag{4.5}$$

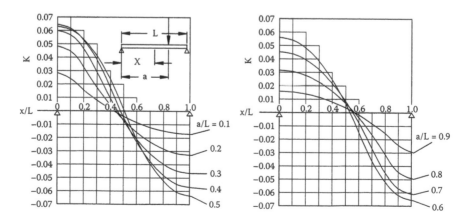

FIGURE 4.16 Angular deformation (slope) of shaft cross sections (radial force applied between the bearings).

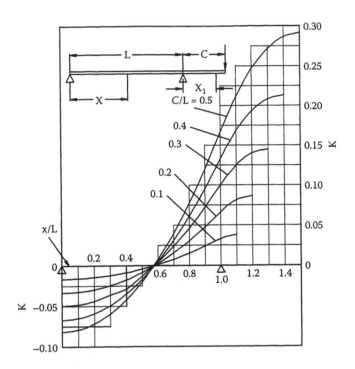

FIGURE 4.17 Angular deformation (slope) of shaft cross sections (overhung radial force).

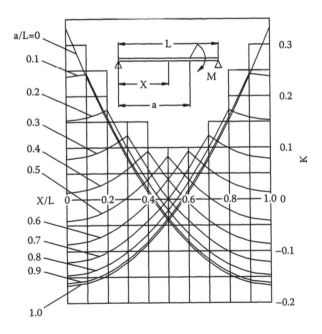

FIGURE 4.18 Angular deformation (slope) of shaft cross sections (bending moment applied between the bearings).

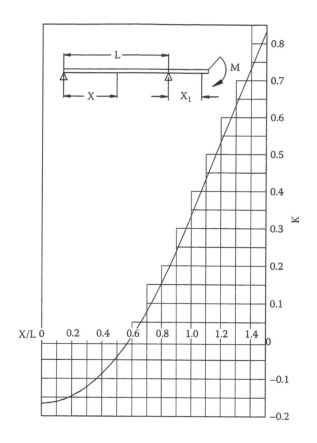

FIGURE 4.19 Angular deformation (slope) of shaft cross sections (overhung bending moment).

For Figure 4.18 and Figure 4.19,

$$\theta = K \frac{ML}{EI} \tag{4.6}$$

where
   F = radial force
   M = bending moment
   E = modulus of elasticity of the shaft material
   I = moment of inertia of the shaft section
   L = shaft length between bearings
   K = value taken from corresponding graphs

Dear reader, if you try to use all the quantities in compatible units (e.g., F in N (newton), M in N·mm, L in mm, E in N/mm² (MPa), and I in mm⁴), the slope angle θ will be obtained in radians. But, if the dimensions are incompatible, the result will have no physical meaning. (See Chapter 13, Subsection 13.4.2 about the choice of units.)

The exact formulas for deformations of shafts with uniform cross section are as follows.

1. A shaft loaded by a radial force F between bearings (see draft in Figure 4.16):

$$y_x = -\frac{Fbx}{6EIL}(L^2 - b^2 - x^2) \quad @\, x \leq a$$

$$y_x = -\frac{Fa}{6EIL}(L - x)(2Lx - x^2 - a^2) \quad @\, x > a$$

$$\theta_x = -\frac{Fb}{6EIL}(L^2 - b^2 - 3x^2) \quad @\, x \leq a$$

$$\theta_x = -\frac{Fa}{6EIL}(2L^2 - 6Lx + 3x^2 + a^2) \quad @\, x > a$$

where $b = L - a$.

2. A cantilever shaft loaded by a radial force F on the overhanging end (see draft in Figure 4.17):

$$y_x = \frac{Fcx}{6EIL}(L^2 - x^2) \quad @\, x \leq L$$

$$y_{x_1} = -\frac{Fx_1}{6EI}(2cL + 3cx_1 - x_1^2) \quad @\, x_1 \leq c$$

$$\theta_x = \frac{Fc}{6EIL}(L^2 - 3x^2) \quad @\, x \leq L$$

$$\theta_{x_1} = -\frac{F}{6EI}(2cL + 6cx_1 - 3x_1^2) \quad @\, x_1 \leq c$$

3. A shaft loaded by a bending moment M between bearings (see draft in Figure 4.18):

$$y_x = \frac{M}{6EI}\left[\left(6a - 3\frac{a^2}{L} - 2L\right)x - \frac{x^3}{L}\right] \quad @\, x \leq a$$

$$y_x = \frac{M}{6EI}\left[3a^2 + 3x^2 - \frac{x^3}{L} - \left(2L + 3\frac{a^2}{L}\right)x\right] \quad @\, x > a$$

$$\theta_x = \frac{M}{6EIL}(6aL - 3a^2 - 2L^2 - 3x^2) \quad @\, x \leq a$$

$$\theta_x = \frac{M}{6EI}\left(6x - 3\frac{x^2}{L} - 2L - 3\frac{a^2}{L}\right) \quad @\, x \geq a$$

4. A cantilever shaft loaded by bending moment M on the overhanging end (see draft in Figure 4.19):

$$y_x = \frac{Mx}{6EIL}(L^2 - x^2) \quad @\, x \le L$$

$$y_{x_1} = -\frac{Mx_1}{6EI}(2L + 3x_1) \quad @\, x_1 \le c$$

$$\theta_x = \frac{M}{6EIL}(L^2 - 3x^2) \quad @\, x \le L$$

$$\theta_{x_1} = -\frac{M}{6EI}(2L + 6x_1) \quad @\, x_1 \le c$$

You should not read at leisure this frightening amount of equations, as you would a Harry Potter Book. But if you really must check your shaft for rigidity, these equations and graphs may be helpful.

EXAMPLE 4.5

If you remember, our intention was to calculate the deformations of some large shaft. Here you are: Figure 4.20 shows the main shaft and part of the drive gear of a subway escalator, which is designed to transport passengers through a height of 65 m. (Dimensions of the shaft and the loading force vectors are given in Figure 4.20 and Figure 4.21a). Forces applied to the hauling chains are

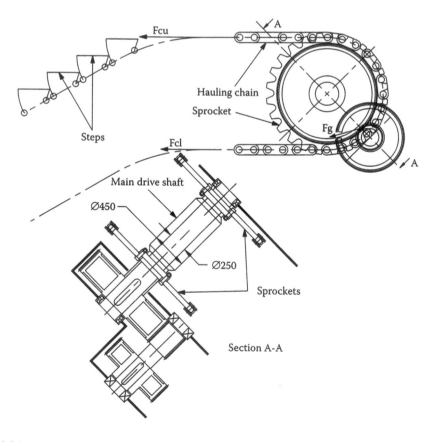

FIGURE 4.20 Escalator main drive.

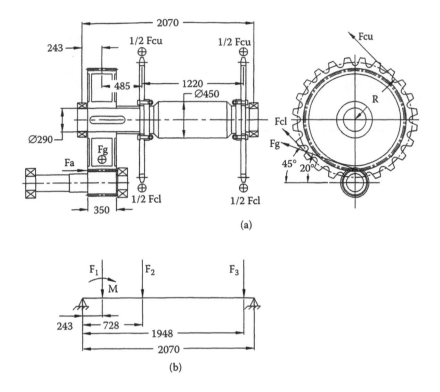

FIGURE 4.21 Escalator main drive shaft with a pinion.

created from below (F_{cl}) by the weight of the chains and steps, and from above (F_{cu}) by the same weight of the chains and steps plus the weight of the passengers. The forces loading the shaft are as follows:

Forces $F_{cu} = 30 \cdot 10^4$ N and $F_{cl} = 10 \cdot 10^4$ N act in the horizontal plane and are distributed equally between the two sprockets.
Gear force $F_g = 24 \cdot 10^4$ N acts in the plane of action of the gear.
Gear axial force $F_a = 5.1 \cdot 10^4$ N.

The directions of the forces are shown in Figure 4.21, where the picture is turned for convenience by 45° counterclockwise.

Of interest to us is the influence of the shaft resilience on gear teeth misalignment (skewness). For that reason, the shaft deformations should be calculated in the plane of action of the gear, and all the forces should be projected onto this plane (i.e., on the plane where force F_g acts). The angle between the forces F_{cl}, F_{cu}, and the gear action plane measures about 25°. The loading layout of the shaft in this plane is given in Figure 4.21b. Here are the forces:

$F_1 = F_g = 24 \cdot 10^4\,\text{N}$
$F_2 = F_3 = 0.5(F_{cl} + F_{cu})\cos 25° = 18.1 \cdot 10^4\,\text{N}$
$M = F_a \cdot 0.5d_g \cdot \sin 20° = 5.1 \cdot 10^4 \cdot 0.5 \cdot 1264 \cdot \sin 20° = 1.1 \cdot 10^7\,\text{N·mm}$

(Here d_g [the gear wheel pitch diameter] = 1264 mm.)

Let's estimate the slope angle of the shaft section in the middle of the gear face width, i.e., at a distance of 243 mm from the left bearing. This will be done using Figure 4.16 and Figure 4.18, as well as Equation 4.5 and Equation 4.6. But these figures and formulas are intended for shafts of uniform cross section. Therefore, we have to determine the uniform diameter of the equivalent shaft, which has nearly the same rigidity as the real one. The real shaft is made up mainly of two segments: diameter $d_1 = 290$ mm (with a total length $L_1 \approx 850$ mm) and diameter $d_2 = 450$ mm ($L_2 \approx 1220$ mm). The longer the segment, the greater its influence on shaft rigidity. For this reason, the short segments are ignored regardless of their diameter.

(Not only the length and diameter of the segment but also its position relative to the bending moment diagram is important; the bigger the bending moment that the segment is exposed to, the greater its contribution to the resilience of the shaft. But for the approximate calculation, this factor may be omitted.)

The deformation of a segment depends on its length and the moment of inertia of its section (I). To take account of these two factors, the moment of inertia of the equivalent shaft I_e is determined as follows:

$$I_e = \frac{I_1 L_1 + I_2 L_2}{L_1 + L_2}$$

$$I_1 = \frac{\pi d_1^4}{64} = \frac{\pi \cdot 290^4}{64} = 3.47 \cdot 10^8 \ mm^4; \quad L_1 = 850 \ mm$$

$$I_2 = \frac{\pi d_2^4}{64} = \frac{\pi \cdot 450^4}{64} = 20.13 \cdot 10^8 \ mm^4; \quad L_2 = 1220 \ mm$$

$$I_e = \frac{3.47 \cdot 10^8 \cdot 850 + 20.13 \cdot 10^8 \cdot 1220}{850 + 1220} = 13.29 \cdot 10^8 \ mm^4$$

(For the curious, the equivalent diameter corresponding to this value of I_e is about 406 mm. We don't need it for the following analysis.)

Now, we go to Figure 4.16 and Figure 4.18. The section in question is placed at a distance of 243 mm from the left bearing, so for this section $x/L = 243/2070 = 0.12$. Coefficients K_{F1}, K_{F2}, and K_{F3} for forces F_1, F_2, and F_3 correspondingly found from Figure 4.19 are as follows:

$$x/L = 0.12, \ a/L = 243/2070 = 0.12, \ K_{F1} = 0.026$$

$$x/L = 0.12, \ a/L = 728/2070 = 0.35, \ K_{F2} = 0.057$$

$$x/L = 0.12, \ a/L = 1948/2070 = 0.94, \ K_{F3} = 0.006$$

For the bending moment M, the K_M value is found from Figure 4.21:

$$x/L = 0.12, \ a/L = 0.12, \ K_M = 0.22$$

Using Equation 4.5 and Equation 4.6 we calculate the following results:

$$\theta_{F1} = K_{F1} \frac{F_1 L^2}{EI} = 0.026 \frac{24 \cdot 10^4 \cdot 2070^2}{2.06 \cdot 10^5 \cdot 13.29 \cdot 10^8} = 9.77 \cdot 10^{-5}$$

$$\theta_{F2} = K_{F2} \frac{F_2 L^2}{EI} = 0.057 \frac{18.1 \cdot 10^4 \cdot 2070^2}{2.06 \cdot 10^5 \cdot 13.29 \cdot 10^8} = 16.15 \cdot 10^{-5}$$

$$\theta_{F3} = K_{F3} \frac{F_3 L^2}{EI} = 0.006 \frac{18.1 \cdot 10^4 \cdot 2070^2}{2.06 \cdot 10^5 \cdot 13.29 \cdot 10^8} = 1.7 \cdot 10^{-5}$$

$$\theta_M = K_M \frac{ML}{EI} = 0.22 \frac{1.1 \cdot 10^7 \cdot 2070}{2.06 \cdot 10^5 \cdot 13.29 \cdot 10^8} = 1.83 \cdot 10^{-5}$$

The *total* slope of the shaft in the middle of the gear width,

$$\theta = \theta_{F1} + \theta_{F2} + \theta_{F3} + \theta_M = (9.77 + 16.15 + 1.7 + 1.83)10^{-5} = 2.945 \cdot 10^{-4} \text{ (rad)}$$

The face width of the gear measures 350 mm, and thus the error in parallelism of the teeth is about 0.1 mm. It is a large error, which results in such an unevenness of load distribution along the teeth that the maximal unit load is more than twice the average. But in this case, the tilt of the gear wheel (along with the adjoined shaft section) is largely compensated by the elastic deformation of the pinion, which tilts in the same direction (as shown in Chapter 1, Figure 1.5e).

Computer calculation of slope angle θ gives $\theta = 3.5 \cdot 10^{-4}$ rad. That means that the error of the rough estimation was 19%. It is too much, but the preliminary estimate lets us know if a more precise deformation analysis should be performed. In this case, the answer is "yes."

The torque, which loads the main shaft between sprockets, is

$$T = 0.5(F_{cu} - F_{cl})R = 0.5(30 \cdot 10^4 - 10 \cdot 10^4)706 = 7.06 \cdot 10^7 \ N.mm$$

(Here, R [the pitch radius of the sprocket] = 706 mm.)

The diameter of the shaft having the required strength may be estimated by Equation 4.1, where the tensile strength of the shaft material S_u is taken equal to 930 MPa:

$$d = (5 \div 6) \sqrt[3]{\frac{7.06 \cdot 10^7}{930}} = 212 \div 254 \text{ mm}$$

If the shaft segment between the sprockets had been chosen with a diameter $d_2 = 250$ mm (instead of 450 mm), the slope angle of the gear wheel would have grown to $\theta = 1.53 \cdot 10^{-3}$ rad. This results in a gear tooth parallelism error of about 0.5 mm, which is absolutely unacceptable.

REFERENCES

1. Timoshenko, S., *Vibration Problems in Engineering,* D. Van Nostrand Company, Toronto, New York, London, 1955.
2. Birger, I.A., Shorr, B.F., and Ioselevitz, G.B., *Calculations for Strength of Machine Elements,* Mashinostroenie, Moscow, 1979 (in Russian).
3. Peterson, R.E., *Stress Concentration Factors,* John Wiley and Sons, New York, London, Sydney, Toronto, 1974.
4. Philimonov, G.N. and Balazky, L.T., Fretting in joints of naval parts, *Sudostroenie,* Leningrad, 1973 (in Russian).

5 Shaft-to-Hub Connections

Parts mounted on a shaft are connected to it by their central portion, which is usually called a *hub*. The connection should be designed so as to

- Fix the part relative to the shaft with needed accuracy
- Ensure load transmission between the part and the shaft

In deciding on a particular type of connection, the magnitude of the torque to be transmitted is the determining factor. The authors chose three types of widely used shaft–hub connections for detailed consideration: interference-fit connection (IFC), keyed joint (KJ), and splined joint (SJ). Each of these connections has merits and demerits that make it preferable or undesirable in certain cases.

5.1 GENERAL CONSIDERATIONS AND COMPARISON

5.1.1 INTERFERENCE-FIT CONNECTIONS (IFCs)

Cylindrical IFCs (Figure 5.1a) have one very important advantage: they are inexpensive. It is enough to machine the diameters of the shaft and hub to a needed size — and the parts are ready for assembly! Other advantages of IFCs are high accuracy of centering the hub relative to the shaft, and the ability to transmit torque, bending moment, radial force, and axial force without any additional parts or devices. Regarding the demerits of IFC, they are not less important:

> High weight: Figure 8.3b and Figure 8.3c give a comparison between SJs and IFCs. For IFCs, the parts must be thick-walled in order to withstand the high pressure between the mated surfaces while keeping the stresses and the interference values within reasonable boundaries (see also Section 5.2).
> Relatively low load capacity: The hub must overcome frictional forces to turn relative to the shaft. To obtain this movement in the SJ, all the teeth must be sheared; the needed torque in this case is larger and usually exceeds the static strength of the shaft in torsion.
> Relatively low reliability: The load in IFCs is transmitted because of frictional forces. In experiments for static axial shift and twisting of IFCs, the friction coefficient f for steel on steel was obtained in the range of 0.08–0.20. At first sight, there is no problem: If we take for the strength calculation $f = 0.01$, the safety factor is 8 at least. Unfortunately, this is not for sure. The authors have successfully used IFCs several times, but one case was unsuccessful. In a reduction gear (described in detail in Example 5.1), the IFC of the first stage wheels with the intermediate shaft slipped under 80% of nominal load. This connection of 40 mm in diameter and 40 mm in length was calculated with a tenfold safety factor and checked under static load of 7.5 times nominal. No slippage was noticed. This experience testifies that the friction coefficient in IFCs may decrease sharply in the presence of vibrations and bending deformations of the shaft.
>
> But this is not the only possible cause of insecurity. The faintest error in the amount of interference or deviation from cylindrical shape may lead to a considerable decrease in surface pressure. This is particularly true in relatively small connections (say, 50 mm diameter and less). In tapered connections, the difference in taper angles of the hub and

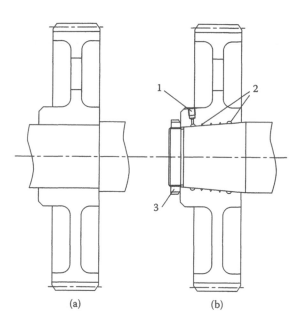

(a)   (b)

FIGURE 5.1 Interference-fit connections.

the shaft presents an additional factor that reduces the load capacity. For these reasons, in important cases (for example, in lifting mechanisms of cranes), it is forbidden to use IFCs without key or other positive locking elements.

Detrimental effect on shaft strength: Elastic deformations under load result in local slippage in the connection and fretting damage of the mated surfaces (see Chapter 2, Section 2.2 and Section 2.3). The fatigue strength of the shaft may be reduced by 70–80%. With higher-strength shaft materials, the detrimental effect of IFCs is greater.

Difficulty of dismantling: Assembly of an IFC is usually very easy: it is enough to warm up the hub or cool the shaft. However, dismantling may be problematic, particularly, when the mating surfaces are of low or middle hardness; there may be seizure and severe damage to the surfaces.

During continuous duty, the condition of the mating surfaces changes (because of the microslippage mentioned previously), and the friction coefficient increases owing to seizure and fretting. Sometimes, pressing out of this connection is impossible and one of the parts must be cut to save the other.

Radical facilitation of dismantling is achieved by using a tapered IFC, which allows hydraulic dismantling (Figure 5.1b). The angle of the taper in this case is usually chosen between 1°54′35″ (taper 1:30) and 1°08′45″ (1:50). Oil is pumped through inlet 1 to spiral grooves 2 under high pressure that must exceed the surface pressure in the connection (the pressure may be in hundreds or even thousands of bar). When this oil pressure separates the hub from the shaft, the friction coefficient becomes very low, and the hub rushes off the shaft with a high speed. To prevent injury to personnel and damage to the parts, safe stoppers must be installed before dismantling. In the design shown in Figure 5.1b, nut 3 slackened by several millimeters may be an ideal stopper.

Tapered connections are more expensive, because machining the tapered surfaces is quite difficult when interchangeability is needed. The next step toward simpler production is separation of the tapered part, and manufacturing a unit at a specialized works equipped for this task. Figure 5.2a shows a rigid coupling for joining two coaxial shafts of the same diameter. Thin-walled sleeve 1 has a cylindrical inner surface and a tapered outer surface. An interference fit is achieved by the axial movement of heavy-walled shell 2. High-pressure oil is led through inlet 3 to spiral grooves

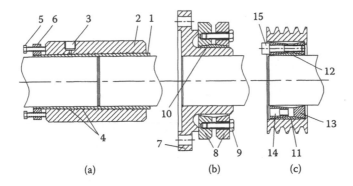

FIGURE 5.2 Easily disassembled interference-fit connections.

4 (to separate the sliding tapered surfaces), and the shell is moved relative to the sleeve by bolts 5 nested in nut 6. (After assembly, bolts 5 must be removed, but the locked nut remains in place.) The contact pressure in the joint is determined indirectly by measuring the increase in the outer diameter of the shell. To dismantle the coupling, the high-pressure oil is led through inlet 3 to spiral grooves 4, and as the oil film separates shell 2 from sleeve 1, the shell gets off the sleeve, and nut 6 serves as a stopper.

Figure 5.2b shows the connection of flange 7 with a shaft by means of radial deformation of the flange hub. The needed surface pressure is created when discs 8 clamped by bolts 9 squeeze double-tapered split sleeve 10, which in turn squeezes the hub of flange 7 and presses it against the shaft. The tapered surfaces of parts 8 and 10 as well as bolts 9 should be coated with MoS_2 grease to decrease the coefficient of friction in these places. That increases the radial squeezing forces that clamp the coupling and facilitates the cone's disengagement while releasing. Mating surfaces of the flange and the shaft should be degreased so as to increase the friction and the load capacity of the IFC.

Another version of such a coupling that attaches a pulley to a shaft is shown in Figure 5.2c. Double-tapered split sleeves 11 and 12 are deformed in the radial direction by means of double-tapered rings 13 and 14 pulled together by bolts 15. In this case, the contacting surfaces of the inner sleeve with the shaft and the outer sleeve with the pulley should be degreased.

The couplings depicted in Figure 5.2 represent versions of IFCs, and all the demerits of IFCs are valid for them, with the exception of "difficulty of dismantling." In some cases, the slippage of IFCs can be considered as a merit: in case of overload, it may play the role of a torque limiter and protect the mechanism from damage. Nevertheless, it is undesirable to use an IFC as a torque limiter for three reasons. First, if you need a torque limiter, it is preferable to use a special device, which enables more exact setting of the maximal torque. Second, the IFC becomes less reliable after slippage because the mating surfaces are subjected to wear, and the coefficient of friction decreases as well. To be on the safe side, the coupling should be dismantled and inspected after the slippage, because the coupling and the shaft may not only be worn out but also overheated, scratched, and possibly cracked. And third, you never know whether there was a slippage in the connection or not, so the reliability of the machine is permanently questionable. Therefore the load capacity of the IFC must be several times greater than the transmitted load to safely prevent slippage.

5.1.2 KEY JOINTS (FIGURE 5.3)

Usually, a KJ is an IFC with the addition of a key that prevents slippage during the transmission of a torque. The KJ is somewhat more expensive than IFC because the keyways must be machined in the hub and on the shaft journal. But KJ is much more reliable and capable of transmitting torque.

In principle, the KJ is also able to transmit a torque when the hub is slide-fitted to the shaft, but the durability of the connection decreases sharply. The permissible load in this case is very

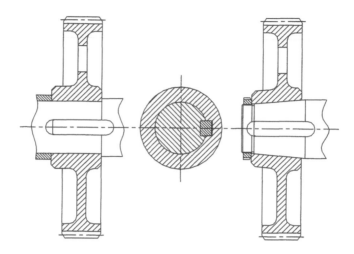

FIGURE 5.3 Key joints.

low, and the key is usually screwed to the shaft to prevent its turning (see Figure 5.10d). When the shaft is rotating, the hub moves cyclically relative to the key and the shaft within the radial clearance, causing wear. Lubrication of such connections is a must.

The aforementioned demerits of IFCs (with the exception of "relatively low load capacity and reliability") are valid for KJs as well. The detrimental effect on the shaft strength is even stronger because the shaft is additionally weakened by the keyway. But the combination of high reliability and load capacity with a moderate cost has motivated the widespread use of KJ in machinery construction, where there is no tight limitation on weight and the needed shaft strength may be achieved by increasing its diameter.

5.1.3 Splined Joints (SJs) (Figure 5.4)

Merits of the SJ:

Highest load capacity when transmitting a torque: As was already said, the static strength of the SJ in torsion is more than that of the shaft. This torsional strength is provided at the moderate joint length required for wear resistance.

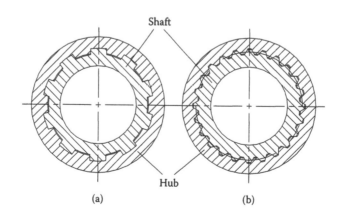

FIGURE 5.4 Splined joints.

Least weight of parts: The thickness of the hub and shaft is dictated by the needed strength for transmission of the load; therefore, the parts may be thin walled (tubular) and relatively lightweight.

Capability of working with clearances and misalignments: These abilities enable an effective use of the SJ for axially shifted parts (for example, for shifting gears in gear boxes). It is also used to compensate misalignment in the connection of two coaxial shafts. Such SJs should be generously lubricated for long-term usage.

Lesser detrimental effect on shaft strength (as compared with IFCs and KJs): Strength decrease in the SJ is caused by geometric stress raisers. Their influence is quite predictable, and it is considerably less than the influence of the stress concentration and fretting in IFCs and KJs.

In SJs, fretting occurs rarely because they are lubricated. But even if fretting occurs, it is limited to the contact surfaces of the teeth, at a certain distance from the most stressed places (which are the tooth fillets) and doesn't affect shaft strength.

Easy assembly and dismantling: SJ is usually designed with a sliding fit, so there is no problem with assembly and dismantling. When the SJ must be tightly fitted, temperature deformation is used for easy assembly.

If there were no demerits in SJs, they would be used everywhere, and we would not be even aware of the other types of the hub–shaft connections. But, fortunately, they have demerits as well:

Relatively high cost: In some cases it is not just expensive but completely impossible to machine splines into the hub or on the shaft because they are too large for the available gear-cutting or broaching machines.

Lower precision of centering the hub relative to the shaft: The simpler the geometry of the mating surfaces, the more precise the machining can be. In IFC and KJ, these surfaces are just cylinders or tapers, and these connections are gapless, so they ensure the best centering precision. In an SJ the spline surfaces are complicated, and machining is less accurate. If high accuracy of centering in the radial direction is needed, auxiliary surfaces — cylindrical or conical — are usually used (see Section 5.4), and that complicates the design.

The areas of application of these connection types may be determined from the features stated previously. Where maximal reliability, minimal weight, or both are required (for example, in aviation and automotive engineering), SJs are usually used. In general engineering, where dimensions and weight are not so restricted, KJs (with a press fit) are prevalent.

The difficulty of dismantling IFCs and KJs doesn't confuse the designer if, in regular service conditions, the connection will not be dismantled. In case of damage, the damaged part may be cut, easily separated from the usable one, and replaced. For example, if the teeth of a gear have been broken, the gear may be cut in radial direction in the area of the keyway to release the surface pressure in the connection. After that, the shaft can be pressed out of the gear, inspected for possible surface damages and fatigue cracks, and then reused if everything is OK.

The IFC (without a key) is used when the torque and the axial force are low relative to the dimensions of the connection, for example, in the connection of railroad car wheels with their axles, the rotor of an electric motor with the shaft, etc. Use of IFCs for transmission of high loads is recommended only on condition that it will be thoroughly tested before being placed in service.

5.2 STRENGTH CALCULATIONS AND DESIGN OF IFCs

5.2.1 CALCULATION FOR TOTAL SLIPPAGE

An IFC transmits load by means of friction. The area of the joint surface is

$$A = \pi d l \ mm^2$$

and the total friction force is

$$F_f = \pi d l \, p f \; N \tag{5.1}$$

where
> d and l are the diameter and length of the connection (mm)
> p = surface pressure (MPa)
> f = coefficient of friction

Equation 5.1 looks extremely logical and easy. The problem is that only the d and l values are known exactly, because these are specified by the designer and they can be measured. Concerning the pressure p and, particularly, the friction coefficient f, the designer is forced to proceed from assumptions and information that may not correspond with reality. But let's consider these two factors in more detail.

5.2.1.1 Surface Pressure

To determine the surface pressure p, the Lame equation (1833) is used. That equation is true for infinitely long and perfectly precise round cylinders, but it is successfully used for real parts with small corrections. The need of corrections arose from the fact that the contact between the hub and shaft takes place not over the entire surface uniformly but first in spots; this is because the parts have surface roughness and waviness, out-of-roundness, and other deviations from the ideal shape. Therefore, the compliance of the parts increases (see the roughness-dependent second member of Equation 6.7 in Chapter 6, Subsection 6.2.5). As the interference grows, the entire surface of the joint comes into contact, and in the spots of initial contact, there may be plastic deformation of the surface layer. Because of this, the influence of the surface deviations is limited. At the relatively high pressures typical of IFCs, it is usually taken into account by reducing the rated radial interference by 60% of the sum of the roughness heights of the mating surfaces R_z.

If the joint diameter is d, the diameter of the central hole in the shaft is d_1 and the outer diameter of the hub is d_2 (Figure 5.5), the surface pressure for materials with the same modulus of elasticity E may be determined from this equation:

$$p = \frac{E\delta}{2d^3} \frac{\left(d^2 - d_1^2\right)\left(d_2^2 - d^2\right)}{\left(d_2^2 - d_1^2\right)} \; \text{MPa}$$

where
> E = modulus of elasticity (MPa) of the hub and the shaft (If the hub and the shaft have different moduli of elasticity, another equation, more complicated, should be used.)
> d_1 and d_2 = diameters (mm)
> δ = rated interference (mm) determined from the equation

$$\delta = (d_S - d_H) - 1.2(R_{ZS} + R_{ZH})$$

where
> d_S and d_H = measured joint diameter of the shaft and the hub
> R_{ZS} and R_{ZH} = roughness heights of these surfaces (all in mm)

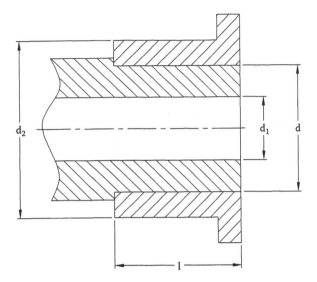

FIGURE 5.5 Interference-fit main dimensions.

If there is no central hole in the shaft ($d_1 = 0$), then the equation for the surface pressure becomes simpler:

$$p = \frac{E\delta}{2dd_2^2}\left(d_2^2 - d^2\right) \tag{5.2}$$

The diameters that appear in these equations may be variable. In such cases, the designers usually simplify the problem. For example, in tapered connections the diameter d may be taken as an average of the minimal and maximal values. The outer surface represented by diameter d_2 may have a flange (as in Figure 5.5) or another stiffener, which increases the surface pressure in the vicinity. Usually these changes in d_2 are neglected. A more precise calculation may be performed using finite element method (FEM).

For tapered IFCs, the equality of the taper angles of the shaft and the hub is very important because it directly influences the amount of interference. Apparently, the maximal divergence of the cones should not exceed 10–15% of the rated interference. The needed interference is achieved by an axial movement of the hub relative to the shaft from the initial position, in which they are pressed against each other with a relatively small force. It is preferable to have the initial contact in the area of the larger diameter of the connection, because there the transmitted torque is maximal.

The real surface pressure may be controlled after assembly by measuring the increase of the outer diameter of the hub (Δ). The increase of the outer diameter for a cylindrical hub may be derived from a formula:

$$\Delta = \frac{d^2 d_2 p}{E\left(d_2^2 - d^2\right)}$$

When E and p are expressed in MPa, d and d_2 are in mm, and so is Δ as well.

If the parts are made of materials with different moduli of elasticity and Poisson's ratio, the formula is somewhat more complicated, and can be found in handbooks.

The calculation of the IFC should take into account the possible differences in temperature during manufacture, storage, and service. During manufacture, the temperature is about 20°C; during storage and use, it may change in the range from, say, −60 to +100°C. If, for example, an aluminum part is mounted with an interference fit on a steel shaft at a temperature of +20°C, the interference will increase by about 0.09 mm for each 100 mm of the joint diameter at −60°C. This may cause some plastic deformation (depending on the initial surface pressure) and reduce the interference. If, in service, the temperature is +100°C, the interference decreases additionally by 0.09 mm for each 100 mm of the joint diameter, and impairs the load capacity of the IFC.

Considerable decrease in interference while in operation may be caused by centrifugal forces. At high rotational speed, the joint may be completely separated, and this may lead not only to slackening of the connection, but also to imbalance of the rotor and increased vibrations. The only remedy for this is to increase the initial interference so as to ensure the needed tightness at maximal working speed.

5.2.1.2 Coefficient of Friction

When the hub is turned or moved in the axial direction relative to the shaft, the force first comes to a maximum (static friction coefficient f_S), and then, it decreases to the amount which conforms to the sliding friction coefficient f (Figure 5.6).

According to experiments,

$$f_S \approx (1.5 \text{ to } 2)f$$

For calculation, it may seem reasonable to use the maximal coefficient of friction f_S, and it is correct when even a small slippage in the connection is inadmissible (for example, in kinematical transmissions). In many cases, a single small slippage on a momentary overload is not dangerous. However, if the connection is once broken, the repeated slippage occurs at a lesser load, nearly conforming to the sliding coefficient of friction f (peaks 1 in Figure 5.6). Therefore, as a rule, the f value is used in calculations for total slippage. The values of f for steel parts in experiments range from 0.08 to 0.18 when assembled by applying an axial force (pressing). When assembled using radial deformation of parts (heating the hub, cooling the shaft, or causing deformation using hydraulic pressure or intermediate tapered parts), the range is from 0.12 to 0.20. It is reasonable that the minimal values should be taken for calculations: $f = 0.08$ for assembly by force and $f = 0.12$ when radial deformation is used. These values are valid for a static loading without any vibrations or cyclic bending. At these conditions, the axial force that the IFC is able to transmit is equal to

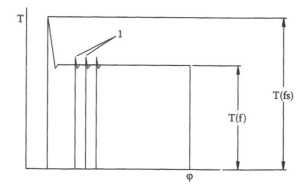

FIGURE 5.6 Torque vs. rotating angle while breaking interference fit.

the friction force F_f and is determined by Equation 5.1. Regarding the torque, it is equal to the product of F_f and $d/2$:

$$T_f = \frac{\pi d^2 l p f}{2} \ N \cdot mm \tag{5.3}$$

Experiments show that during cyclic bending of the shaft, the slip resistance of IFC decreases. The reason for this is that the bending is accompanied by a local microslippage, and in these areas, the friction coefficient decreases almost to zero. In addition, if the bending stress on one side of the connection is larger than on the other side, there arises an axial force aimed to press the shaft out of the hub toward the larger bending stress. (See Chapter 2, Section 2.1.)

According to experimental data,[1] when the IFC is loaded with an axial force or a torque, and the shaft is simultaneously loaded with a cyclic bending moment, the dynamic coefficient of friction f_{dyn} may be obtained from the formula

$$f_{dyn} = f - \frac{\beta \sigma_a}{p} \frac{d}{l} \tag{5.4}$$

where
  f = static coefficient of friction
  σ_a = bending stress amplitude
  p = surface pressure in the connection
  d and l = diameter and length of the connection
  $\beta = 0.08 \pm 0.008$ is an experimentally determined factor

It was noticed that the mean bending stress doesn't influence the load capacity of the IFC. Only the amplitude σ_a affects it.

When the bending stress becomes so large that $f_{dyn} = 0$, the hub begins moving relative to the shaft (in the axial direction) without any external force. If the value of f_{dyn} obtained from Equation 5.4 is negative, some form of stopper must be used to prevent the self-dismantling of the connection.

Equation 5.4 is valid when the bending stress on one side of the connection is zero. If there are bending stresses on both sides, the experiments mentioned previously give the following relation:

$$f_{dyn} = f - \frac{\beta_1 (\sigma_{a1} - \sigma_{a2}) + \beta_2 \sigma_{a2}}{p} \frac{d}{l} \tag{5.5}$$

In this formula, σ_{a1} is the larger bending stress, and σ_{a2} is the lesser bending stress (on the other side of the IFC); $\beta_1 = 0.08$ and $\beta_2 = 0.05$.

It was noted that if the external force (load) tended to move the hub relative to the shaft in a direction opposite to the direction of self-dismantling, then $\beta_1 = \beta_2 = 0.05$.

The substitution of $\sigma_2 = 0$ into Equation 5.5 transforms it into Equation 5.4. As is seen from these equations, the longer the connection, the less the influence of bending on the load capacity of the IFC.

In contrast to the transmission of axial forces, the torque transmitting capacity never decreases to zero, but it may be as little as 15% or less of the static load capacity. The data cited previously is based on extensive experiments with IFCs of 30, 70, and 160 mm in diameter and with d/l ranging from 0.1 to 1.5.[1]

EXAMPLE 5.1

Let's analyze the case of IFC slippage mentioned in the beginning of Section 5.1 regarding "relatively low reliability." The case in point is the intermediate shaft of a double-reduction gear

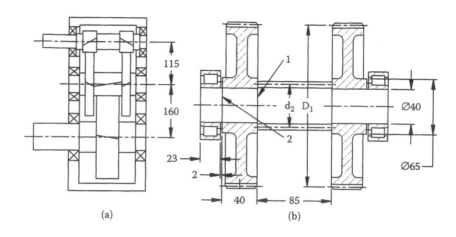

FIGURE 5.7 Intermediate shaft of a double-reduction gear.

shown in Figure 5.7a (sketch of a gear) and Figure 5.7b (the shaft to be investigated). The center distance of the input stage $C_1 = 115$ mm, gear ratio $i_1 = 5$, so the gear diameter $D_1 = 191.7$ mm. The output stage center distance $C_2 = 160$ mm, $i_2 = 5$, so the pinion diameter $d_2 = 53.3$ mm. The output torque of the gear is 2000 N·m, so at the intermediate shaft, each gear transmits a nominal torque of 200 N·m to the pinion. The IFC strength calculation was performed without taking into account dynamic factors and unevenness of load distribution between the gears of the input stage, but the safety factor was taken as $k = 10$. The needed surface pressure was found from Equation 5.3:

$$p = \frac{2T_f}{\pi d^2 l f} k \qquad (5.6)$$

Substitution of $T_f = 2 \cdot 10^5$ N·mm, $d = l = 40$ mm, $f = 0.12$, and $k = 10$ gives $p = 165.8$ MPa. The manufacturing tolerances of the shaft and gears were chosen so as to ensure this pressure at minimal interference.

Now it is time to do a more accurate calculation of the IFC of the gear to the pinion shaft. First, consider the possibility that the load distribution may be uneven between the two halves of the "herringbone." The helix angle of the gear teeth is $\psi = 30°$, and for this angle the load factor is $K_{Wh} = 1.32$ (see Chapter 7, Figure 7.22). The dynamic factor may be approximately taken as $K_d = 1.35$. Hence, the total torque transmitted by one gear equals

$$T = 200 \cdot 1.32 \cdot 1.35 = 356.4 \; Nm$$

(Admittedly, there is some exaggeration in such use of the dynamic factor K_d. The dynamic load has been mostly taken by the mass of the gear, and only part of it results in increased torque. But we try to find out the reasons for the IFC slippage; so we take the maximum possible overload.)

A calculation of the bending stresses in the pinion shaft gives the following: in section 1, $\sigma_1 = 121.2$ MPa; in section 2, $\sigma_2 = 29.2$ MPa. From Equation 5.5, the coefficient of friction equals

$$f_{dyn} = 0.12 - \frac{0.08(121.2 - 29.2) + 0.05 \cdot 29.2}{165.8} = 0.067$$

Thus, we see that owing to the bending stresses, the coefficient of friction may degrade to almost half of its static value. But it doesn't explain the observed slippage of the IFC. To transmit the 354.6 N·m torque, it is enough to have coefficient of friction (from Equation 5.3)

$$f = \frac{2T}{\pi d^2 l p} = \frac{2 \cdot 354600}{\pi \cdot 40^2 \cdot 40 \cdot 165.8} = 0.021$$

So, the obtained coefficient of friction $f_{dyn} = 0.067$ is enough to transmit even threefold torque. But because there was a slippage, the real coefficient of friction was less than 0.021. Consequently, our calculation was based on wrong assumptions about the f value. Taking into consideration that the connection has passed the static test under 7.5-fold load, an additional reduction in the coefficient of friction owing to vibrations should be supposed. In experiments with bolt connections, a reduction in the coefficient of friction (because of vibrations) from 0.13–0.14 to 0.01, and even to 0.005, was noticed. Obviously, this was the second cause of the slippage; low accuracy of the gears set up vibrations, and the friction in the IFC became too low.

When IFCs are not exposed to heavy vibrations and bending loads, the rated safety factor $k = 3$ is usually considered as sufficient (once again, when the possible slippage is not dangerous).

5.2.2 Design of IFCs

Every design begins with something known; in this case, it is the approximate diameter d (reasoning from the diameter of the shaft) and load to be transmitted. The designer chooses the length of the connection (usually $l = (1-1.5)d$), the coefficient of friction f, and the safety factor k. This is enough to determine the needed surface pressure p, the needed amount of interference, and then the stresses that the IFC induces within the parts.

The maximal stresses take place on the joint surfaces. On the shaft,

$$\tau_{max}^S = p \frac{d_1^2}{d^2 - d_1^2} \tag{5.7}$$

and on the joint surface of the hub,

$$\tau_{max}^H = p \frac{d_2^2}{d_2^2 - d^2} \tag{5.8}$$

These stresses are constant and they should not, under any circumstances (such as change of temperature, influence of load and other), exceed the yield point S_y of the material, otherwise the real interference and surface pressure may become less than the rated ones. So,

$$\tau_{max} \leq 0.5 S_y$$

These stresses should be calculated for the minimal rated value of p. The sufficient safety factor in this case is 1, because if there is a plastic deformation, the surface pressure decreases automatically to the admissible value, and the plastic deformation ceases.

The sum of the radial deformations of the hub (outward) and the shaft (inward) represents the interference value δ that is needed to achieve the required surface pressure p. The larger the interference, the less the influence of manufacturing tolerances on the load capacity of IFC. But on the other hand, the large interference may cause problems when assembling.

Example 5.2

A shaft transmits torque of $T = 10$ kN·m. On its end is attached a flange by means of IFC. The bending moment is negligible. The tensile strength of the shaft material is $S_u = 950$ MPa. Roughness height of the mating surfaces $R_z = 10$ mkm.

The diameter of the shaft may be approximately obtained from Equation 4.1 because the shaft has a sharp stress raiser (an IFC):

$$d \approx (5 \text{ to } 6) \cdot \sqrt[3]{\frac{10000 \cdot 10^3}{950}} = 109 \text{ to } 130 \text{ mm}$$

We take $d = 110$ mm, and the length of the connection $l = 150$ mm. Assuming $f = 0.12$ and the safety factor $k = 3$, the required surface pressure p derived from Equation 5.6 is

$$p = \frac{2 \cdot 10000 \cdot 10^3 \cdot 3}{\pi \cdot 110^2 \cdot 150 \cdot 0.12} = 87.7 \text{ MPa}$$

First, take the shaft without a central hole ($d_1 = 0$) and the outer diameter of the hub $d_2 = 1.6d \approx 175$ mm. Now, let's determine the deformations of the hub and the shaft under this pressure. Increase of the hub inner diameter is

$$\Delta d_H = \frac{dp}{E}\left(\frac{d^2 + d_2^2}{d_2^2 - d^2} + \mu\right) = \frac{110 \cdot 87.7}{2.06 \cdot 10^5}\left(\frac{110^2 + 175^2}{175^2 - 110^2} + 0.3\right) = 0.122 \text{ mm}$$

Decrease of the shaft outer diameter is

$$\Delta d_S = \frac{dp}{E}\left(\frac{d^2 + d_1^2}{d^2 - d_1^2} - \mu\right) = \frac{110 \cdot 87.7}{2.06 \cdot 10^5}(1 - 0.3) = 0.033 \text{ mm}$$

To determine the required value of the interference δ, we should sum up the deformations of the hub and the shaft and add 60% of the total roughness height:

$$\delta = \Delta d_H + \Delta d_S + 1.2(R_{Z1} + R_{Z2}) = 0.122 + 0.033 + 1.2(0.01 + 0.01) = 0.179 \text{ mm}$$

This value of interference should be considered as minimal. Taking into account the tolerance of the shaft neck (0.022 mm) and that of the hub inner diameter (0.035 mm), the maximal interference in this connection will be $0.179 + 0.022 + 0.035 = 0.236$ mm.

To ensure some assembly clearance by heating the hub, the radial deformation caused by heating should be at least 0.3 mm. Considering the linear expansion factor of steel $\alpha = 12 \cdot 10^{-6}$ 1/°C, the required difference between the temperatures of the hub and the shaft is

$$\Delta t = \frac{\delta}{d \cdot \alpha} = \frac{0.3}{110 \cdot 12 \cdot 10^{-6}} = 227°\text{C}$$

That means that if the shaft temperature is 20°C, the hub should be heated to 250–270°C. If the shaft is chilled in liquid nitrogen to −70°C, the hub needs be heated only to 160–180°C.

Let's calculate this connection with thin-walled parts. Say, the shaft has a central hole $d_1 = 70$ mm, and the outer diameter of the hub $d_2 = 140$ mm. Because the torque transmitted is the same, the required surface pressure is also the same. In this case, the changes in the diameters of the hub and of the shaft are as follows:

$$\Delta d_H = \frac{110 \cdot 87.7}{2.06 \cdot 10^5} \left(\frac{110^2 + 140^2}{140^2 - 110^2} + 0.3 \right) = 0.212 \text{ mm}$$

$$\Delta d_S = \frac{110 \cdot 87.7}{2.06 \cdot 10^5} \left(\frac{110^2 + 70^2}{110^2 - 70^2} - 0.3 \right) = 0.097 \text{ mm}$$

The minimal interference (the difference of the diameter d values measured on the shaft neck and inside the hub) is

$$\delta = 0.212 + 0.097 + 1.2(0.01 + 0.01) = 0.333 \text{ mm}$$

Taking into account the manufacturing tolerances, the maximal interference may be as large as $0.333 + 0.022 + 0.035 = 0.390$ mm. For sufficient radial deformation (by about 0.45 mm) using heating or chilling, the difference in temperatures of the hub and shaft should be about 340°C.

Now it is worth checking the stresses in the parts caused by the interference fit. As mentioned, we check it at a minimal required surface pressure. Using Equation 5.7 and Equation 5.8, we obtain the maximal shear stresses in the shaft and the hub:

$$\tau_{max}^S = 87.7 \frac{70^2}{110^2 - 70^2} = 59.7 \text{ MPa}$$

$$\tau_{max}^H = 87.7 \frac{140^2}{140^2 - 110^2} = 229.2 \text{ MPa}$$

The stress in the hub is quite big, and so the material should have a yield point of at least 500 MPa.

The design of IFCs is extremely simple. Measures to diminish the detrimental effect of this connection on the fatigue strength of the shaft have been discussed in Chapter 4, Section 4.2.

5.3 DESIGN AND STRENGTH CALCULATION OF KEY JOINTS

5.3.1 ROLE OF THE IFC IN THE KEY JOINT

The main condition that should be met to ensure durable serviceability of KJ is a complete (as far as possible) prevention of any relative movement in the connection. This is achieved by prestressing the joint using an IFC, and the friction force should transmit the entire torque with a safety factor of 1.3 to 1.5.

The needed prestressing may be made also by taper keys or tangent keys, but they move the hub in radial direction relative to the shaft, causing a certain radial runout of the hub. Therefore, such connections are used rarely, usually for low-speed shafts, and when a high accuracy is not required. But they are not discussed here.

In connections with a large taper (say, 1:10, where the angle of the taper is numerically larger than the coefficient of friction), the needed prestress can't be achieved by just assembly, because the hub will slip off the shaft. There must be an axial tightening using a nut or an end plate (Figure 5.8). The needed tightening force is obtained from the equation

$$F_{tight} = \frac{2T(\sin \alpha + f \cos \alpha)}{d_m f} k \, N \qquad (5.9)$$

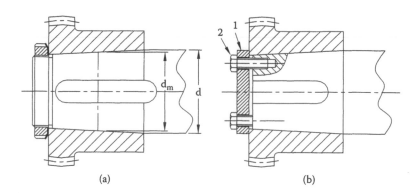

FIGURE 5.8 Tightening of tapered joints by axial force.

where

α = half-angle of the taper
f = coefficient of friction
d_m = mean diameter of the taper (in the middle of its length) (mm)
T = magnitude of the transmitted torque (N · mm)
k = safety factor

EXAMPLE 5.3

Let's calculate (just for fun!) a tapered connection with some parameters taken from Example 5.2: torque T = 10 kN·m and diameter of the shaft d = 110 mm. The length of the connection l = 120 mm, taper 1:10 (α = 2°51′45″). The connection is tightened with a nut as shown in Figure 5.8a. The material of both the shaft and nut: steel SAE 4340 heat-treated to HRC 28–35, tensile strength S_u = 880 MPa, and yield point S_y = 725 MPa.

The minimal diameter of the taper is 110 − 0.1 · 120 = 98 mm; hence, the mean diameter of the taper is

$$d_m = \frac{110 + 98}{2} = 104 \text{ mm}$$

Setting the values of k = 1.4 and f = 0.12, we obtain from Equation 5.9 the needed axial force of the nut:

$$F_{tight} = \frac{2 \cdot 10000 \cdot 10^3 (\sin 2°51′45″ + 0.12 \cos 2°51′45″)}{104 \cdot 0.12} 1.4 = 2.72 \cdot 10^5 \ N$$

Because the minimal diameter of the taper is 98 mm, we choose the standard thread M95 × 2 and the standard locknut. The outer diameter of the locknut is 125 mm and its length, 17 mm. The needed tightening torque of the nut may be determined from the following equation:[2]

$$T_{tight} = F_{tight}\left[\frac{d_t}{2}\left(\frac{s}{\pi d_t} + 1.15\mu_t\right) + \mu_f R_f\right]$$

where
  s = thread pitch (here s = 2 mm)
  d_t = thread diameter (d_t = 95 mm)
  μ_t and μ_f = coefficients of friction in the thread and between the hub and the face
        of the nut, correspondingly (we take $\mu_t = \mu_f = 0.14$)
  R_f = average radius of the nut face (R_f = 52 mm)

Inserting these values into the preceding formula gives us the nominal tightening torque:

$$T_{tight,nom} = 2.72 \cdot 10^5 \left[\frac{95}{2} \left(\frac{2}{\pi \cdot 95} + 1.15 \cdot 0.14 \right) + 0.14 \cdot 52 \right] = 4.147 \cdot 10^6 \; N \cdot mm$$

When tightening is performed using a manual torque wrench or precision screwdriver, the possible deviation of the tightening force from its calculated value may be about ±23% (according to the Bossard estimation[3]). It depends on friction fluctuations. Hence, to ensure the needed axial force at a higher friction, the tightening torque should be increased by 23%:

$$T_{tight} = 4.147 \cdot 10^6 \cdot 1.23 = 5.1 \cdot 10^6 \; N \cdot mm$$

But at a lower friction, the axial force may be greater than needed by 23% because of lesser friction and by another 23% because of increased tightening torque. That means,

$$F_{tight,max} = 2.72 \cdot 10^5 \cdot 1.23^2 = 4.12 \cdot 10^5 \; N$$

What do you think about this force? Imagine, it is 40 tons! Surely, the shaft and the nut should be checked for strength. The inner diameter of the thread is about 92 mm, so the diameter of the circular groove between the thread and the tapered part of the shaft may be equal to 90 mm. The area of this section A and the tension stress σ are as follows:

$$A = \frac{\pi \cdot 90^2}{4} = 6362 \; mm^2; \; \sigma = \frac{4.12 \cdot 10^5}{6362} = 64.8 \; MPa$$

No problem. What about the thread strength? It looks too weak for the huge axial load. The shear stress in the thread can be determined from the formula[2]

$$\tau = \frac{F_{tight,max}}{\pi d_i h k}$$

where
  d_i = inner diameter of the thread ($d_i \approx d - 1.3p = 95 - 1.3 \cdot 2 = 92.4$ mm)
  h = engaged length of the thread (h = 17 mm)
  k = coefficient that takes into account the thread vees (k = 0.87 for triangular thread)

$$\tau = \frac{4.12 \cdot 10^5}{\pi \cdot 92.4 \cdot 17 \cdot 0.87} = 96 \; MPa$$

The yield stress in shear

$$\tau_y \approx 0.6 S_y = 0.6 \cdot 725 = 435 \ \text{MPa}$$

Thus, the safety factor for shear strength equals 435/96 = 4.5.

Let's calculate the strength of the nut. It looks like a ring; Figure 5.9 shows (a) the radial forces in the threads and (b) axial forces evenly distributed over the circumference. Because the axial forces and contraforces are distanced in the radial direction (C, Figure 5.9b), the ring is loaded by a shear force, which is equal to $F_{tight, \ max}$, and a twisting moment. Both the shear force and twisting moment are evenly distributed over the circumference. Now, as we have analyzed the kinds of loading forces, we are ready to calculate the stresses.

The mean diameter of the thread $d_{mt} = 93.7$ mm, and the unit axial force is

$$q_a = \frac{F_{tight,max}}{\pi \cdot d_{mt}} = \frac{4.12 \cdot 10^5}{\pi \cdot 93.7} = 1400 \ N/mm$$

The angle of the thread profile is 60°, so the unit radial force in the thread equals

$$q_r = q_a \tan 30° = 1400 \cdot 0.5773 = 808 \ N/m.$$

The tension stress in the ring caused by these radial forces is (approximately)

$$\sigma_1 = \frac{q_r d_{mt}}{2 A_r}$$

Here, A_r is the area of the ring section. The dimensions of the nut are as follows: in the axial direction it measures 17 mm, and in the radial direction, 11 mm (from the outer diameter of the

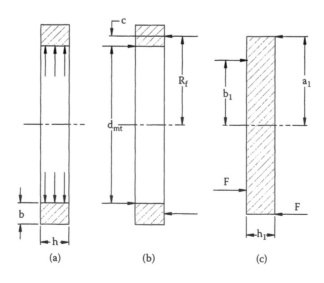

FIGURE 5.9 Sketch to strength calculations.

thread to the bottom of the slot on the outer diameter of the nut). So the area $A_r = 11 \cdot 17 = 187$ mm^2, and the tension stress equals

$$\sigma_1 = \frac{808 \cdot 93.7}{2 \cdot 187} = 202 \text{ MPa}$$

The maximal stress caused by the distributed twisting moment may be determined from the following equation[4]:

$$\sigma_2 = \frac{3F_{tight,max}C}{\pi bh^2}$$

where
 b = radial dimension of the cross section ($b = 11$ mm)
 h = axial dimension of the cross section ($h = 17$ mm)
 C = radial distance between the distributed axial forces applied to the nut

$$C = R_f - \frac{d_{mt}}{2} = 52 - \frac{93.7}{2} = 5.15 \text{ mm}$$

$$\sigma_2 = \frac{3 \cdot 4.12 \cdot 10^5 \cdot 5.15}{\pi \cdot 11 \cdot 17^2} = 637.4 \text{ MPa}$$

The total tension stress in the nut is $\sigma = \sigma_1 + \sigma_2 = 202 + 637.4 = 839.4$ MPa. The shear stress

$$\tau = \frac{q_a}{h} = \frac{1400}{17} = 82.4 \text{ MPa}$$

(Please don't forget to watch the dimensions! In the last equation, q_a was expressed in N/mm and h, in mm, so the result was in N/mm^2, which is MPa.)
 The equivalent stress equals

$$\sigma_{eq} = \sqrt{\sigma^2 + 4\tau^2} = \sqrt{839.4^2 + 4 \cdot 82.4^2} = 855.4 \text{ MPa}$$

This stress is quite high and exceeds the yield point of the material. Besides, the stress calculation was approximate, and some sufficient safety factor (say, 2) should be taken; so the cross section of the nut should be considerably increased. This is easy to do if there is enough space.
 But first let's think over how to tighten this nut. If we take a usual wrench for this kind of nuts (with four jaws engaging with four slots on the outer diameter of the nut), the length of the lever should be 10 ft and two sturdy guys of 200 lb each should hang by the end of it to create the needed tightening torque. It seems a bit exotic, does it not? For this purpose a mechanical screwdriver should be used (if it is available), or the design should be changed. Figure 5.8b shows a design with end plate 1 and bolts 2 (instead of a nut). Six bolts M14, property class 10.9, provide the needed axial force. The maximal tightening torque for this bolt (at $\mu_t = \mu_f = 0.14$) equals 148 N·m, and at this, the tightening force is 74100 N, and the stress is 90% of the yield point. But the real coefficient of friction may be smaller by 23%; so the tightening torque should be decreased correspondingly so as to avoid yielding of the bolt. However, the friction may be higher by 23%

as well, and the tightening force (already decreased by 23% because of decreased tightening torque) will decrease additionally. So the minimal tightening force is

$$F_{tight,min} = \frac{74100}{1.23^2} = 48980 \; N$$

The total minimal tightening force is 48980 x 6 = 2.94 · 10^5 N, and this is rather more than we need (2.72 · 10^5 N).

The end plate should be checked for strength and rigidity under a maximal tightening force (74100 N on each bolt). For the approximate calculation, the end plate may be represented as a round plate simply supported around the periphery. It is loaded by an axial force F evenly distributed over the circumference passing through the centers of the bolts (b_1 = 38 mm, Figure 5.9c). To begin with, let's take the thickness of the plate to be about equal to the bolt diameter: h_1 = 15 mm.

The evenly distributed bending moment in the plate at a radius b_1 is determined by the formula[5]

$$M = \frac{F}{4\pi}\left[\frac{(1-v)\left(a_1^2 - b_1^2\right)}{2a_1^2} - (1+v)\ln\frac{b_1}{a_1}\right]$$

where F = 74100 x 6 = 4.45 · 10^5 N, a_1 = 52 mm, b_1 = 38 mm, and v = 0.3; thus,

$$M = \frac{4.45 \cdot 10^5}{4\pi}\left[\frac{(1-0.3)(52^2-38^2)}{2\cdot 52^2} - (1+0.3)\ln\frac{38}{52}\right] = 2.02 \cdot 10^4 \; N \cdot mm/mm$$

The bending stress in the plate equals

$$\sigma = \frac{6M}{h_1^2} = \frac{6 \cdot 2.02 \cdot 10^4}{15^2} = 538.6 \; MPa$$

Now we should take into account that the real end plate has holes for the bolts on this diameter. Six holes of 15 mm diameter take 37.7% of the circumference of the bolt circle of 38 mm radius. Hence, the real length of the plate section that takes the bending moment is only 62.3% of the circumference. Thus, the calculated stress in the real plate equals 538.6/0.623 = 864.4 MPa. This stress is too high. If we made the plate from a high-strength steel with the yield point of 900 MPa and take a safety factor of 1.5 (considering that our calculation is approximate), the admissible calculated stress is 600 MPa. The stress is related inversely to the square of the plate thickness h_1; so the needed thickness should be

$$h_1 = 15 \; mm \cdot \sqrt{\frac{864.4}{600}} = 18 \; mm$$

Approximating to the nearest standard thickness of the rolled plate, we choose 20 mm.

The end plate bends elastically under load, and there will be some angle between the plate and the bearing surface of the bolt head. Angular deflection of the plate surface at the radius b_1 is determined as follows:[5]

$$\Theta = \frac{Mb_1}{D(1+v)}; \quad \text{where}$$

$$D = \frac{E\,h_1^3}{12(1-v^2)} = \frac{2.06 \cdot 10^5 \cdot 20^3}{12(1-0.3^2)} = 1.51 \cdot 10^8 \; N \cdot mn$$

From here,

$$\Theta = \frac{2.02 \cdot 10^4 \cdot 38}{1.51 \cdot 10^8 (1 + 0.3)} = 3.91 \cdot 10^{-3} \; rad$$

Usually, an angular misalignment of 0.01 rad is reckoned as admissible for statically loaded bolts. In our case, the angle is less than that value.

Concerning the precision of this calculation, it should be noticed that the formulas used here are exact for thin plates, but our plate is quite thick. Besides, the load from the bolts is not distributed evenly but concentrated around the holes, which are stress raisers. This may give you the impression that our calculation is just a simulation of mental work. To disprove this impression, we have determined the stresses in the plate, 15 mm thick, using FEM. According to this exact calculation, the maximal stress on the contour of a hole was 1070 MPa. Recall that in our approximate calculation, the stress was 864.4 MPa; i.e., the error was 25%. So this estimative calculation, supplemented with a safety factor of 1.5 enabled us to do correct design.

Experience leads us to conclude that estimative calculations are very helpful, even if you use FEM; they help you find the error in exact calculations, which may occur because of false insertion of boundary conditions and other data.

5.3.2 STRENGTH OF KEYS

Figure 5.10a represents a standard key joint with a radial clearance s between the key and the hub. This clearance is quite large. For a 100-mm shaft, the nominal clearance is 0.4 mm, but taking into account standard tolerances, it may come to 1 mm. Under load, the key turns under the action of a couple of forces (from the shaft and the hub), which are not coaxial (Figure 5.10b). Therefore, the load distribution over the side surfaces of the key is highly uneven. If the radial clearance is decreased to a minimum, the load distribution will be more uniform (Figure 5.10c). But typical tolerance values dictate the need for a certain clearance.

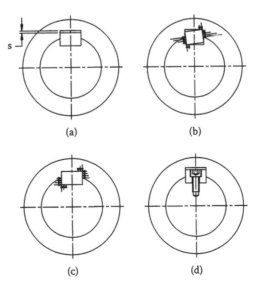

(a) (b)

(c) (d)

FIGURE 5.10 Key displacement and load distribution.

The bearing stress on the sides of the key is usually calculated under the assumption that the entire torque is transmitted through the key, and the pressure is distributed evenly over the key's side surfaces:

$$\sigma = \frac{4T}{dlh} \qquad (5.10)$$

where
  T = torque
  d = shaft diameter
  l = length of the working surfaces of the key (without roundings)
  h = height of the key (deducting chamfers)

As stated previously, in reality, the torque shall be transmitted mainly through the IFC between the hub and shaft. In addition, the load distribution over the height of the key obviously can't be even. So this calculation is just a matter of convention. The admissible stress is usually equal to the yield stress of the weakest material (shaft, hub, or key) divided by 2 or 3.

The reason for such a rough estimation of the bearing stress is that it is a hard task to determine the actual load distribution on the key working surfaces. Besides, it is not so important because some plastic deformation of the surfaces may lead to better load distribution. More important is the shear stress calculation because, in this case, an overstress leads to failure of the joint (again, assuming that the key is the only part that transmits torque). The shear stress equals

$$\tau = \frac{2T}{dlb}$$

Here, b is the width of the key. Usually, $b = (1.5–1.8)h$, so if the bearing stress doesn't exceed the allowable value, the shear stress is safely less than the allowable.

If it is desired that the hub move along the shaft, the admissible bearing stress calculated using Equation 5.10 can be about 10 to 15 MPa. The key in this case should be attached to the shaft by bolts to prevent its turning (Figure 5.10d). Such a design is very archaic. Modern connections with a movable hub are generally splined.

5.3.3 STRENGTH OF THE SHAFT NEAR THE KEYWAY

In the cross section of the shaft shown in Figure 5.11a, the loaded part of the keyway looks like a tooth with a load applied mainly to its tip. In the root of the "tooth," there are considerable bending (tension) stresses, especially because the fillet radius is quite small, and the stress concentration factor is correspondingly high. When the IFC is too weak or doesn't exist (for instance, when there is a slide fit), the variable load is mainly or entirely taken by the key. This may cause a fatigue crack to originate in the root of the keyway as shown in Figure 5.11a. This crack propagates nearly parallel to the shaft surface. But if the IFC takes almost the entire torque (except rare peaks), the key works under a low load, and no fatigue cracks can begin in the root of the tooth.

The maximal von Mises stresses in the root of the keyway tooth and at the end of the keyway (point 1, Figure 5.12a) were calculated using FEM for the following conditions:

- Two patterns of load distribution over the working surface of the key: triangular (Figure 5.11a) and uniform (Figure 5.11b)
- Two types of hub-to-shaft fits: interference fit and slide fit

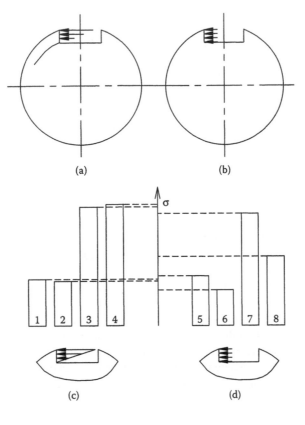

FIGURE 5.11 Tension stress (von Mises) in the shaft keyway.

The bar charts in Figure 5.11c and Figure 5.11d show the stresses for the following:

- The interference fit — bars 1, 2, 5, 6
- The slide fit — bars 3, 4, 7, 8
- The root of the tooth — bars 1, 3, 5, 7
- Point 1 (Figure 5.12a) — bars 2, 4, 6, 8

As is seen from the charts, the stress level depends only slightly on the load distribution pattern, but it grows sharply when the fit changes from interference to slide. The reason for this dramatic effect is that the hub, when fitted with interference, blocks the possible deformation of the tooth

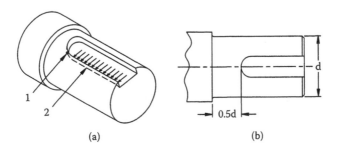

FIGURE 5.12 Stress concentration at the keyway end.

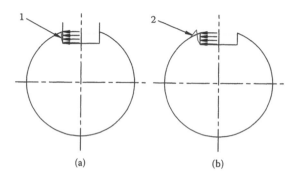

FIGURE 5.13 Keyway deformation under load.

in a radial direction (deformed contour is shown in Figure 5.13a, curve 1). When the hub is slide fitted, the deformation of the tooth is not restricted by the hub (see curve 2 in Figure 5.13b), and the deformation at the same load is much bigger.

When the tooth deforms, it pulls the end fillet of the keyway. In Figure 5.12a, the deformed contour of the keyway is represented by dashed line 2. The aforementioned point 1, where the tension stress is maximal, belongs to the keyway contour and the shaft neck surface, so the material at this point suffers from both high stress and surface damage caused by IFC (fretting and others; see Chapter 2). These damages are related to the microslip in the connection and concentrated at the end area of the connection where the transmitted torque is maximal. If the end of the keyway is moved away from the end of the connection (Figure 5.12b), the fatigue strength of the shaft may be increased by 50–70%.[6]

5.3.4 STRENGTH OF HUB NEAR THE KEYWAY

Undoubtedly, strength problems are usually concerned with the shaft, but sometimes hub failures also are observed. Figure 5.14 shows the von Mises and maximal principal stresses in the keyway fillet calculated using FEM.

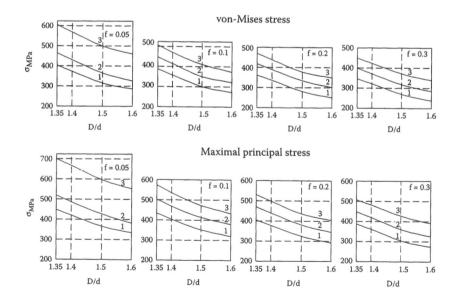

FIGURE 5.14 Stress in the keyed hub.

TABLE 5.1
Parameters of Calculated KJs

Hub diameter D, mm	135	140	150	160
Interference (diametral), mm	0.094	0.088	0.076	0.070
Surface pressure (by Lame), MPa	41.6	42.3	41.4	41.8

The parameters of the key joints are as follows: shaft diameter $d = 100$ mm, hub diameter D is variable (see Table 5.1), key dimensions 28×16 mm, fillet radius $r = 1.5$ mm. The length of the connection $L = 120$ mm. The connection is press-fitted, and the interference is changing depending on the hub diameter, so as to achieve nearly the same surface pressure in all cases.

The stresses are calculated for different values of the friction coefficient ($f = 0.05, 0.1, 0.2$, and 0.3) and for three loading conditions: torque transmitted $T = 0$ (the hub is stressed by the press fit only, curves 1), 3750 N·m (curves 2), and 7500 N·m (curves 3).

From these charts one can see not only that the outer diameter of the hub influences significantly the stress level (this was clear *a priori*), but also that the friction coefficient is a very strong factor, because it characterizes (together with the surface pressure) the ability of the friction forces to transmit the entire torque or most of it. Figure 5.15 shows the stresses at $T = 7500$ N·m. It is seen that at $f < 0.1$, where the load capacity of friction forces becomes less than the torque applied, the stress in the keyway fillet grows quickly. This demonstrates once again, as stated already, the necessity of press fit in KJs able to transmit the entire torque by friction forces; some safety margin is welcome.

Let's compare the fatigue strength of the hub with that of the shaft.

EXAMPLE 5.4

We take for this analysis one of the cases calculated previously: $d = 100$ mm, $D/d = 1.6$, $f = 0.2$ (after some on-period), and $T = 7500$ N·m. The material of the hub: SAE 4130, normalized, tensile strength $S_u = 622$ MPa, and yield point $S_y = 484$ MPa. The von Mises stress in the hub equals $\sigma_1 = 250$ MPa when unloaded and $\sigma_2 = 350$ MPa when transmitting the torque. If the torque (because of torsional vibrations) varies cyclically by, say, 15%, the stress amplitude equals

$$\sigma_a = (\sigma_2 - \sigma_1)0.15 = 15 \text{ MPa}$$

The maximal stress equals $\sigma_{max} = 350 + 15 = 365$ MPa $< S_y$. Thus, the requirement for static strength of the hub is fulfilled.

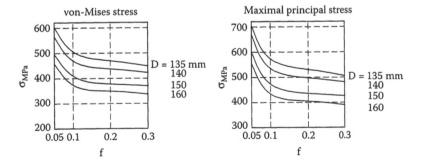

FIGURE 5.15 Stress in the keyed hub ($T = 7400$ N · m).

What about the fatigue? We can determine the fatigue limit of this steel approximately as follows (see Chapter 12, Subsection 12.1.2):

$$S_{-1} = 0.454\,S_u + 8.4\,(\text{MPa}) = 0.454 \cdot 622 + 8.4 = 291\,\text{MPa}$$

The fatigue limit of the part equals (see Chapter 12, Subsection 12.2.4):

$$S_{-1,p} = S_{-1}\frac{K_S K_d}{K_e}$$

where
  K_S = surface quality factor = 0.8 (see Chapter 12, Subsection 12.2.1)
  K_d = dimension factor = 0.9 (see Equation 12.58 in Chapter 12, Subsection 12.2.2)
  K_e = effective stress concentration factor ($K_e = 1$ in this case because the stress is the real maximal stress determined using FEM)

$$S_{-1,p} = 291\frac{0.8 \cdot 0.9}{1} = 210\,MPa$$

If we neglect the mean stress (for the sake of simplicity and because its influence is weak; see Chapter 12), the safety factor of the hub against fatigue failure equals

$$n_H = \frac{S_{-1}}{\sigma_a} = \frac{210}{15} = 14$$

So the hub strength doesn't worry us. What about the shaft strength? If the shaft is loaded with the torque only, the shear stress equals

$$\tau = \frac{16T}{\pi d^3} = \frac{16 \cdot 7500 \cdot 10^3}{\pi \cdot 100^3} = 38.2\,MPa$$

(Note: $T = 7500$ N·m $= 7500 \cdot 10^3$ N·mm. We had to convert the torque units into N· mm, because the diameter d is expressed in mm.) Taking the stress amplitude as 15% of the mean stress, we obtain the shear stress amplitude: $\tau_a = 0.15 \cdot 38.2 = 5.73$ MPa.

Let's assume that the shaft is made of SAE 4340 heat-treated to HRC 28–35. For this steel $S_u = 880$ MPa and $S_y = 725$ MPa. The fatigue limit equals

$$S_{-1} = 0.383\,S_u + 94\,(\text{MPa}) = 0.383 \cdot 880 + 94 = 431\,\text{MPa}$$

For the shaft, $K_S = 0.9$ (ground surfaces), $K_d = 0.9$, and $K_e = 3.5$ (stress concentration factor for the press-fit connection). Thus, the fatigue limit for the shaft equals

$$S_{-1,p} = 431\frac{0.9 \cdot 0.9}{3.5} = 99.7\,\text{MPa}$$

Because

$$\frac{S_{-1}^2}{S_{-1,\tau}^2} = 3$$

where $S_{-1,\tau}$ = fatigue limit in shear (see Chapter 12, Subsection 12.1.2), and

$$S_{-1,\tau} = \frac{99.7}{\sqrt{3}} = 57.6 \text{ MPa}$$

Taking the amplitude of alternating shear stress as 15% of the mean stress, we obtain

$$\tau_a = 38.2 \cdot 0.15 = 5.73 \text{ MPa}$$

From here, the safety factor of the shaft against fatigue failure equals

$$n_S = \frac{S_{-1,\tau}}{\tau_a} = \frac{57.6}{5.73} = 10$$

Hence, though the shaft is largely overdesigned, its safety factor is still less than that of the hub. But in most cases, the shaft is loaded not only by a torque, but also by a bending moment simultaneously. Let us assume that the hub and the shaft are parts of the gear output shaft shown in Figure 4.1 (Chapter 4). The force developed in the gear mesh loads the shaft (in addition to the torque) with a bending moment $M = 5.2 \cdot 10^6$ N·mm and produces in it the bending stresses σ_a that change cyclically at each turn of the shaft:

$$\sigma_a = \frac{32M}{\pi d^3} = \frac{32 \cdot 5.2 \cdot 10^6}{\pi \cdot 100^3} = 53 \text{ MPa}$$

The equivalent stress (see Chapter 12, Subsection 12.1.2),

$$\sigma_e = \sqrt{\sigma_a^2 + 3\tau_a^2} = \sqrt{53^2 + 3 \cdot 5.73^2} = 54 \text{ MPa}$$

And then the safety factor of the shaft equals

$$n_S = \frac{S_{-1,p}}{\sigma_a} = \frac{99.7}{54} = 1.85$$

The safety factor of the hub remains unchanged ($n_H = 14$).

From this example, we clearly see why we don't usually have problems with the hub strength: it is loaded nearly statically. But sometimes, for example, when the torque transmitted is reversible and the press fit is not so tight as to transmit most of it by friction forces, the stress amplitude in the hub increases immensely, and its strength may become insufficient.

In some cases, the keyway is placed in the nearest vicinity of the application of a cyclic load as shown in Figure 5.16a. In this case, the round pins shown in Figure 5.16b can be preferable.

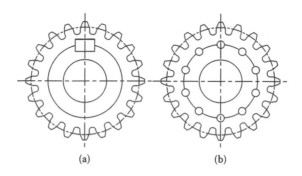

FIGURE 5.16 Key joints of thin-walled parts.

5.3.5 ROUND KEYS (FIGURE 5.17)

Round keys are used for permanent joints, because the holes are drilled and reamed in the shaft and hub together. The connection with round keys is less detrimental to the strength of the shaft and hub and may be used for thin-walled parts, provided that the needed load capacity of the IFC may be practically achieved.

The diameter of keys d_k is recommended within $(0.08–0.15)d$, and the length $l_k = (3–5)d_k$. The maximal bearing stress equals

$$\sigma_{max} = \frac{16T}{\pi d\, d_k\, l_k\, n} \tag{5.11}$$

where n is the number of keys.

You can ask, where is this formula from? Why 16 and not 17 or, say, 24? Oh, the derivation of the formula (Equation 5.11) is really simple. First, look at Figure 5.17b. Hub 1 rotates relative to

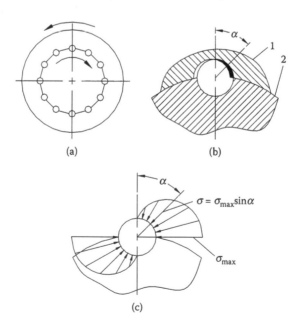

FIGURE 5.17 Round pins.

shaft 2 under load (owing to elastic deformations) by a very small angle. This angle is so small that the movement in the vicinity of one key can be considered as translational displacement in a tangential direction (assuming also that the key diameter is small as compared to the diameter of the connection). This displacement measured in tangential direction is nearly the same at all points, but the bearing deformation in contact of the hub and the key, which is measured perpendicular to the key surface (in Figure 5.17b, this area is blackened), is obviously proportional to $\sin \alpha$. Because the stress is proportional to the deformation, we can write the formula for the bearing stress (see Figure 5.17c):

$$\sigma = \sigma_{max} \sin \alpha$$

Thus, at $\alpha = 0$, $\sigma = 0$; at $\alpha = \pi/2$, $\sigma = \sigma_{max}$.

Force is the product of pressure by area. The infinitesimal area on the key surface is

$$dA = r \cdot d\alpha \cdot l_k = \frac{d_k}{2} l_k \cdot d\alpha$$

From here, the infinitesimal force (applied to the infinitesimal area perpendicular to the surface of the pin)

$$dF = \sigma \cdot dA = \sigma_{max} \sin \alpha \frac{d_k l_k}{2} d\alpha$$

The tangential component of this force, dF_t, which opposes the tangential force created by the torque, equals

$$dF_t = dF \cdot \sin \alpha = \sigma_{max} \sin^2 \alpha \frac{d_k l_k}{2} d\alpha$$

The tangential force taken by one pin equals

$$F_0 = \frac{2T}{n}$$

The obvious equilibrium condition is

$$F_0 = \int_0^{\pi/2} dF_t = \sigma_{max} \frac{d_k l_k}{2} \int_0^{\pi/2} \sin^2 \alpha \cdot d\alpha = \frac{\pi}{8} \sigma_{max} d_k l_k = \frac{2T}{n}$$

From here, Equation 5.11 is obtained directly. In reality, the pressure distribution is not exactly sinusoidal because the radial components of the key pressure deforms the hub in the radial direction, and the pressure concentrates more in the vicinity of the joint diameter. Well, it is too complicated to get into any deeper.

5.4 SPLINED JOINTS

Splined connections are made in two types: straight (Figure 5.18a) and involute (Figure 5.18b). The straight spline has less friction force when moving the hub axially under load. The involute spline is stronger (because of the larger number of teeth) and less detrimental to the strength of

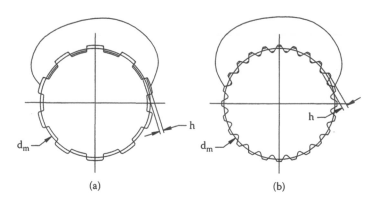

FIGURE 5.18 Spline shapes.

the shaft. The effective stress concentration factor for the involute spline is about 1.7 times less than that for the straight spline.[4]

Performance capabilities of SJs depend strongly on the kind of load transferred — whether it is a torque only or torque with a radial force. The two kinds of loading are considered separately in the following subsections.

5.4.1 SJs LOADED WITH TORQUE ONLY

Figure 5.19 shows several examples of tightened SJs that are loaded (at first sight) with a torque and a radial force. But notice the following: in all these connections, the centering of the hub relative to the shaft is made using additional surfaces, so the spline doesn't participate in the transmission of the radial force.

In Figure 5.19a, flange 1 is centered on the left directly on shaft 2 and, on the right, through insert 3 pressed into flange 1. The insert is used to ease manufacturing by allowing the inner spline to be broached. Tightened nut 4 is locked with special washer 5, which is provided with inner and outer protuberances engaged with corresponding slots in the shaft and nut. Snap ring 6 holds the washer in place.

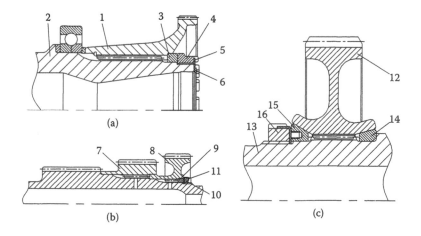

FIGURE 5.19 Centering of spline connections using additional surfaces.

Figure 5.19b represents a shaft of a triple-speed drive. In this design, pinion 7 is centered on the left directly on the shaft and, on the right, through the hub of pinion 8. The latter is centered on the right through insert 9. All these parts are tightened with nut 10 that is locked with a deformable washer 11.

In the design shown in Figure 5.19c, gear 12 is centered relative to shaft 13 using split cones 14 and 15. To make dismantling easier, the angle of the cones is made larger than the coefficient of friction. Usually, the vertex angle is 30° and 60° for cones 14 and 15, respectively. The cones may be made of bronze or brass to discourage fretting. To reduce to a minimum the possible microslip in the connection under load, nut 16 should be tightened strongly.

Figure 5.20 shows the intermediate shaft of a tractor gear box. Gears 1, 2, and 3 are mounted on shaft 4 through rolling bearings, so that they can rotate independently of the shaft. When one of these gears is brought into operation, it is made to rotate with the shaft by coupling 5 or 6 so as to transmit torque. These couplings are connected to shaft 4 through splined sleeves 7 and 8, and they can be engaged or disengaged by shifting their axial positions.

In all, there are seven SJs loaded only with torque:

- Shaft 4 with sleeves 7 and 8
- Coupling 5 with sleeve 7 and gear 1 (when engaged)
- Coupling 6 with sleeve 8 and gears 2 and 3 (when engaged)

The connections of sleeves 7 and 8 are centered relative to the shaft by the outer diameter of the splines, tightly fitted and tightened with nut 9. But in the rest of the connections listed, there must be sufficient clearances to enable free axial movement.

SJs usually fail because of tooth wear; therefore, the strength of SJs is checked by calculating the bearing stress. The tangential force is

$$F_t = \frac{2T}{d_m}$$

where
T = torque transmitted (N · mm)
d_m = mean diameter of the SJ (mm)

The working area of the teeth that take this force is

$$A = hln$$

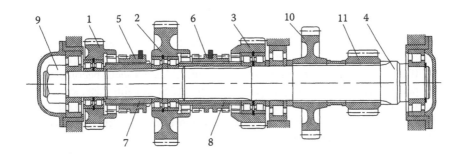

FIGURE 5.20 Intermediate shaft of tractor gear box.

where

> h = height of the active surface of the tooth minus the chamfers (measured in radial direction)
> l = engaged length of the teeth
> n = number of teeth

From here, the nominal bearing stress equals

$$\sigma_{nom} = \frac{F}{A} = \frac{2T}{d_m hln} \tag{5.12}$$

The authors can't refrain from reminding the reader of the necessity to keep compatible dimensions for all the homogenous terms of an equation. In particular, in Equation 5.12, if you choose newtons and millimeters, all the dimensions (d_m, h, and l) must be expressed in millimeters (mm), and torque, T, in newton millimeters (N · mm). The result will be obtained in newton per millimeter, squared (N/mm² = MPa). If the torque is given in newton meters (Nm) and you don't want to change this, all the dimensions listed previously must be given in meters, and the result will be expressed in newton per meter, squared (N/m² = Pa). If you choose pounds and feet — well, you know what to do!

Involute splines usually have a 30° pressure angle and $1.2m$ tooth height (m is the *module* of the teeth, given in millimeters; conformably to the AGMA system, $m = 25.4/P$ where P is the diametral pitch). The engaged height of the teeth $h = m$, and the mean diameter $d_m = m \cdot n$, so the number of teeth $n = d_m/m$. Substitution of these data into Equation 5.12 gives the expression for the nominal bearing stress:

$$\sigma_{nom} = \frac{2T}{d_m^2 l} \tag{5.13}$$

Sometimes, gear teeth are used to form a spline joint. The pressure angle of these teeth is 20°, and the tooth height, $2.25m$. The engaged height $h = 2m$, i.e., twice as much as in spline teeth. Therefore, this joint is more wear resistant (because the bearing stress is half as large as in the usual SJs), but these teeth are more "breakable" and more sensitive to misalignment. Therefore, when there is considerable angular misalignment in the connection, this option (with 20° teeth) is used with crowned (barrel-shaped) teeth (see Figure 5.8). Teeth with engaged height $h = 1.6m$ are also in use.

Equation 5.12 and Equation 5.13 are derived with the assumption that the contact pressure is distributed uniformly over the working surfaces of the teeth. To find the maximal bearing stress, some correction factors are used:

$$\sigma_{max} = \sigma_{nom} k_T k_L \tag{5.14}$$

Factor k_T is used to consider the unevenness of load distribution between the teeth caused by errors of shape and the angular position of the teeth. For nonwearing SJs at the usual accuracy, $k_T = 1.3–1.4$. If there is some wear of the spline, the more loaded teeth become more worn, and after a run-in period, all the teeth will be loaded uniformly. So for this case, $k_T = 1$.

Load distribution may also improve in nonwearing joints because of local plastic deformation of the overloaded surfaces. Therefore, in all cases, it is worth marking the parts so as to assemble the spline always in the same position.

Factor k_L is intended to consider the unevenness of load distribution along the teeth caused by different twisting of the shaft and the hub. (This phenomenon is similar to that in IFC; see Chapter 2.) The larger the length of the SJ, the greater is the k_L value.

The k_L value depends on the rigidity of the parts and the teeth. For example, if the number of teeth is 6, their compliance is greater, and k_L is less than if the number of teeth is, say, 16. For involute splines that have a relatively large number of teeth, the factor k_L may be determined approximately from this equation:

$$k_L \approx 3\frac{l}{d} \tag{5.15}$$

For example, $l/d = 1$, $k_L = 3$. More exact calculation can be done using FEM.

In wearing joints that transfer a constant torque, the teeth will run in, and for them, $k_L = 1$. If the torque value changes considerably, the load distribution along the teeth will never become uniform. Half of the k_L value obtained from Equation 5.15 may be taken in this case just by guesswork.

5.4.2 SJs Loaded with Torque and Radial Force

Examples of this type of SJ are given in Figure 5.20; gears 10 and 11 are connected with shaft 4 by splines that transmit both the torque and the force applied to the teeth.

Let's imagine that a spline is manufactured and assembled with absolute accuracy. If this spline transmits only torque, the load is distributed uniformly between the teeth (Figure 5.21a). In this case, the shaft and the hub are coaxial and the vector sum of all forces applied to the teeth equals zero. If the SJ is also loaded with a radial force, the load distribution between the teeth must be nonuniform, because the vector sum of the tooth forces must equal the loading radial force. (Nobody has cancelled Newton's 3rd Law!) Figure 5.21b shows how the picture changes when the radial force is applied. The hub shifts relative to the shaft in the direction of the force by some small amount ε. At this force, the load of teeth 2 and 3 increases (in the figure, it looks as an interference of the teeth surfaces that were before in touch, but, in reality, it symbolizes increased contact deformation). The load of teeth 5 and 6 decreases because their working surfaces move

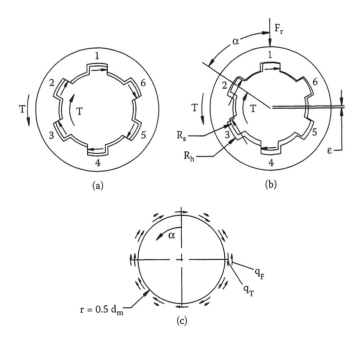

FIGURE 5.21 Load distribution in spline joints.

apart (in the figure, a gap appears between the surfaces; in reality, it is not necessarily a gap, but it may be just a decrease in contact deformation; but a gap may arise as well, if the radial force is large enough). The load of teeth 1 and 4 should not change because the distance between the working surfaces doesn't change as a result of radial displacement of the hub.

As the joint rotates relative to the vector of the radial force, the center of the hub describes a circle of radius ε relative to the shaft, and the working surfaces slip cyclically. Each turn of the joint conforms to one cycle of slippage, and each cycle is accompanied by a certain wear of the working surfaces.

The aforesaid is wholly correct when the hub is centered relative to the shaft by the working (side) surfaces of the teeth. But if the hub is centered by any additional surfaces (for example, by the outer diameter of the teeth, when $R_s = R_h$ [see Figure 5.21b], which excludes any radial displacement of the hub), the spline transmits only the torque, and the radial force is taken by the centering surfaces. In practice, there is often a mixed design: the hub is centered by the outer (or sometimes by the inner) diameter of the teeth, but there is a small clearance (R_h is a bit bigger than R_s) to ensure easy assembly and, if needed, easy axial movement of shifted gears. If there is a radial force, the hub can move in a radial direction within the clearance, and the greater the clearance, the larger the radial displacement, leading to slippage and wear. Therefore, it is desirable to decrease the clearance between the centering surfaces. It should be noticed that the cyclic motion of the hub relative to the shaft brings about wear to the centering surfaces as well. This means that the initially small clearance between the centering surfaces will increase with time and the wear process will accelerate exponentially.

But what can one do! Everything has its working life. And if we sometimes feel sorry in this "connection," the SJ is not the object we usually grieve for.

It is seen from Figure 5.21b that the "change in distance" between the working surfaces of the mating teeth associated with the hub radial displacement is proportional to sin α. (For teeth 1 and 4, sin α = 0, so their load doesn't change.) To derive the equations for the maximal and minimal tooth force, let's assume that there are infinitely many teeth. Then we consider the load distribution from torque T and from radial force F_r separately. The unit tangential force from the torque (per millimeter of the SJ's circumference) assuming its uniform distribution (see Figure 5.21c) is

$$q_T = \frac{T/r}{2\pi r} = \frac{T}{2\pi r^2} \ N/mm$$

The unit tangential force from the radial load is

$$q_F = q_{F\max} \sin \alpha \ N/mm$$

The vector sum of the q_T forces is zero; so the equilibrium condition lies in the fact that the scalar sum of the q_F forces must be equal to F_r. The resolution of forces q_F by vertical and horizontal components shows that the horizontal components on the right and on the left get balanced. The sum of the vertical components $q_{F vert}$ shall counterbalance the force F_r:

$$q_{F vert} = q_F \sin \alpha = q_{F\max} \sin^2 \alpha$$

The infinitesimal force equals

$$df = q_{F vert} \, ds = q_{F vert} \, r d\alpha = q_{F\max} \, r \sin^2 \alpha \cdot d\alpha$$

From the equilibrium condition,

$$F_r = \int_0^{2\pi} df = \int_0^{2\pi} q_{F\max} \, r\sin^2\alpha \cdot d\alpha = \pi r q_{F\max}$$

$$q_{F\max} = \frac{F_r}{\pi r} \; N/mm$$

So the maximal and the minimal unit tangential loads are

$$q_{\max} = q_T + q_{F\max} = \frac{T}{2\pi r^2} + \frac{F_r}{\pi r};$$

$$q_{\min} = q_T - q_{F\max} = \frac{T}{2\pi r^2} - \frac{F_r}{\pi r};$$

(5.16)

Let's denote k_R as the factor that considers the unevenness of load distribution between the teeth that is caused by radial force. Because $r = 0.5d_m$,

$$k_R = \frac{q_{\max}}{q_T} = \frac{\dfrac{T}{2\pi r^2} + \dfrac{F_r}{\pi r}}{\dfrac{T}{2\pi r^2}} = 1 + \frac{2F_r r}{T} = 1 + \frac{F_r d_m}{T}$$

(5.17)

If we would like to know the maximal tangential force applied to one tooth, we can calculate it from Equation 5.16, taking into consideration that the force is equal to the product of the unit load and the length of an arc. The length of arc per tooth

$$s = \frac{\pi d_m}{n}$$

Thus, the max tooth force equals

$$F_{\max} = q_{\max} s = \left(\frac{T}{2\pi r^2} + \frac{F_r}{\pi r}\right)\frac{\pi d_m}{n} = \frac{2}{n}\left(\frac{T}{d_m} + F_r\right)$$

An additional unevenness of load distribution along the teeth is caused if the radial force vector is displaced from the middle of SJ by some amount e, (Figure 5.22a). Such a design should be avoided because uneven wear of the spline will result in misalignment of the hub relative to the shaft. If the hub belongs to a gear, the misalignment of the gear teeth may cause its premature failure. It may be preferable to make the spline shorter but symmetrical relative to the radial force (Figure 5.22b).

For a spur gear that is mounted on a splined shaft (such as gears 10 and 11 in Figure 5.20), radial force F_r is a force F_g applied to the gear teeth. This force equals

$$F_g = \frac{2T}{d_g \cos 20°} \approx \frac{2T}{d_g}$$

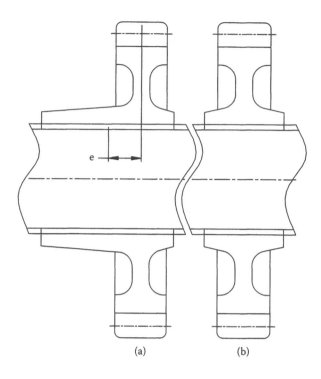

FIGURE 5.22 Influence of toothed rim dissymmetry on a spline connection.

(Here d_g is the pitch diameter of the gear.) From the previous equation, also taking into consideration Equation 5.17,

$$k_R = 1 + 2\frac{d_m}{d_g} \tag{5.18}$$

From Equation 5.18, we can see that the bigger the ratio d_m/d_g, the greater is the unevenness of load distribution between the teeth of the SJ.

5.4.3 ALLOWABLE BEARING STRESSES IN SJs

Prestressed SJs that practically don't wear are calculated for maximal bearing stress. Because in these joints, the centering is usually performed using additional surfaces (see Figure 5.19), the spline transmits the torque only, and the bearing stress may be determined using Equation 5.12 to Equation 5.14. This stress should not exceed the yield point of the material divided by some safety factor of about 1.2–1.4. It is clear that for this calculation, the maximal torque, including the possible vibrations and momentary overloads, must be taken into account.

If the hub may move relative to the shaft, the spline is calculated both for the maximal bearing stress and for wear. The first one is performed just as for the prestressed joints, but if the SJ is loaded by a radial force as well, the factor k_R should be taken into account.

For the wear calculation, the momentary loads should not be considered. Usually, the calculation is based on the maximal of the long-acting loads typical of the work cycle. The contact stress in this case may be determined using Equation 5.12 to Equation 5.14, where $k_L = k_T = 1$. But if there is a radial force applied to SJ, the factor k_R must also be included.

The allowable bearing stress for wearing joints depends on the wear resistance of the teeth and the needed service life. The determining factor for the wear resistance is the hardness of the working surfaces of the teeth (i.e., their flanks). These surfaces are often treated to produce a hard layer (as in induction hardening, carburizing, or nitriding) and/or coated with materials that decrease friction and wear. Besides, the wear rate depends on the amplitude of the cyclic displacement of the hub relative to the shaft. It may depend on the d_m/d_g ratio (for gears), or it may be determined by functional needs (e.g., SJs of vehicle cardan shafts or SJs used to connect shafts with a certain misalignment; see Figure 5.23 and Chapter 6, Figure 6.43). Wear resistance is also greatly dependent on lubricant quality and cleanness.

Usually, calculation of the allowable bearing stress is based on experimental and service data. As a guide, for tempered steels (HRC 30–35), the allowed bearing stress may be within the limits of 15 to 30 MPa, and for case hardened teeth (HRC 56–61), 40 to 60 MPa. It doesn't mean that the stress can't be 100–150 MPa, or even higher at certain conditions, but everything should be checked experimentally.

5.4.4 Lubrication of SJs

The mechanism of lubrication of metals is described in Chapter 6, Subsection 6.3.2, in which the absolute necessity of lubrication, at least before assembly, becomes completely clear. For heavily loaded prestressed joints, where micromovements are expected, a permanent access to liquid lubricant may be needed (Figure 5.19b).

If the hub is expected to move relative to the shaft, the spline may be greased or lubricated with liquid oil. In joints placed outside a mechanism (for example, in cardan shafts), the splines are generally greased for life or with a greaser that enables periodical relubrication.

Grease lubrication simplifies the design of the joint: there is no need for an oil tank, or oil supply devices; the seals are much simpler and, even if the seal is worn a bit, no leakage can be observed. But the heat removal in this case is relatively small. Therefore, if heat release in the joint

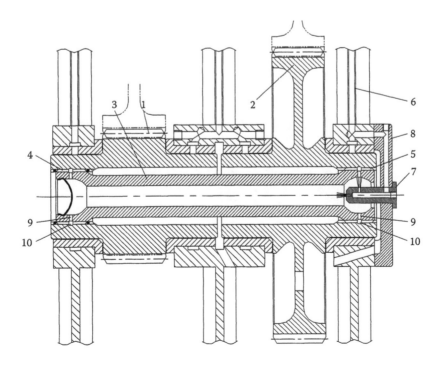

FIGURE 5.23 Lubrication of loose spline joints.

is high (because of high speed, increased cyclic movement in the joint, or both), the liquid lubrication system should be used to take the heat off effectively.

When placed inside an oil-lubricated mechanism, the SJ is lubricated with oil as well. Figure 5.23 shows a typical connection of two gears, 1 and 2, by torsion bar 3 with involute splines 4 and 5 on its ends. Gears 1 and 2 are not concentric because the radial forces applied to the gears are directed oppositely, and the radial displacement of the gears in their sliding bearings is opposite. As a result, there is a misalignment and cyclic movement with the frequency of rotation in the SJs. The lubricating oil comes from oil duct 6 to the sliding bearing of the gear and at the same time to oil sprayer 7 (through the ducts in the housing and cover 8). The sprayer has two nozzles: one, central, jets oil to spline 4; the other, a side nozzle, supplies oil to spline 5. Under centrifugal forces, the oil fills up the annular spaces formed inside the torsion bar and then through radial holes 9 flows to circular grooves 10. These grooves are turned both in the inner and outer splines. The oil fills up the grooves and flows out from there through the backlashes in the splines, taking off the heat and washing out all the wear debris.

REFERENCES

1. Grechistchev, E.S. and Iliashenko, A.A., *Interference Fit Connections*, Machinostroenie, Moscow, 1981 (in Russian).
2. Birger, I.A. and Iosilevich, G.B., *Threaded Joints*, Machinostroenie, Moscow, 1973 (in Russian).
3. Bossard, *Metric Fasteners for Advanced Assembly Engineering*, Bossard Ltd. Fasteners, 1990.
4. Birger, I.A., Shorr, B.F., and Iosilevich, G.B., *Calculation of Machine Elements for Strength*, Mashinostroenie, Moscow, 1979 (in Russian).
5. Timoshenko, S. and Woinowsky-Krieger, S., *Theory of Plates and Shells,* McGraw-Hill, New York, Toronto, London, 1959.
6. Kogaev, V.P., *Calculations for Strength at Stresses Variable with Time*, Mashinostroenie, Moscow, 1977 (in Russian).

6 Supports and Bearings

At the beginning of this chapter, two technical terms should be defined: *bearing* and *support*. Here, a bearing is considered a part or a unit that serves directly as an interface between the rotating and stationary parts (say, between the shaft and the housing). A support is considered a unit that includes at least one bearing (maybe two or more), adjoining part of a housing, and possibly other parts needed for the axial fixation of the bearings, end play adjustment, etc. No other terms need any elucidation.

6.1 TYPES AND LOCATION OF SUPPORTS

Usually, shafts have two supports. Each support fixes the adjoined part of the shaft in a radial direction. Additionally, the support may fix this part in one axial direction (type F1), in both directions (type F2), or not to fix in the axial direction at all (type NF).

For shafts fixed in the axial direction, a combination of supports, F2 + NF (Figure 6.1a) or two F1 supports (Figure 6.1b), is used. In the latter case, the axial play that is set when assembling must be greater than the probable difference in thermal deformations of the shaft and the housing; otherwise, the supports may become jammed and damaged. This difference in thermal expansions can't be exactly determined because it depends on the power loss in the mechanism, environment temperature, cooling conditions, and so on. As a rule, if the shaft needs an exact axial fixation, such a design is not acceptable.

A shaft may have two NF supports if it is joined to another part, which is fixed axially, for example, by a flange connection to another shaft (Figure 6.1c) or by the engagement of two double-helical gears (Figure 6.1d). Error of axial location of such a shaft includes axial play of the axially fixed part plus the possible clearance in the connection of the two parts. Specifically, in the flange connection, the clearance is zero. But in the engagement of double-helical gears (Figure 6.1d), the situation is more complicated. The nominal position of teeth 1 (belonging to the pinion) relative to teeth 2 of the gear wheel is symmetric (Figure 6.1e). In this position, which is realized under load, both the right and the left teeth of the pinion are in contact with the corresponding teeth of the gear. If the gear is unloaded, pinion 1 may be moved relative to gear 2 in an axial direction by amount X (Figure 6.1f). In this position, pinion teeth 1 and gear teeth 2 are in contact from above on the left and below on the right side. Obviously, the pinion may be moved by the same distance in the opposite direction. The total axial movement of the pinion is given by

$$2X = \frac{Z}{\sin \psi \cos \varphi}$$

where
Z = backlash magnitude in the engagement (mm)
ψ = helix angle
φ = pressure angle

For example, if the backlash magnitude $Z = 0.5$ mm, $\psi = 30°$, and $\varphi = 23°$, the possible axial movement of the pinion with respect to the gear equals 1.09 mm. Under load, axial forces F_a move the pinion to a symmetric position depicted in Figure 6.1e. In this position, the teeth on the left

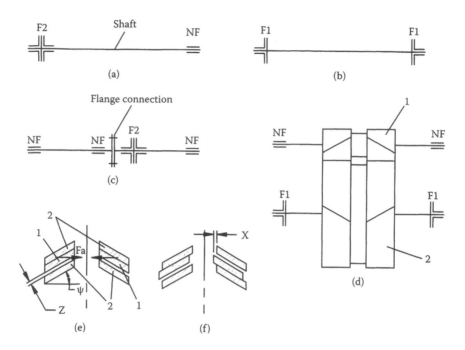

FIGURE 6.1 Examples of axial fixation of shafts.

and on the right are transmitting the same load, and forces F_a are even on both sides. Gear tolerances (such as radial and lateral runouts, variations in tooth thickness and helix angle, and others) make the pinion oscillate with small amplitude during rotation to keep the symmetric position. This movement should not be hindered, because it is aimed at achieving equality of forces F_a and, consequently, of even load distribution between the left and right teeth.

To put it more precisely, the complete equality of forces F_a is not achieved because of frictional forces in the teeth, bearings, and couplings. This point is discussed more comprehensively in Subsection 6.2.3 (friction in roller bearings; see also Example 6.3) and Chapter 7, Section 7.8.

A supported section of a shaft may move relative to the housing within limits determined by clearances in the support, runout of the bearings, and elastic deformations of bearings and other parts of the supports. Therefore, selection of a distance between supports and their location relative to shaft-mounted elements (gears, cams, pulleys, and so on) may have a dramatic effect on the shaft loads and on the shaft's actual position (alignment) relative to the housing. Let's explore this in more detail.

The alignment of a shaft in radial supports, which have clearances and certain compliances, depends on the magnitude and direction of the radial forces applied to the supports. In the simplest case, when the shaft rotates in radial bearings and the reactions at the supports, R_1 and R_2, have the same direction (Figure 6.2a), the angle of misalignment of the shaft relative to the housing, Θ, is determined by the difference in radial clearances and elastic deformations of the supports:

$$\Theta = \frac{\Delta_1 - \Delta_2}{L} \ (rad)$$

where Δ_1 and Δ_2 are the radial displacements of the bearing necks of the shaft in supports 1 and 2, respectively, under the action of forces R_1 and R_2, and L = spacing of the bearings.

If the supports are made with plain bearings, the inequality of vectors R_1 and R_2 may cause an additional misalignment of the shaft because of the different eccentricity of the shaft journals in the bearings and the different angular position of the point of minimal clearance.

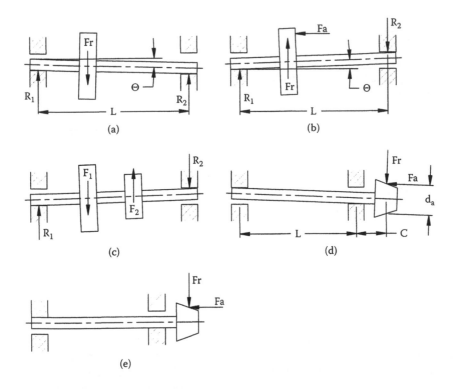

FIGURE 6.2 Shaft misalignment due to displacement in bearings.

If the shaft is loaded as shown in Figure 6.2b to Figure 6.2d, the reactions at the supports may have an opposite direction, and in these cases, the angle of misalignment will be much bigger:

$$\Theta = \frac{\Delta_1 + \Delta_2}{L}$$

To decrease the misalignment, the clearances in the supports should be made as small as possible, and the stiffness of the supports and spacing L should be increased as much as practical.

Not always is the increased spacing of supports welcomed. If the shaft is exposed to bending moments, its increased length may impair the alignment of machine elements mounted on it. For example, the misalignment angles Θ_b in the middle of the bevel pinion, which is caused by bending of the shaft by forces F_r and F_a, is given by

$$\Theta_{b,Fr} = -\frac{F_r C}{6EI}(2L+3C); \quad \Theta_{b,Fa} = \frac{F_a d_a}{6EI}(L+3C)$$

(The dimensions are shown in Figure 6.2d; angles Θ_b are not signed.)

The final misalignment of the pinion results from the vector summation of these bending-conditioned angles with the supports-conditioned angle Θ, and the optimal ratio of dimensions should be found in each case, separately.

EXAMPLE 6.1

Figure 6.3a shows a double-geared planet wheel with straight teeth. As applied to the topic under discussion, this planet wheel represents a shaft, which has two supports of type F1 placed inside

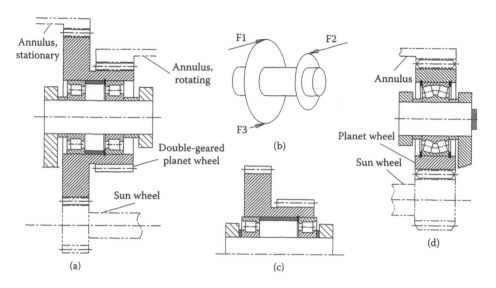

FIGURE 6.3 Planet wheel bearings.

it and two toothings from the outside. This planet wheel doesn't need an exact axial fixation. The direction of forces applied to the teeth is shown in Figure 6.3b (to make the scheme clearer, only the tangential components are depicted). Force F_3 is relatively small, and the planet wheel is mainly loaded by forces F_1 and F_2. The designer of this gear had placed the bearings so that each of them is symmetric about one of the toothings (Figure 6.3a). During the test, a strange phenomenon was found: after short-term work under heavy load, the planetary gear appeared seized up, so that it was impossible to turn it by hand. (However, a good blow by heavy hammer would unlock it.) Dismantling revealed that the roller bearings of the planet wheels had severe smearing on their raceways concentrated on a third of their width. In addition, some of the bearings had broken shoulders of the inner rings. These damages were brought about by angular misalignment (skewing) of the planet wheel relative to its axle, which led to the skewing of outer rings of the bearings relative to their inner rings. This trouble was eliminated by increasing the bearing spacing (Figure 6.3c), when not only the angle of skew but also the load of the bearings was markedly decreased. The change was made within the existing width of the planet carrier openings.

Now, as we have made certain how important it is to decrease the misalignment of a part, let's consider (just for a change) another example, in which, in contrast, the part should have the freedom to skew freely in any direction. Figure 6.3d shows a spur planet wheel supported by a single spherical bearing installed symmetrically about the middle of the tooth. This design has a unique property: self-adjustment of the planet wheel relative to the mating gears (sun wheel and annulus) so that the load is evenly distributed along the teeth. This method may be used not only for planet wheels, but also for any idler. Uniform load distribution along the teeth, which is assured by this design, enables the use of gears with an increased face width and correspondingly decreased diameters. This may be important when the radial dimensions of the drive are limited.

The question is why this design is not used for all spur planet wheels and idlers? There can be several reasons for this. First, a spherical bearing of needed load capacity can't always be installed inside the gear. Quite often, the idler should be installed on an axle with the bearings placed on both sides of the gear, where their diameters can be made much greater. Second, spherical bearings are less suitable for high-speed applications than, for example, straight roller bearings. And third, perhaps, a certain inertia of the designer's thinking and distrust of new, untried solutions should be mentioned. This approach is positive and helps prevent mistakes, but sometimes it hampers progress. Indeed, if some design is not used in such a trustworthy field as aviation or transportation engineering, the designer doesn't feel confident enough to use it. At the same time, the design with idlers or planet spur gears supported by a single spherical roller bearing has been successfully proved in low- and middle-speed drives.

Let's come back to the usual design of shafts supported in two points. With respect to shaft misalignment, one of the most unfavorable variants is an overhung loading (Figure 6.2d). Roller bearings, when properly spaced ($L \geq 2C$ is recommended[1]), are able to position the shaft with reasonable accuracy; see Figure 6.15 (input shaft 6) and Figure 6.20a. Nevertheless, for bevel gears, one usually uses special tooth geometry with localized contact (for example, spiral bevel gears), which are less sensitive to misalignment. In addition, the gears should be adjusted during assembly in such a way as to displace the contact pattern toward the vertex of the cone as shown in Figure 6.20a. Under load, the shaft becomes bent and displaced in the supports as shown in Figure 6.2d, and then the real contact pattern is expected to move toward the base of the cone and to take up position in the middle of the tooth length.

Shafts with an overhung load supported by sliding bearings may keep an aligned position under load if the sleeve bearings are not concentric (Figure 6.2e). The magnitude and direction of the eccentricity should be calculated so as to provide the needed location of the shaft. Such a design requires nice calculation of the plain bearings, and the partial load conditions should be taken into account as well. An additional point to emphasize is that the bearings must not have any wear while in service; otherwise, all this elaborate work becomes meaningless.

Overhung loads are usually avoided in heavily loaded mechanisms. In Figure 6.16c and Figure 7.31 (see Chapter 7), the gear is so designed that the bevel pinion is placed between two bearings.

As noted in the preceding text, the supports may be made of rolling bearings (RBs) or sliding bearings (SBs). The SBs are not as good at a low speed when the machine is starting or reciprocating, because in these cases, there is mixed or boundary friction, which is accompanied by wear and increased heat generation.

"Less good" doesn't always mean "bad." For example, the piston pin bearing in piston engines is usually made of a bronze sleeve and works successfully, in spite of relatively low speed and reciprocating motion.

SBs are more sensitive to the quality of the lubricating oil and to the mode of lubrication. In many cases, they need a pressure lubrication system with filtration to maintain lubricant cleanliness. At the same time, RBs content themselves at low and moderate speeds with splash lubrication or with grease applied during assembly. For these reasons, small and midsize mechanisms are usually provided with RBs, which have very low friction starting from zero speed.

Of fundamental importance is that RBs can work with very small clearances or even with prestressing. That allows the shaft to be positioned more exactly. RBs are not as good at high speed; their load capacity decreases (and service life is shorter because of the large number of cycles). In addition, the running temperature is higher. The reliability of RBs at a high speed is more influenced by errors of manufacturing and assembly, such as runouts and waviness of races, imbalance or poor centering of cages, runouts of the shaft necks and shoulders, fit tolerances, accuracy of axial play adjustment, and so on. If the speed is more than two thirds of the catalog limit, considerable attention should be given to the working conditions of RBs; such factors as vibrations, poorly purified oil, insufficient (and also excessive) lubrication, and nonoptimal oil viscosity may cause the failure of a bearing. In these cases, it is worth asking the bearing manufacturer for recommendations.

As opposed to RBs, the load capacity of SBs increases when the speed rises. In SBs, the continuous oil film that separates the shaft journal from the bearing is achieved safely at a high speed. Under such conditions, the SB can work over a long period of time practically without any wear. Therefore, in large high-speed machines, in which heavy load and high speed are combined with a long service life, the SB is quite often the only acceptable type of bearing.

One of the controlling considerations is cost. Large SBs are much cheaper than RBs both in production and in maintenance. Early detection of an SB's damage can be easily accomplished by measuring its temperature or by sampling metal particles in the oil. This enables the machine to be stopped before significant damage occurs. The repair is usually cheap and consists of replacing the damaged bearing sleeve. In contrast to SBs, the failure of RBs often comes with little warning and is accompanied by damage to other parts. The small radial dimensions and weight of SBs and their ability to dampen impact loads give them wide application in combustion engines, including car motors.

6.2 ROLLING BEARINGS (RBs)

Catalogs of RBs are provided with brief information concerning design, technical data, and directions for use. This information must be taken into consideration. In the following text are discussed some important factors for the reliability and durability of RBs that will help designers orient themselves in the information given by the manufacturers.

6.2.1 Design of RBs

Deep groove ball bearings (Figure 6.4a) consist of two rings, inner and outer, and a set of balls between them, separated by a cage. Raceways are shaped closely to the balls; a cross section of the raceway represents an arc of circle with radius larger by 3% than that of the ball. The number of balls depends on the possibility to put them in place. Figure 6.5 shows the stages of assembly: placement of the balls inside the "crescent" (Figure 6.5a), shifting of the inner ring to the central position (Figure 6.5b), and equal distribution of the balls around the circumference (Figure 6.5c). After that, the cage should be installed in place to keep the distribution of the balls even. The cage is made of two halves, put in from both sides of the balls and riveted together within intervals between the balls.

If for some reasons the cage is destroyed in service, the assembly process may be reversed: the balls may gather and fall out of the bearing. In this case, the shaft will dangle with big amplitude and destroy everything on its way. This feature of such bearings should be taken into consideration when thinking over possible emergency conditions. From this point of view, bearings with greater quantity of balls or rollers are preferable; when the cage fails, the quantity of rollers keeps the inner ring concentric to the outer ring, and the shaft may remain in place until the bearing failure is revealed.

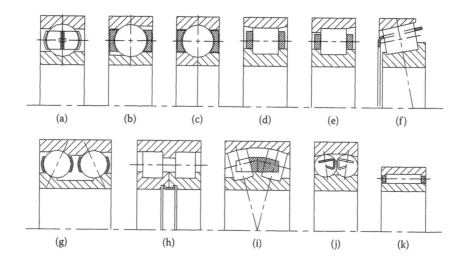

FIGURE 6.4 Basic types of rolling bearings commonly used in machines.

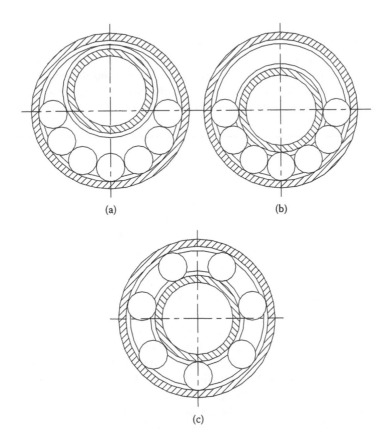

FIGURE 6.5 Assembling a radial ball bearing.

Single-row angular contact ball bearings (Figure 6.4b) and four-point bearings (Figure 6.4c) have more balls because the assembly conditions are better. The number of balls here is limited by the strength of the cage between the balls. The main responsibility of the cage, which may be made solid for these bearings, is to prevent contact between the rolling elements, because the high sliding velocity in their contact may result not only in high power losses but also in burns and scoring of their surfaces.

Bearings with straight or tapered rollers (Figure 6.4d to Figure 6.4f) have cylindrical or tapered raceways, respectively. These surfaces are usually modified to reduce the sensitivity of these bearings to the misalignment of the rings. The modification consists of small removal (about several microns) of metal at the ends of the raceways or the rollers. Nevertheless, roller bearings are very sensitive to angular misalignment of the shaft, which should not exceed 0.001 rad.

The function of the cage in roller bearings is to separate the rollers and to prevent skewing of the rollers relative to the bearing axis. (The latter function is also performed by the shoulders of the rings, and thus they help each other.) In separable bearings (Figure 6.4d to Figure 6.4f and Figure 6.4k), the cage holds the set of rolling elements together. It saves us from crawling on all fours and trying to find these balls and rollers that have fallen out of the bearing.

Full complement cylindrical roller bearings are used for particularly heavy loads. They are designed as shown in Figure 6.4d and Figure 6.4e but without a cage. This enables installing additional rollers, and owing to that, the service life of the bearing may be increased by a factor of 1.5–2.5. Because the rollers are rubbing against each other, the allowable speed of a full complement bearing is nearly half that of a similar bearing with a cage.

Further increase in the load capacity of roller bearings may be achieved by increasing the number of rows. (It is clear that, in this case, the axial dimensions of the bearing are increased.) Figure 6.4g and Figure 6.4h show two-row RBs. A serious disadvantage of such bearings is that they are extremely sensitive to angular misalignment. In modern machinery, the shafts are mainly made of high-strength materials. Both the shafts and the housings are highly stressed and hence have significant elastic deformations. This restricts the field of application of bearings shown in Figure 6.4g and Figure 6.4h.

For the same reason, a two-row spherical roller bearing (Figure 6.4i), which combines very high load capacity with insensitivity to skewing, has been very successful. The two-row self-aligning ball bearing (Figure 6.4j) also has the property of insensitivity to skewing, but the geometry of the outer raceway is very unfavorable with respect to contact stresses, so that its load capacity is even less than that of the regular ball bearing shown in Figure 6.4a.

It should be pointed out that bearings with spherical outer raceways are not indifferent to which ring — inner or outer — is rotating. If the inner ring rotates, the normal operation of the bearing is not influenced by the misalignment angle until the rollers or balls overstep the limits of the outer raceway. On the other hand, if the inner ring is stationary and the outer ring rotates (as in Figure 6.3d), there is a lateral sliding in the contact of the rollers or balls with the outer raceway, and the speed of sliding is in direct proportion to the skewing angle. Friction forces generated in sliding contact may be large enough to have a detrimental effect upon bearing operation and reliability. In this case, it is worth asking for a recommendation from the bearing manufacturer. A more detailed discussion of the friction forces in rolling and sliding contact is presented in Subsection 6.2.3.

Another favorite type of RB is the needle bearing, depicted in Figure 6.4k. It has relatively small radial dimension and is suitable for cramped places. It is also advantageous for oscillatory motion. Owing to the relatively small diameter of the rollers, a minor oscillation of the ring results in a sizable rotation of the rollers, which makes them less prone to the formation of "false brinelling" (see the following text). But, again, because the rollers are longer, these bearings are very sensitive to misalignment.

The long rollers of a needle bearing can't be directed by the shoulders of the rings (unlike regular roller bearings with the length to diameter ratio between 1 and 1.5); therefore, it is desirable to provide them with a cage. The roller of a cageless needle bearing, when it comes to an unloaded zone, may skew relative to the generating lines of the raceways within the radial gap. When such a skewed roller comes to the loaded zone, it is bent and may be broken and cause distortion to the bearing.

The simplest cages are formed by stamping from sheet metal (Figure 6.4a). Such cages are less strong, being supported directly by the rolling elements. It is advisable to avoid using RBs with stamped cages at increased vibration and speed, and in the field of centrifugal forces (for example, in the supports of planet wheels). Under such working conditions, solid cages of copper or aluminum alloys, centered by one of the rings, are preferred. Centering by the outer ring provides a better effect, because the centrifugal force presses the cage to the centering surface by a heavier side. So the wear takes off material from the heavier part of the cage, i.e., decreases its imbalance. (Because of the same mechanism, the wear of a cage centered on the inner ring increases its imbalance.) In addition, the area of the centering surface is larger on the outer ring (because of the larger diameter), and consequently, the surface pressure and the wear are less.

The design of the cage has a strong influence on the lubrication conditions of RBs and is of critical importance at a high speed. Figure 6.6 gives the results of an experimental investigation of three versions of a straight roller bearing, undertaken with the purpose of increasing the speed limit of the bearing.[2] Version 1 has a solid cage centered by the lips of the inner ring. The cage has slots on its centering surface to let the oil in, but still the oil entrance is hindered. In version 2, the cage is centered by the lips of the outer ring, and it is provided with slots on its centering surface for the oil outlet. Between the cage and the inner ring, there is a considerable circular gap, through which the oil enters freely into the bearing. But outlet of the oil is hindered, because it is blocked

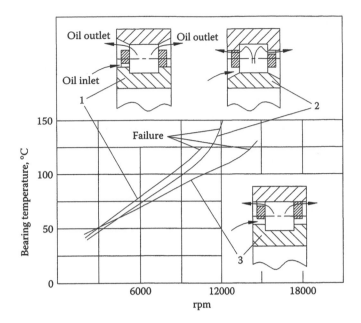

FIGURE 6.6 Comparison of high-speed roller bearings.

between the lips of the outer ring. Nevertheless, the speed limit of this bearing is higher. In version 3, the favorable centering of the cage by the outer ring is retained, but the outer ring is made without a lip. Now the conditions of inlet and outlet of oil are the best, and the speed limit is greater than either version 1 or version 2.

Some types of RBs have cages of composite materials, which are more sensitive to high temperature and certain oil additives. When using such bearings, the corresponding information of the manufacturer should be studied thoroughly. Rings and rolling elements of RBs are usually made of low-alloy, high-carbon steel that reaches a hardness of 59–63 HRC after hardening and low-temperature tempering. Tempering temperature is about 150–160°C, and most of the residual stresses due to quenching remain in the steel, which becomes brittle. Therefore, RBs must not be subjected to impact during installation; this may release part of the quenching stresses and change the geometry of the raceways. And what is more, the rings may crack and even break to pieces, which may fly apart and injure personnel. Generally, machines like to be treated nicely and gently. Mounting and dismounting of RBs should be performed using special devices or temperature deformation (usually heating in oil with controlled temperature). It is clear that the heating temperature must not exceed 150°C (usually, the recommended temperature limit is 120°C), because it may lead to a decrease of hardness and to a change in the geometry of parts. No doubt, they would become less brittle if they were heated up with eagerness, but this scrap would not be a bearing anymore.

6.2.2 Stresses and Failures in RBs

The most strained element of RBs is the rolling contact where the stresses may be as much as 2500–5000 MPa. When the bearing rotates, each point on the surfaces of the rolling elements and the raceways is exposed to the cyclic load, which changes from zero to maximum. This results in fatigue distortion of the surfaces in the course of time. In appearance, this distortion looks like a separation of small rounded particles from the surfaces and formation of pits with dark, lusterless surfaces. This form of surface deterioration is called *pitting*. It is accepted that the rolling contact surfaces of RBs don't have a fatigue limit, and they should be calculated for a certain service life. The spread in service life of RBs can be large (tenfold and more), even if they are taken from the same lot and tested under the same conditions on a laboratory stand. Usually, the service life calculation is based on 90% probability. This probability could be increased by reducing the load.

When pitting is considered, the meaning of *service life* is rather vague. The distortion of working surfaces may develop slowly (especially when the speed of rotation is not too high), and the degradation of functional performance of the bearing may be very slow as well. Therefore, in many cases the real service life of a bearing is finally established on the basis of experiments and observations in service. It may appear much longer than calculated.

The load capacity of an RB depends not only on its geometry, but also on the quality of the metal and heat treatment, on the accuracy of machining and surface conditions, on technological subtleties, lubrication, and so on. That is why the formulas for contact stresses given in the following text are intended for understanding purposes only. The load capacity of RBs should be calculated in accordance with the recommendations and data of the manufacturers given in the catalogs. Let's examine the contact stresses by the simplest example of a cylindrical roller bearing. Figure 6.7a shows a bearing loaded with radial force R. The most loaded roller is pressed between two rings with force F that can be approximately derived from the equation

$$F \approx \frac{5R}{z} \tag{6.1}$$

where z is the number of rolling elements in the bearing.

Contact of the roller with the inner ring represents a contact of two cylinders with radii r and r_i (see Figure 6.7). Owing to elastic deformation, the line contact changes under load into area contact of width b_C (Figure 6.7b). (Length L of the area is the length of the contacting cylinders.) The compressive stress is distributed in the cross section in a semielliptical form. The maximum magnitude of this stress σ_H may be determined using Hertz's formula:

$$\sigma_H = 0.418\sqrt{\frac{FE}{r_e L}} \tag{6.2}$$

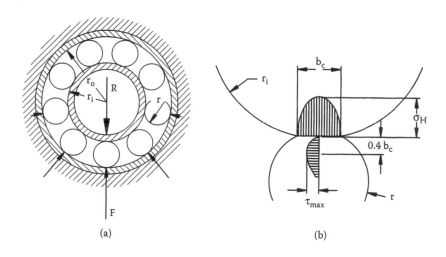

(a)   (b)

FIGURE 6.7 Stresses in two-cylinder contact.

where
    F = load, N
    E = modulus of elasticity (MPa)
    L = length of contact line (mm)
    r_e = equivalent radius of curvature (mm) defined as

$$r_e = \frac{r_i r}{r_i + r}$$

The width of the contact area is

$$b_C = 3.04 \sqrt{\frac{F r_e}{LE}} \tag{6.3}$$

The elastic deformation in the radial direction Δ_C (reduction of distance between the centers of cylinders) is given by

$$\Delta_C = \frac{2F \cdot (1 - v^2)}{\pi \cdot L \cdot E}\left(\frac{2}{3} + \ln\frac{4r}{b_C} + \ln\frac{4r_i}{b_C}\right) \tag{6.4}$$

For the contact of the roller with the outer ring, which is concave, Equation 6.2 to Equation 6.4 are applicable as well, but the r_e value must be derived from another formula:

$$r_e = \frac{r_o r}{r_o - r}$$

Inasmuch as the denominator in this case is less (difference of radii instead of sum), the r_e value is larger, and the stress σ_H is less than it was in contact with the inner ring.

EXAMPLE 6.2

Let's calculate the stresses and deformations in a straight roller bearing NU310E. The diameter and length of the rollers are equal to 16 mm; hence, $r = 8$ mm, and $L = 16$ mm. The radius of the inner raceway $r_i = 32.5$ mm, that of the outer raceway $r_o = 48.5$ mm. The number of rollers $z = 12$. The radial load $R = 10{,}000$ N.

The maximal force on a single roller according to Equation 6.1 is

$$F = \frac{5 \cdot 10000}{12} = 4167\ N$$

The equivalent radius of curvature in contact with the inner ring

$$r_{ei} = \frac{32.5 \cdot 8}{32.5 + 8} = 6.42\ mm$$

The maximum Hertzian stress in this contact region

$$\sigma_{H,i} = 0.418 \sqrt{\frac{4167 \cdot 2.06 \cdot 10^5}{6.42 \cdot 16}} = 1208 \; MPa$$

Width of the contact area is given by

$$b_{C,i} = 3.04 \sqrt{\frac{4167 \cdot 6.42}{16 \cdot 2.06 \cdot 10^5}} = 0.274 \; mm$$

Elastic deformation in the radial direction is given by

$$\Delta_{C,i} = \frac{2 \cdot 4167(1 - 0.3^3)}{\pi \cdot 16 \cdot 2.06 \cdot 10^5} \left(\frac{2}{3} + \ln \frac{4 \cdot 8}{0.274} + \ln \frac{4 \cdot 32.5}{0.274} \right) = 0.0085 \; mm$$

In the contact region of the roller with the outer ring,

$$r_{eo} = \frac{48.5 \cdot 8}{48.5 - 8} = 9.58 \; mm$$

$$\sigma_{H,o} = 0.418 \sqrt{\frac{4167 \cdot 2.06 \cdot 10^5}{9.58 \cdot 16}} = 989 \; MPa$$

$$b_{C,o} = 3.04 \sqrt{\frac{4167 \cdot 9.58}{16 \cdot 2.06 \cdot 10^5}} = 0.335 \; mm$$

$$\Delta_{C,o} = \frac{2 \cdot 4167 \cdot 0.91}{\pi \cdot 16 \cdot 2.06 \cdot 10^5} \left(\frac{2}{3} + \ln \frac{4 \cdot 8}{0.335} + \ln \frac{4 \cdot 48.5}{0.335} \right) = 0.0085 \; mm$$

As we see from this example, the stress in contact with the inner ring is greater; therefore, the pitting begins here (and on the rolling elements, certainly). This is true for almost all RBs because the inner raceway is convex and the outer raceway is concave. The sole exception is provided by the self-aligning ball bearing (Figure 6.4j), in which the shape of the outer raceway is not close to the ball, and the maximal Hertzian stress takes place where the ball contacts the outer ring.

Rupture sources in rolling contact are not completely studied yet. There are a lot of factors influencing fatigue crack origination: tension stress on the contact area boundaries, shear stress at depth, plastic deformation of the surface layer and transformations in its microstructure, hydraulic and chemical effects of the lubricant, and so on. So, the Hertzian stress is used as an acceptable value for comparative strength calculations.

It is known for a fact that sometimes the fatigue cracks originate at a depth. The most shear stress (for roller bearings) is located at a depth of 0.4 b_C, and its magnitude is

$$\tau_{max} = 0.304 \sigma_H$$

If the contacting parts have a hardened layer and the core (sap) is not strong enough, the fatigue crack may develop under the hardened layer and then move outward. As a result, relatively large fragments of the hardened layer may detach from the surface (spalling). Elements of RBs are usually hardened through, but when case hardening is used, the thickness of the hardened layer and the strength of the sap should be sufficient to avoid subsurface cracking. (For more details, see Subsection 6.2.8.)

One of the typical modes of failure of RBs is *abrasive wear*, which occurs when abrasive contaminants get into the lubricant. In such cases, pitting may not happen, but the clearances in the bearing grow, and it may become unsuitable for its main role, i.e., to keep the adjoined part in a certain position relative to the housing. To prevent lubricant contamination, better seals and oil filters should be employed. But in hard usage (such as in agriculture and construction site machinery, in mining, and so on), it is quite difficult to completely prevent abrasives from entering the lubricant.

Strength of the cage is one of the important contributors to a bearing's reliability, especially at high speeds. The cage is exposed to centrifugal forces and to loads from vibrations and imbalance. In addition, the cage interacts with the rolling elements: in the loaded zone of the bearing, the cage is a driven part, and in the unloaded zone it is a driving part. In the unloaded zone, the rolling elements are separated from the cage by an oil film. In this way, the rolling elements may continually correct their position relative to the cage so as to interact with it predictably. If the rolling elements are clamped between the inner and the outer rings (for instance, because of a too tight fit or as a result of a shaft's overheating), the unloaded zone may disappear. In this case, the rings move the rolling elements forcibly and after some period of time, their spacing may change and become uneven. But the spacing of the cage pockets remains the same, and then the rolling elements stress the cage. This may cause distortion of the cage, especially when running idle.

The rolling elements may have slightly dissimilar speeds. This phenomenon is caused by small differences in diameters and by different degrees of slippage. Usually, these differences are microscopically small and negligible in typical cases; the rolling elements can take up the needed position relative to the cage with a clearance equal to the oil-film thickness. But if there is no unloaded zone, the difference in speed will gradually accumulate and result in macroscopic differences in spacing of the rolling elements. But the cage doesn't allow the rolling elements to change their spacing as they want to, and the interaction between them may break the cage.

The risk of cage breakage under load is less, because the rolling elements and raceways get elastically deformed in the loaded zone, where the inner ring shifts toward the outer ring. Owing to this displacement, a small clearance may arise in the unloaded zone, enabling the rolling elements to match the cage pockets' positions. Even if there is no clearance, the clamping of the rolling elements in the unloaded zone is considerably weaker, and their resistance to the cage may be less than the cage's strength.

When the rings of the bearing are misaligned, the cage is subjected to additional loads, including lateral bending. In two-row RBs, which have a common cage for both rows, unequal loading of the rows may cause an additional load on the cage, for example, when one row takes both the radial and the axial load, while the other takes only a radial load of lesser magnitude.

If RBs are working under scarce lubrication or if the lubricant is contaminated, the cage is the first element to suffer. There is a sliding contact in its interaction with the rolling elements, and it is prone to wear and overheating.

Sometimes, the failure of RBs is related to damage of the working surfaces caused by plastic deformation (due to overload), if an electric current is present, or with prolonged exposure of the nonrotating bearing to variable load or vibrations (false brinelling). There can also be chipping of the ring lips caused by excessive skewing (see Example 6.1).

The replacement of RBs is indicated either according to a maintenance schedule (after a certain period of work) or by the physical conditions of the bearing, such as increased noise and vibration, overheating, loss of accuracy and other causes.

6.2.3 Design of Supports with Rolling Bearings

Figure 6.8 shows the prevalent supports of type F2 fixing the shaft axially in both directions. That means, the F2 support has to include either one bearing that is able to take axial force in both

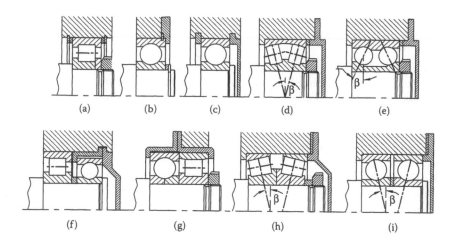

FIGURE 6.8 Supports of type F2.

directions (Figure 6.8a to Figure 6.8g), or two bearings with the ability to take the axial force in one direction only (Figure 6.8h and Figure 6.8i).

The supports shown in Figure 6.8 differ from each other both by their load capability and by the achievable accuracy of axial positioning. Supports according to Figure 6.8a are mostly used for shafts that are not loaded in an axial direction (but for small and accidental forces); e.g., shafts of spur gears. The use of such a support for bearing a continuous axial load is usually not recommended.

Figure 6.9 shows a straight roller bearing loaded with axial force F_a. This force is taken in the contact of the roller faces with the lips of the rings (in Figure 6.9b, these areas are hatched). In these places, there is sliding friction, but the oil wedge can't be formed there, so these locations have boundary-layer lubrication. Therefore, in spite of high hardness of the contacting surfaces,

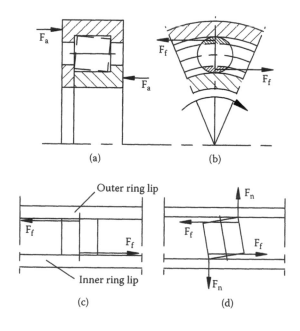

FIGURE 6.9 Cylindrical roller bearing under axial load.

they wear out in course of time (the period of time depends on load magnitude, speed, conditions of lubrication, and heat dissipation).

Axial forces F_a applied to the roller tend to skew it relative to the bearing axes (see dashed contour in Figure 6.9a), causing uneven distribution of radial load along the roller. So the radial load is needed to prevent skewing of the rollers as the dashed contour depicts.

Friction forces F_f applied to the roller (Figure 6.9b and Figure 6.9c) tend to skew it perpendicularly to the previous direction. When the gap between the roller and the lips of the rings increases (because of wear of both rollers and lips), the roller becomes turned as shown in Figure 6.9d, and it contacts the lips by the edges, not by areas as it did in the beginning. As a result, the contact stress increases, as do the friction coefficient and wear. Such conditions may lead to seizing, causing the roller to be jammed between the lips. Forces F_n in this case grow intolerably, and the lips may be broken.

Thus, straight roller bearings are not intended for axial loads, because in such a case, they are not only RBs. Partly, they become sliding bearings with all their drawbacks, such as increased heat generation, sensitivity to the viscosity and cleanliness of oil, to heat dissipation, and so on. Nevertheless, under certain conditions, straight roller bearings are able to take some continuous axial load in accordance with recommendations of the manufacturer. A short-term axial load may even be much larger.

Supports shown in Figure 6.8b to Figure 6.8e can take both radial and axial (bidirectional) loads. These supports are intended basically for radial load, particularly the version in Figure 6.8d, because, under high axial load, only one row of the rolling elements participates in the load transmission.

Supports shown in Figure 6.8b to Figure 6.8d are not that effective in axial fixation of the shaft; because the angle β is relatively small (see Figure 6.10a and Figure 6.10b); a small radial clearance C_r results under axial load in a considerable axial movement S_a. Elastic deformation of rolling elements is an aggravating factor.

The double-row angular contact ball bearings (Figure 6.8e) provide an exact and rigid axial fixation, because they have small clearances (between the rolling elements and the rings) and big pressure angles ($\beta = 25–45°$). But these bearings are very sensitive to misalignment.

The supports shown in Figure 6.8b to Figure 6.8e have the common property that under combined load (radial + axial) the inner ring moves in both axial and radial directions within the radial clearance until it becomes self-centered relative to the outer ring. At the same time, the other support, which is of NF type, is loaded with radial force only. In this support, the inner ring shifts from the center of the outer ring by a half of the radial clearance. In that way, an additional shaft misalignment is created, which may have a negative effect on the reliability of the mechanism. A combination of two bearings, one of which takes only the radial component of the load and the

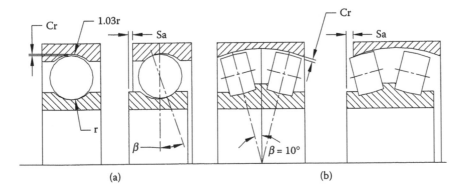

(a)   (b)

FIGURE 6.10 Radial clearance C_r and axial play S_a in the radial bearings.

other, only the axial component (Figure 6.8f, Figure 6.8g, Figure 6.16b, Figure 6.16c and Figure 6.16d), may ensure minimal misalignment of the shaft.

Supports F2, made of two tapered roller bearings or two angular contact ball bearings (Figure 6.8h and Figure 6.8i), allow infinite adjustment of clearance after the bearings are mounted, and the fit tolerances don't influence the clearance. This is a crucial advantage, when the axial play should be very small or when a preload is to be applied. But this is also a disadvantage, because the axial play adjustment needs practical skills and instruments. In service, when the bearings must be replaced, lack of these elements may lead to incorrect adjustment of the axial play (or preload) and to unsatisfactory operation of the mechanism. From this standpoint, bearings that don't need any adjustment in service are preferable.

Figure 6.11 shows supports of type F1. Here, in Figure 6.11a and Figure 6.11b, we can see the same bearings, able to take axial load in both directions, but they are fixed relative to the housing in one direction only. Figure 6.11c and Figure 6.11d show straight roller bearings, designed to transmit an axial load in one direction only. They are also fixed relative to the housing in one direction. (All the earlier statements about the load capability of such bearings under axial load are valid in this case as well.)

Supports with angular contact ball bearings and tapered roller bearings (Figure 6.11e and Figure 6.11f) are used as a rule with pairs of bearings. This means that a bearing of the same type (often the same bearing) forms the second support as well, being placed in a mirror position (see, for example, shaft 3 in Figure 6.15). If these supports are loaded unequally (for example, because of axial force or considerable difference in magnitude of radial load), a noticeable misalignment of the shaft relative to the housing may occur. The cause of such misalignment has been described in the preceding text: the shaft moves in the direction of the larger axial force until it becomes centered in one support, while in the other support it moves radially as much as the radial clearance allows (Figure 6.12b and Figure 6.12c). It is clear that increased axial play results in increased misalignment of the shaft, so the axial play should be as small as possible. While specifying the axial play, all the influencing factors, such as temperature and elastic deformations of the shaft, the housing, and the bearings under load should be carefully taken into account (see also Chapter 9, Section 9.4).

A typical support with a tapered roller bearing is shown in Figure 6.13a. The axial play is adjusted using shim 1. Sometimes, supports can be seen in which the adjustment of axial play is performed by a central screw 2 with a disk 3 (Figure 6.13b). The assumption that the wobbling

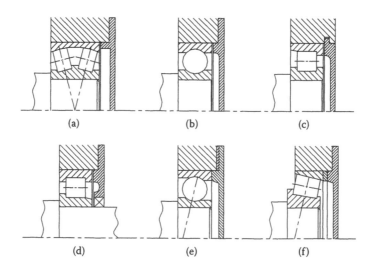

FIGURE 6.11 Support of type F1.

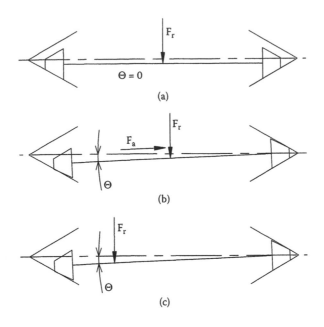

FIGURE 6.12 Shaft misalignment caused by its axial movement under load.

disk 3 enables the bearing to self-adjust is true only for purely axial loading. If the bearing is loaded with a radial force F_r, the load is transmitted only in a part of the bearing, and a corresponding axial force F_a loads its outer ring in this local area. Under this load and enabled by the axial clearance, the outer ring tilts (together with disk 3) around the tip of bolt 2 (see Figure 6.13c). Therefore, the outer ring of the bearing becomes misaligned relative to the inner ring. Obviously, such a design must not be used when the bearing is exposed to radial or combined loads.

Figure 6.14 shows supports of type NF. A support of this type exists with any radial bearing that is not fixed relative to the housing in the axial direction (Figure 6.14a and Figure 6.14b). Such a support allows the adjoined part of the shaft to move in either axial direction. Of course, the

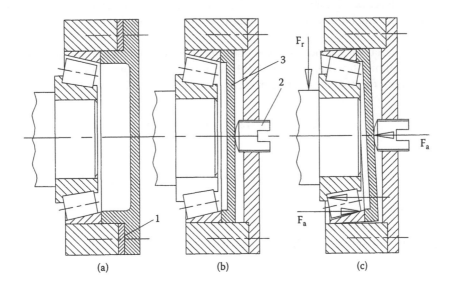

FIGURE 6.13 Supports with tapered roller bearings.

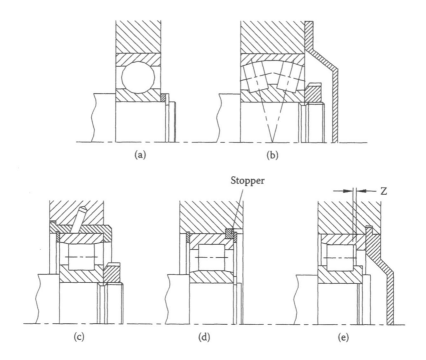

(a)   (b)

(c)   (d)   (e)

FIGURE 6.14 Supports of type NF.

outer ring of the bearing should be installed in the housing with a sliding fit; otherwise, support NF becomes similar to F2. The consequences of this transformation depend on the magnitude of friction forces between the outer ring and the housing and on the axial force value. In the case of bad luck, one of the bearings may be damaged causing failure of other parts of the mechanism.

But sometimes, even a sliding fit with assured clearance may be problematic; for instance, a heavy shaft and heavy radial loads may cause friction forces between the bearing and the housing, which are comparable (by magnitude and the effect created) with an interference fit. Therefore, a more reliable variant is the NF support with a cylindrical roller bearing (Figure 6.14c and Figure 6.14d). Here, the outer ring can be installed with tight fit as long as there is still a clearance between the rollers and rings.

However, a certain resistance to axial movement exists in this variant as well. From our experiments[3] it follows that the magnitude of the force of resistance to axial movement of the *rotating* inner ring of a straight roller bearing, F_S, can be described by the following formula:

$$F_S = f_0 F_r [0.042 + arc\tan(500 V_S / V_\Sigma)] \qquad (6.5)$$

where
   f_0 = equivalent coefficient of friction when the *nonrotating* inner ring is moved under radial load in axial direction
   F_r = radial load
   V_S = speed of axial movement
   V_Σ = summary rolling speed in the bearing determined from the formula

$$V_\Sigma = \frac{\omega d_2}{2} \cdot \frac{d_2 + 2d_r}{d_2 + d_r}$$

ω = angular speed of the inner ring (rad/sec)
d_2 = inner race diameter
d_r = roller diameter

The experiments revealed decreasing f_0 during running-in of the bearings. For example, after a 10-h run under load, the friction coefficient decreased from 0.14 to 0.09. For practical calculations, it can be taken that $f_0 = 0.11$.

In the same research, the axial force created by a straight roller bearing with an angular misalignment of rings was measured. The magnitude of this force can be determined using Equation 6.5, where

$$\frac{V_S}{V_\Sigma} = 0.5 \tan(\gamma_k \sin \Theta)$$

Here

γ_k = angle between the axes of the inner and outer rings
Θ = angle between the vector of radial load and the plane that passes through the axes of the rings

(Please be so kind as to present the angles in radians only, because degrees belong to the weather as well, and you never know whether the forecast is true.)

According to the experiments, at $V_S / V_\Sigma < 0.0002$ the resistance to axial movement of the ring becomes negligible.

EXAMPLE 6.3

Cylindrical roller bearing NU310E, $d = 50$ mm, $D = 110$ mm, $b = 27$ mm, $d_2 = 65$ mm, $d_r = 16$ mm. Radial load $F_r = 10{,}000$ N, rotational speed $n = 500$ r/min. The shaft oscillates axially at the frequency of rotation and with an amplitude of $a = 0.1$ mm. What is the axial force of resistance?

$$\omega = \frac{\pi n}{30} = \frac{\pi \cdot 500}{30} = 52.4 \, s^{-1}; \quad V_\Sigma = \frac{52.4 \cdot 65}{2} \cdot \frac{65 + 2 \cdot 16}{65 + 16} = 2038 \, mm/s$$

Maximum speed of axial movement of the shaft:

$$V_S = a\omega = 0.1 \cdot 52.4 = 5.24 \, mm/s$$

Next we find,

$$\frac{V_S}{V_\Sigma} = \frac{5.24}{2038} = 0.00257;$$

$$F_S = 0.11 \cdot 10000[0.042 + arc \tan(500 \cdot 0.00257)] = 1046 \, N$$

This is the maximum magnitude; the minimum is zero when the shaft passes the return points.

The design shown in Figure 6.14e is used when the axial movement is relatively small. Gap Z is installed to ensure the needed motion.

6.2.4 CHOICE AND ARRANGEMENT OF SUPPORTS

Let's consider several options of laying out supports. Figure 6.15 shows the top view of a three-stage reduction gear (the cover is removed). Shafts 1, 2, and 3 have two supports of type F1 each.

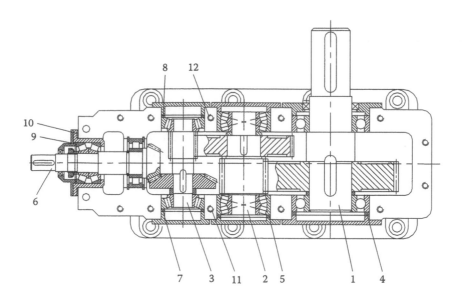

FIGURE 6.15 Bearing arrangement in a three-stage gear.

Such a design is the simplest. The inner rings of the RBs are mounted against the shoulders of the shaft. As a rule, no additional fastening of the inner rings but the tight fit is needed. End caps bolted to the gear housing restrict the possible motion of the outer rings. Shafts 1 and 2 don't need exact axial positioning, and their axial play is limited using shims 4 and 5 to within 0.5 to 1 mm.

If the gear reciprocates, as a travel mechanism of a crane does, every change in direction of the torque transmitted leads to axial displacement of the shafts within existing axial play. To decrease impacts and wear of the housing in such cases, the axial play is to be minimized.

The axial play of the shaft 3 tapered-roller support bearings needs to be precisely adjusted, and also the shaft must be positioned axially so as to superpose the gear 5 cone vertex with the axis of pinion shaft 6. Therefore, shaft 6 is provided with two shims, 7 and 8, which first allow adjustment of the axial position of the shaft (for example, by changing the thickness of shim 7) and then that of the axial play of the bearings, using shim 8.

The design is more complicated when a combination of supports F2 + NF is used (compare shafts 1 and 2 in Figure 6.15 and their variants in Figure 6.16a and Figure 6.16b). In this case, the bearing installed in support F2 must be fixed relative to the shaft in both directions, and this requires an additional component, such as nut or snap ring. The outer ring of this bearing also must be fixed in both directions relative to the housing, and for this purpose, additional components are required: cups, retaining rings, etc. Although complicated in design, which is certainly undesirable, this combination of types F2 + NF yields some important advantages. Influence of thermal and elastic deformations on axial play is completely avoided, and axial positioning of the shaft neck adjoined to the F2 support can be made as exact as needed. Furthermore, the misalignment of the shaft caused by taking up the radial clearances unevenly (as shown in Figure 6.12) is avoided as well.

Support F2 of shaft 6 in Figure 6.15 is made up of two tapered roller bearings. Shim 9 is required to adjust the axial play of these bearings, and shim 10 to adjust the axial position of the bevel pinion so as to superpose the vertex of its base cone with the axis of shaft 3. This support is designed according to Figure 6.8h, but it may be designed by variants shown in Figure 6.8e, Figure 6.8g, and Figure 6.8i as well. Other options from Figure 6.8 are not suitable for different reasons: the support as in Figure 6.8a is not suitable for axial loads, and supports shown in Figure 6.8b to Figure 6.8f don't ensure the needed positioning precision because of increased axial movement (see explanations in the preceding text).

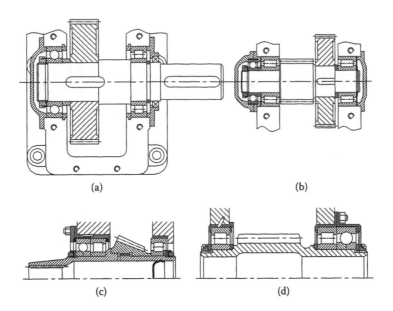

(a)   (b)

(c)   (d)

FIGURE 6.16 Bearing arrangement F2 + NF.

The shaft depicted in Figure 6.16a represents a version of shaft 1 in Figure 6.15. In this version, a more durable cylindrical roller bearing is installed on the output-end side. Because this bearing is not designed for taking axial loads, this function is imposed on the ball bearing, which forms support of type F2.

In the version shown in Figure 6.16b, in comparison with the shaft 2 in Figure 6.15, there is no need to adjust bearings, and owing to separate taking of radial and axial loads, possible misalignment of the shaft (such as depicted in Figure 6.12) is avoided. The design of supports shown in Figure 6.16c and Figure 6.16d has been taken from aircraft reduction gears. Here, the axial force is taken by a four-point ball bearing. Its outer ring is installed with a radial clearance and can't take any radial load.

6.2.5 FITS FOR BEARING SEATS

In operation, the inner ring of an RB is similar to a part of the shaft and should work as an integral part of it. Similarly, the outer ring should be similar to an integral part of the housing. In other words, the connection of the bearing rings with the shaft and the housing should be as durable as possible.

From this point of view, it would be ideal to use RBs without rings, produce raceways directly on the shaft and in the housing, and put between them a set of rolling elements (balls or rollers) with or without a cage. This ideal, as are other ideals, is very costly or unattainable at all. Sometimes, when there is no place for the rings, ringless RBs are indeed used (see Subsection 6.2.8), but as a rule, RBs are produced separately from the other parts of the mechanism. This allows product standardization and mass production of RBs, which reduce their price by far. Besides, when the service life of an RB is over, it can be easily replaced by a new one irrespective of the other parts. Therefore, even in aviation engineering, where reliability and decreasing weight are of paramount importance, RBs are mostly used as separable units.

Generally, the connection of the rings with the shaft and the housing should prevent their turning relative to the mating parts, because it may result in wear of the fitted surfaces. The main, and also

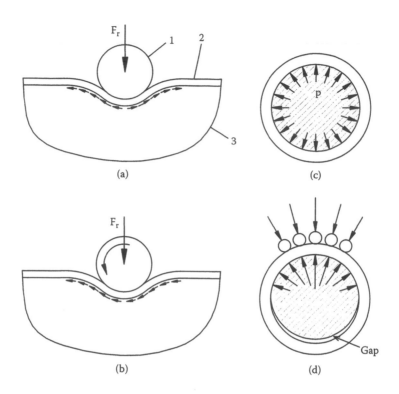

FIGURE 6.17 Interaction of rollers, ring, and shaft.

the cheapest, way to achieve this goal is to use interference fits. To define the magnitude of interference needed to prevent turning of the rings, the interaction between the ring and the adjoined part is considered. The mechanism of this interaction (as researchers from the Odessa Polytechnic University have proposed it*) is as follows: If a cylindrical roller 1 presses on a strip 2 lying freely on an elastic foundation 3 (see Figure 6.17a), the strip indents the foundation, and during the indentation, there occurs some sliding between them. The friction forces are symmetrical about the vertical axis of the roller and directed as shown in Figure 6.17a. If the roller is rolling (Figure 6.17b), the pressure between the strip and the foundation increases in front of the roller. Behind the roller, the pressure decreases, and the friction forces reverse their direction. In this way, all the friction forces between the strip and the foundation, as well as the friction in rolling contact, act in the same direction and move the strip relative to the foundation in the direction of rolling. As applied to RBs, strip 2 is a ring of a bearing, elastic foundation 3 is a shaft or housing, and roller 1 is a rolling element of a bearing.

The mechanism described in the preceding text provides an explanation for the well-known phenomenon of turning of the stationary outer rings of RBs relative to the housing when slide-fitted. Similarly, the inner ring may turn relative to the shaft if it is fitted with insufficient interference. For an approximate determination of the interference magnitude needed to prevent turning of the ring, the friction in the rolling contact is neglected. With this assumption, the tangent force moving the ring may be estimated from a formula:

$$F_m = F_r \cdot f_C$$

---

* Unfortunately, the authors didn't succeed in establishing the names of the researchers. They are asked to send CRC Press a copy of their article for reference in the following edition.

where

F_r = radial load on the rolling elements

f_c = friction coefficient in the areas of interaction between the rolling elements, ring, and shaft (or housing)

Circumferential friction force, which prevents the ring from moving, is estimated from the interference fit:

$$F_h = p\pi db f_P$$

where

p = surface pressure caused by the interference fit

d = diameter of the mounting surface (say, the shaft neck)

b = width of the ring

f_P = friction coefficient in the interference fit

Putting F_m equal to F_h and f_c equal to f_P, we obtain the needed minimal magnitude of the surface pressure p:

$$p = \frac{F_r}{\pi db} \tag{6.6}$$

The magnitude of the interference δ (that means the difference between the diameters of male and female parts), which is required to achieve the needed pressure p, may be determined from an equation:[4]

$$\delta = pd \cdot 10^3 \left[\frac{1}{E_1}\left(\frac{1+\xi_1^2}{1-\xi_1^2} - v_1 \right) + \frac{1}{E_2}\left(\frac{1+\xi_2^2}{1-\xi_2^2} + v_2 \right) \right] + 2Cp^{0.5} \; \mu m \tag{6.7}$$

(Here p is expressed in MPa, and d in mm.) and

$$\xi_1 = \frac{d_1}{d}; \quad \xi_2 = \frac{d}{d_2}$$

where

d_1 = diameter of the central hole in the male part (shaft or outer ring of the bearing) (mm)

d_2 = outer diameter of the female part (inner ring of the bearing or boss of the housing) (mm)

E_1 and E_2 = modulus of elasticity for the male and female parts, respectively (MPa)

v_1 and v_2 = Poisson's coefficients for the male and female parts, respectively

By the term $2C \cdot p^{0.5}$, the increased compliance of the surface layer has been taken into account. It depends on the surface roughness. For the roughness height $R_a = 1.25$ μm, $C = 1.4$; for $R_a = 0.63$ μm, $C = 1.1$; for $R_a = 0.32$ μm, $C = 0.8$.

EXAMPLE 6.4

Calculation of the interference required between the inner ring of the bearing and the shaft: Cylindrical roller bearing NU310E ($d = 50$ mm, $b = 27$ mm) is loaded with a radial force $F_r = 10,000$ N.

The needed pressure in the interference fit can be found from Equation 6.6:

$$p = \frac{10000}{\pi \cdot 50 \cdot 27} = 2.35 \, MPa$$

Both the bearing ring and the shaft are made of steel, so $E_1 = E_2 = 2.06 \cdot 10^5$ MPa, $v_1 = v_2 = 0.3$. The shaft doesn't have a central hole, so $d_1 = 0$ and $\xi_1 = 0$. The outer diameter of the inner ring $d_2 = 65$ mm, so $\xi_2 = 50/65 = 0.769$. Roughness of the fitted surfaces $R_a = 0.63$ μm, so $C = 1.1$. Now we put all these values into Equation 6.7:

$$\delta = 2.35 \cdot 50 \cdot 10^3 \frac{1}{2.06 \cdot 10^5} \left(1 + \frac{1 + 0.769^2}{1 - 0.769^2}\right) + 2 \cdot 1.1 \cdot \sqrt{2.35} = 6.2 \, \mu m$$

Thus, the minimum interference should be 6.2 μm. To fulfill this requirement, the m5 fit should be chosen for the shaft. This fit gives interference within 9–32 μm.

Now, we have to check the bearing clearance in the case of maximum interference, which is 32 μm. Increase in diameter of the inner raceway may be found as follows:

- First from Equation 6.7 can be found by iterations the pressure p at an interference of 32 μm: $p = 19$ MPa.
- Then the increase of the inner raceway diameter d_2 can be found using the Lame formula.

$$\Delta d_2 = \frac{2d^2 d_2 p}{E\left(d_2^2 - d^2\right)} = \frac{2 \cdot 50^2 \cdot 65 \cdot 19}{2.06 \cdot 10^5 (65^2 - 50^2)} = 0.0173 \, mm = 17.3 \, \mu m$$

Radial clearance in this bearing before mounting (for normal clearance group) lies within 30–60 μm. That means that even if the fit interference is maximal, the bearing clearance is still 12.7 μm, at least.

There is another, old-stager explanation of the phenomenon of the turning of bearing rings fitted with insufficient interference. Without a radial load, the pressure in the fit is distributed evenly (Figure 6.17c), but under load, the pressure in the loaded part increases, and on the opposite side, it decreases to the point of disappearance (Figure 6.17d).

Imagine that you have a ring on your finger sitting tight. But if you press on it in the radial direction (especially from the inner side of the palm), on the opposite side, there will appear a gap between the ring and the finger.

If reality conforms to what is shown in Figure 6.17d, rotation of the shaft will certainly lead to turning of the ring relative to the shaft. Investigation has revealed[5] that the minimum interference at which there is no gap between the ring and the shaft in the unloaded zone can be determined from the equation

$$\delta = 0.0043 \frac{F_r}{b} \, \mu m$$

where
  F_r = radial load (N)
  b = width of the ring (mm)

Taking into account the roughness of the surfaces, geometrical errors, and the necessary factor of safety, the author of the investigation recommends increasing the interference more than threefold:

$$\delta = 0.0143 \frac{F_r}{b} \, \mu m$$

For bearing NU310E (from Example 6.3), the interference calculated by this formula should be

$$\delta = 0.0143 \frac{10000}{27} = 5.3 \, \mu m$$

This result is close to what was obtained in Example 6.4 (6.2 μm), but Equation 6.6 and Equation 6.7 are more physical as they take into account real dimensions of the parts and roughness of the contacting surfaces.

Not always can RBs be installed with interference that ensures a completely immovable connection. For example, sometimes, it is desirable to use slide fit to make assembly and dismantling easier or to enable axial movement of the bearing relative to the housing to compensate for temperature expansion of the shaft. But what if the RB is fitted with a clearance? As experience shows, the behavior of the slide-fitted connections depends mostly on ring motion relative to the radial load vector. If the ring rotates relative to the load vector (so-called *circumferential load* of the ring), the clearance connection will be soon damaged. Depending on the hardness of the shaft neck (or the housing socket), the rotational speed, and the load magnitude, there may be observed fretting, wear, scoring, and even local welding and tearing-out of metal particles. But if the ring is nominally stationary relative to the load vector (*point load*), the ring will slowly slip (creep) relative to the mating part. Thus, wear must certainly occur, but if the mating part is made of wear-resistant material, the wear could be of low intensity and quite acceptable for the needed service life. If the wear is too intense, the ring should be secured against slippage. (For example, the stopper in Figure 6.14d is put into a slot cut in the bearing's outer ring and in the housing. This stopper should be strong enough to take the tangential force, which is the product of radial load by friction coefficient, and to withstand wear, because the ring is slightly moving.)

In most cases, the shaft with the inner ring rotates, and the radial load vector is constant in direction (for example, this condition exists for all bearings in Figure 6.15 and Figure 6.16). Thus, the inner rings are exposed to a circumferential load and the outer rings, to a point load. Under these conditions, the inner ring should have interference-fit connection with the shaft, and for the outer rings, slide fit is acceptable. It is especially important because the housing is split through the axes of gears, and the interference fit here is impracticable.

In the cases shown in Figure 6.3, the load vector is also constant in direction, but the inner rings of the bearings are nominally stationary (point load) and the outer rings are rotating (circumferential load). Thus here, the outer rings should have a tight fit, and the inner rings can be slide-fitted.

Another typical case is that of an eccentric weight attached to a rotating shaft (for example, in mechanical vibrators). Here, the inner rings of the bearings are exposed to a point load and the outer rings, to a circumferential load. Therefore, the inner rings may be slide-fitted, but the outer rings should be fitted tightly.

The slide fit for the "point-loaded" rings is only optional, and it is desirable to make it tighter. But the interference fit for the "circumferentially loaded" rings is mandatory, with very rare exceptions, which need special consideration.

On this basis, the designer should choose the fit taking into consideration the following:

1. Circumferentially loaded rings should be installed with interference fit. The minimal magnitude of the interference may be determined using Equation 6.6 and Equation 6.7.
2. The interference should not be too much to keep a certain clearance in the bearing; bearings with increased clearance (C3 and C4) can be used when needed.

It is appropriate here to discuss the effect of clearances on the load capacity of RBs. Figure 6.18a shows the load distribution between the rolling elements under the assumption that before load application, all clearances were zero. All rolling elements placed in the loaded half of the bearing take part in the load transmission. If there was some initial clearance in the bearing (Figure 6.18b), some of the rolling elements (at the ends of the loaded half of the bearing) would not be able to transmit load because of the clearance. The neighboring elements come into contact, but they are loaded less than before, when the clearance was zero. And the rolling elements in the middle of the loaded area are overloaded. If there is a clearance between the bearing and the housing (Figure 6.18c), the outer ring is too pliable to transfer the radial load; it becomes elastically deformed, and the clearance of the fit has nearly the same effect as the internal clearance of the bearing.

So, it is desirable to keep the clearances as small as possible, and in the first place, by installing rings with a tight fit. In practice, there may be problems caused by different limitations. They can be related to the assembly method, dismantling, or the requirement for free axial motion of a loose bearing, as mentioned in the preceding text. It is not recommended that considerable interference is used when the shaft neck is slotted or splined, because the initially round ring may become polygonal. In the latter case, the needed strength of the connection may be achieved by axial tightening with a nut (Figure 6.19a).

For the same reason (uneven deformation of the rings), a tight fit should be avoided when mounting RBs into a split housing. If the outer ring in a split housing is loaded circumferentially,

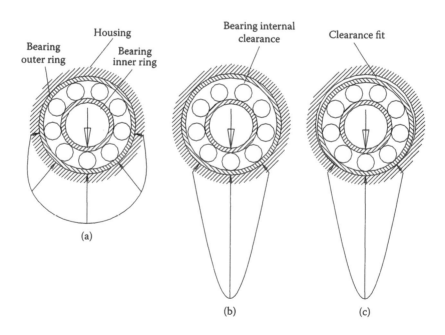

FIGURE 6.18 Load distribution in a bearing.

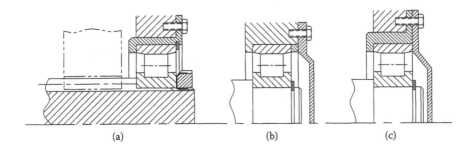

FIGURE 6.19 Fastening of bearing rings.

it should be fixed additionally by axial tightening with bolts (Figure 6.19b), or it can be installed with interference fit in a thick-walled sleeve (Figure 6.19c), which is in turn slide-fitted and bolted to the housing.

When using decreased bearing clearance, determination of temperature deformation and machining of the parts must be performed with higher accuracy. Interference within the bearing may lead to high contact stresses, which are greater than that from the working load. Therefore, prestressing (i.e., preloading) of RBs is used only for more exact positioning of the shafts, and not to increase their load capacity.

For prestressing, adjustable bearings are usually employed: angular contact ball bearings and tapered roller bearings (Figure 6.4b and Figure 6.4f). In these bearings, the prestress is applied after they have been mounted on the shaft and in the housing. So the tolerances of the fitting surfaces don't affect the prestress value, which may be adjusted independently and infinitely using nuts, shims, or axial forces created by springs. The preferable arrangement of such bearings (especially when the shaft is overhanging) is shown in Figure 6.20a. It has two advantages:

- Better ratio between bearing spacing L and overhanging length C.
- When the bearings are prestressed, thermal expansion of the shaft in axial and radial directions compensate each other, so that the prestress doesn't change a lot. In the design shown in Figure 16.20b, both the radial and the axial expansions of the shaft sum up and increase the initial prestress.

The ring exposed to a point load may be installed with a tight fit or a sliding fit. The latter is needed for some supports of type NF (as in Figure 6.14a and Figure 6.14b) and in cases where the bearing can't be tightly fitted (some of them have been discussed in the preceding text).

Possibly you ask yourself: who needs all these complicated things, all these formulas? In all catalogs, we can find recommendations on how to choose bearing fits, depending on the type of the bearing, mode of load, and load level. Perhaps they are false? God forbid! The catalog recommendations are perfectly right, but sometimes not detailed enough. For example, the scale of load levels often looks as follows: low load, normal load, high load, and high load with shocks. It is reasonable that designers dealing with, for example, ship gears (30,000 h service life), helicopter gears (500 h), and racing motorcycle gears (designed for only one lap) have different ideas about what is normal load. Besides, short-term overloads may occur in service, associated with planned tests or with emergency operation. They don't influence the service life of the bearings (as long as the loads don't exceed the permitted static load), and it is not clear whether they should be taken into account while choosing the bearing fits. Let's consider one practical case.

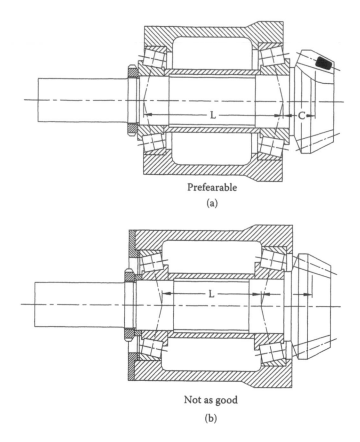

Prefearable

(a)

Not as good

(b)

FIGURE 6.20 Arrangement of tapered roller bearings.

Example 6.5

The main shaft of the escalator presented in Figure 4.20 and Figure 4.21 (see Chapter 4) rotates in two spherical roller bearings 23056 with dimensions $d = 280$ mm, $D = 420$ mm, and $b = 106$ mm. The radial load on the bearing near the gear is about $35 \cdot 10^4$ N, and on the other bearing, $30 \cdot 10^4$ N. Twice a year, a test of the emergency brake is performed, and during the test, the radial load of the latter bearing comes up (for several turns of the shaft) to $50 \cdot 10^4$ N. Of course, this overload has not been taken into account when calculating the service life of the bearings.

Angular speed of the shaft $n = 11.2$ r/min, loading rating (dynamic) $C = 1.56 \cdot 10^6$ N, and for the radial load of $35 \cdot 10^4$ N, it was found, $L = 200,000$ h. The escalator is in service about 20 h a day, 300 d a year; thus, the service life is more than 30 years. When choosing the fit, the designer had opened the manufacturer's recommendations and started to think over the load level: is it normal or high with shocks. If you were going by escalator, you would just feel that there are no shocks. And taking into account the enormous service life, you can't say that the load is heavy! It is low, or normal. For the normal load, the recommended fit is m6, which results in interference of 20 μm minimum and 87 μm maximum. This fit was chosen by the designer. And then the escalator was built, installed in the metro station, and tested. When testing the emergency brake, it was revealed that the inner ring of the bearing turns relative to the shaft neck. The shaft is made of mild steel, so the relative motion would be able to wear out the neck very soon. The way out of this situation was found in additional axial fastening of the bearing by an end washer with plenty of bolts. (Imagine what a pleasure it was to machine all these dozens of threaded holes on the shaft by hand tools!)

Let us determine the needed fit of the inner ring under radial load of $50 \cdot 10^4$ N, using Equation 6.6 and Equation 6.7. The surface pressure in the connection needed to prevent turning of the ring,

$$p = \frac{50 \cdot 10^4}{\pi \cdot 280 \cdot 106} = 5.36 \ MPa$$

The outer diameter of the inner ring (on average) $d_2 = 326$ mm; thus,

$$\xi_2 = \frac{d}{d_2} = \frac{280}{326} = 0.859$$

Minimal interference of the fit (at roughness of the surfaces 1.25 μm) is

$$\delta = 5.36 \cdot 280 \cdot 10^3 \frac{1}{2.06 \cdot 10^5} \left(1 + \frac{1 + 0.859^2}{1 - 0.859^2}\right) + 2 \cdot 1.4 \cdot 5.36^{0.5} = 54.7 \ \mu m$$

Hence, the inner ring should be mounted with the fit p6, which results in interference of 56 μm minimum and 123 μm maximum. According to the catalog, this fit is recommended for high load with shocks. If this fit had been used initially, the turning of the ring would have been prevented.

6.2.6 Requirements for Surfaces Adjoined to RBs

Geometric accuracy is the most important requirement. Rings of RBs are relatively thin. Being pressed to the mating surfaces by the interference fit or by the load, they duplicate (partly) errors such as out-of-roundness and taper. These errors may cause increased stresses in the bearings, vibrations, dynamic loads, and premature failure of the bearing.

Bearings mounted with a tight fit are always pressed up to the shoulder, which must be perpendicular to the shaft axis. Error in perpendicularity (measured as face runout) automatically skews the ring. If the bearing is mounted with heating, there can be a gap between the bearing and the shaft shoulder formed after cooling down. To ensure exact position of the ring, this gap should be taken up by pressing.

Technical requirements of geometric accuracy and roughness are given in the catalogs, but they should be given in the machine drawings as well.

Geometry of the mounting surfaces achieved by machining may be changed because of elastic deformations of the parts after assembly with interference fit or under load. These changes may be useful or harmful depending on design. Figure 6.21b represents an RB mounted on a hollow shaft. Under load, the thin-walled shaft and the inner ring of the bearing become ovalized, so that the clearances in the loaded zone are taken up and the load is distributed more uniformly between the rolling elements (compare with Figure 6.21a). The thickness of the shaft wall may be as small as $0.10d$ to $0.15d$, provided that the strength and the rigidity of the shaft are guaranteed. But usually the shaft wall should not be thinner than the inner ring of the bearing.

Notice that use of hollow shaft leads to fit with an increased interference. This is automatically considered when Equation 6.7 is used to determine the needed interference. But while choosing the fit on the basis of the catalog recommendation, the increased compliance of the hollow shaft should be taken into consideration as well.

Figure 6.21c shows a bearing seat supported by a rib. Because the rigidity of the housing is increased in the area of maximum load, the load distribution between the rolling elements is worsened. Ribs shown in Figure 6.21d improve the load distribution and result in increased service life of the bearing.

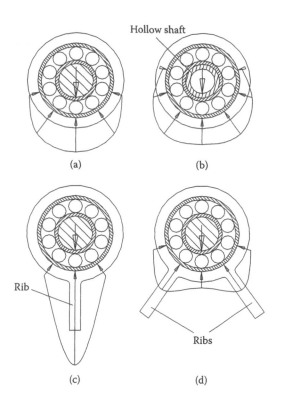

FIGURE 6.21 Flexibility of shaft and housing and load distribution in the bearing.

Strength and wear resistance of the surfaces adjoined to the RBs are of secondary importance if the connection between the bearing ring and the seat is motionless. This immobility should be ensured not only at normal conditions, but also while starting, on trials, in emergency operation with overload and vibrations (because of out-of-balance), and so on. If the wear resistance of the shaft is low, even momentary turning of the bearing ring about the shaft neck may decrease the interference and result in distortion of the connection. That's why it is advisable to ensure wear resistance of RB seats. Sometimes, the shaft necks are surface-hardened by carburizing or nitriding, especially when this surface treatment is needed for the strength of the shaft itself or of other elements such as splines, gear teeth, cams, etc. But, usually, there is no need for surface hardening. When the bearing ring is slide-fitted, the wear resistance of the surfaces is of primary importance. Seats in light metal housings should be shielded with sleeves made of heat-treated steel. The sleeves are press-fitted into the housing so as to retain the interference fit despite the greater expansion of the housing under working temperature. Additionally, the sleeves are usually fixed with pins or bolts (Figure 6.14c, Figure 6.16c, and Figure 6.16d). Holes for the pins are drilled after the sleeves are installed in place.

The height of the shaft shoulders may be important, particularly when the shoulder transmits an axial force from the bearing (when using tapered roller bearings, this force can be created as a component of radial load). The authors have seen shoulders worn out in contact with a common ball bearing. If the bearings are of adjustable type (such as angular contact ball bearings or tapered roller bearings), the gradual wear of the shoulders increases their axial play and leads to increasing radial displacement of the shaft. High wear resistance is needed not only for the shaft shoulders, but also for support elements that fix the bearing ring axially: end cups, spacers, shims, etc.

The designer is interested in making the shaft shoulder height as small as possible; it enables decreasing of the shaft diameter and reduces stress concentration. However, there can be negative consequences of decreasing the shoulder height, as illustrated by the following bitter experience.

EXAMPLE 6.6

In Figure 6.22 is shown the output shaft assembly of a reduction gear. The designer was aspiring to decrease the diameter and the cost of the shaft. Diameters d_1 and d_3 can't be decreased, because they are limited by the strength of the shaft. Diameter d_2 serves as a shoulder for the gear and the right-hand bearing. The shoulder height is dictated by the bearing catalog recommendations, but there is a well-known ruse: the shoulder can be made small, but between it and the bearing a spacer ring can be installed, which has the height as recommended by the catalog. This method is excellent on condition that the compressive resistance of the small shoulder is sufficient to take the axial forces originated from the load transferred by a mechanism. This aspect was not taken into consideration in this case. The designer didn't find a suitable steel tube for sleeve 1, so he made it of low-quality gray iron. In operation, the axial force is not distributed uniformly over the entire circumference, but concentrated on a part of it (see Figure 6.13c). Taking into account that the maximal radial load on a rolling element is given by

$$R_{max} \approx \frac{5R}{z\cos\beta}$$

where z is the number of rolling elements in the bearing, it can be assumed that the load concentration factor is about 5 in the axial direction as well. This loaded area in contact between sleeve and shoulder moves circumferentially with the shaft rotation speed. Sleeve 1 was mounted on the shaft with a clearance, so its motion relative to the shaft shoulder was unhindered. This cyclic loading resulted in wear and squash of the contacting surfaces 2, which became tapered. After that, sleeve 1 cracked along its generatrix, moved toward the gear (together with the bearing, certainly) and got on the d_2 part of the shaft. The shaft became terribly skewed, and severe noise and vibration in the gear suggested to the maintenance staff that something must be wrong with it. For the renewal, a new shaft was made, with increased diameter between the gear and the bearing, d_2, without any sleeve. The gear and the pinion were replaced as well, because they were severely damaged. In short, this is a story about unlucky economization. As they say, "the stingy pay twice."

6.2.7 ELASTIC DEFORMATION OF RBs UNDER LOAD

As was already said, the possible movement of the shaft in the support involves clearances in the bearing and in the fits, resilience of the bearing, and housing deformation where the bearing is installed. In this section, we will study the resilience of the bearings, which consists mostly of deformation where rolling elements contact the rings (δ_1), and also includes deformation of the housing (δ_2) and the shaft (δ_3) under tight fits. Deformations of the bearings have been investigated in detail. In the following text are given simplified formulas to calculate them.[4]

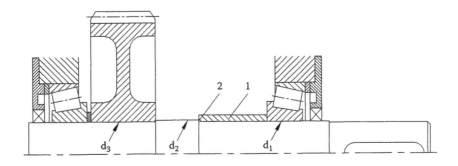

FIGURE 6.22 Height of the shaft shoulder.

For approximate determination of δ_1, the following averaged equations can be used. For single-row ball bearings:

$$\delta_1 = k_b F_r^{2/3} \ \mu m \qquad (6.8)$$

For roller bearings:

$$\delta_1 = k_r F_r \ \mu m \qquad (6.9)$$

Here

F_r = radial load of the bearing (N)

k_b, k_r = factors that depend on type and size of bearing

For bearings with inner diameter d of 40–100 mm, the factors k are as follows:

- Single-row ball bearings: $k_b = 0.153–0.00044d$
- Regular tapered roller bearings: $k_r = 0.052/d$
- Wide tapered roller bearings: $k_r = 0.033/d$
- Regular straight roller bearings: $k_r = 0.065/d$
- Wide straight roller bearings: $k_r = 0.043/d$

(Here, d shall be substituted in millimeters.)

These formulas are correct when the initial (before loading) clearance in the bearing is zero. If there is an initial clearance, a lesser number of rolling elements will take the load, and their deformation will increase. In the case of preload, the result will be opposite. The recommended method of taking the clearance or the preload into account is as follows:

- If the ratio of the initial clearance (after the bearing is mounted on the shaft) to deformation δ_1 rated from Equation 6.8 and Equation 6.9 is 1:1, 2:1, or 4:1, δ_1 shall be increased by 10, 20, or 30%, correspondingly.
- If the bearing is preloaded, its deformation δ_1 should be divided for roller bearings by 2–2.5 and for ball bearings by 2–5, depending on the preload magnitude.

The resilience of the tightly fitted connections (δ_2 and δ_3) is relatively small. In total they come to 25–30% of δ_1 for roller bearings and 20–25% of δ_1 for ball bearings.

EXAMPLE 6.7

It is desired to define radial deformation of a straight roller bearing NU310E loaded with a radial force $R = 10,000$ N. The bearing is mounted on a shaft with a tight fit m5, and the internal clearance as mounted measures 12.7–42.7 µm (see Example 6.4).

$$\text{Coefficient } k_r = 0.065/d = 0.065/50 = 0.0013$$

From Equation 6.9, deformation of a bearing with zero clearance,

$$\delta_1 = 0.0013 \cdot 10,000 = 13 \ \mu m$$

The ratio of the internal clearance to δ_1 equals 0.98 at a smaller clearance and 3.28 at a greater one. Hence, the calculated deformation should be increased by 10 and 26%, correspondingly.

Thus, the calculated deformation in contact of the rollers with the rings comes to 14.3–16.4 μm. This result corresponds with that of Example 6.2, where the total deformation of the most loaded roller was obtained as 8.5 + 8.5 = 17.0 μm.

To compare the rigidity of roller bearings with that of ball bearings, let's calculate the deformation of a ball bearing 6310, which has the same dimensions as the aforementioned NU310E. Because we take the same radial load, the same fit m5 should be chosen. As we know from Example 6.4, at maximum interference, the inner clearance will be decreased by 17.3 μm; therefore, the bearing should be of clearance group C3, with an initial clearance of 18–36 μm. Thus, the internal clearance in the bearing mounted on a shaft will be within 0.7–18.7 μm.

$$\text{Coefficient } k_b = 0.153 - 0.00044 \cdot 50 = 0.131$$

From Equation 6.8, deformation of a bearing with zero clearance,

$$\delta_1 = 0.131 \cdot 10000^{2/3} = 60.8 \, \mu m$$

The influence of clearance can be neglected here. From these examples, we can see the difference in rigidity between ball bearings and roller bearings. In this case, radial deformation of the 6310 bearing is about fourfold that of the NU310E bearing.

6.2.8 RBs with Raceways on the Parts of the Mechanism

As mentioned earlier, RBs without rings is a very expensive luxury. Nevertheless sometimes, when the support must be installed in a restricted place, it becomes necessary to use a rolling bearing with only one ring (Figure 6.23[2]) or even completely ringless (Figure 6.24). Usually, they are straight roller bearings, which have cylindrical races and, in respect to technology, are relatively easy to produce. At the same time, the raceways should meet some requirements concerning strength and quality of machining. Commonly, these parts are made of high-strength alloy steel, carbonized, and hardened. The hardness of the surface layer should be HRC 59–63, and the thickness (according to FAG recommendations) within $(0.07-0.12)d_r$, but not less than 0.3 mm (d_r is the diameter of rollers). The recommended surface roughness $R_a < 0.2$ μm.

To prevent underlayer failure, the hardened layer should be thicker than the location of maximal shear stress, which is located at a distance of $0.4b_C$ from the surface (Figure 6.7). Let's test the recommended thickness of the hardened layer from this point. To ease our work, let's suppose that the Hertzian stress $\sigma_H = 5000$ MPa (see this evaluation in the beginning of Subsection 6.2.2), the

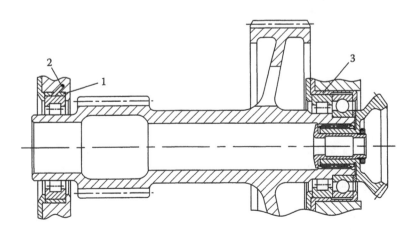

FIGURE 6.23 Roller bearings without inner rings.

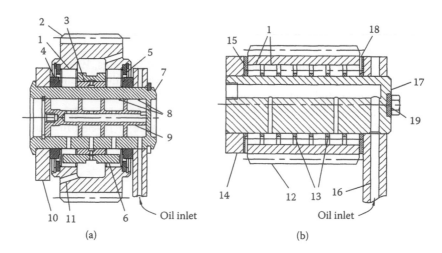

FIGURE 6.24 Self-made roller bearings.

length of the roller $L = 1$ mm, and the roller contacts a flat surface (in this case, $r_e = 0.5d_r$). Now, from Equation 6.2 we can obtain the force F:

$$F = \frac{\sigma_H^2 r_e L}{0.418^2 E} = \frac{5000^2 \cdot 0.5 \cdot d_r}{0.418^2 \cdot 2.06 \cdot 10^5} = 347.3 d_r \ N/mm$$

The width of the contact area is obtained from Equation 6.3:

$$b_C = 3.04 \sqrt{\frac{347.3 \cdot d_r \cdot 0.5 \cdot d_r}{2.06 \cdot 10^5}} = 0.088 d_r \ mm$$

As we can see, the recommended thickness of the hardened layer approximately equals the width, b_C, of the contact area, i.e., 2.5 times deeper than the location of the maximal shear stress.

It should be noticed that the part containing the raceways may also have other elements limiting the thickness of the hardened layer. For example, gear teeth need the carbonization depth to be $(0.28–0.007m) \ m \pm 0.2$ mm plus grinding allowance.

(Here, m is the module of the gear, in mm. In the American system of terms,

$$m = \frac{25.4}{P} \ mm$$

where P is the diametral pitch.

The carbonization depth is uniform on all surfaces of a part, and it should be suitable to all "interested parties."

Figure 6.23 shows a double-geared shaft with a bevel gear attached to it. It has one NF support (straight roller bearing without inner ring) and one F2 support. The F2 includes an identical roller bearing and a four-point ball bearing, the latter taking only the axial force generated in the gears. The inner raceways of the roller bearings are made directly on the shaft ends. A bearing manufacturer supplies the outer rings with a set of rollers and a cage.

In supports depicted in Figure 6.24, rollers 1 are the only parts received from a bearing manufacturer, and the manufacturer of the gear makes the rest. Support of a planet wheel as shown in Figure 6.24a is classic. Planet wheel 2 runs on two rows of rollers that are separated by rings 3 and 4. Two rings 5 fixed by snap rings prevent the axial motion of the planet wheel. Cage 6 is centered by the middle ring 3. Lubrication oil enters the unloaded part of the bearing through holes in axle 7. The oil ducts are organized in such a way, that if there is some solid particle 8 in the oil, it will be separated by centrifugal forces and will remain in the labyrinth of insert 9. Axle 7 should be safely fixed against rotation to keep the oil ducts in the axle and in planet carrier 10 aligned. The oil flows out of the planet wheel through holes 11.

The planet wheel support shown in Figure 6.24a is formed, in effect, by a single two-row cylindrical roller bearing. Axle 7 and two rings 4 form (together with planet carrier 10) the inner ring with two lips. The outer ring with two lips is formed by planet wheel 2 and two rings 5. Ring 3 serves as a floating center lip that separates two rows of rollers.

Rings 3, 4, and 5 can be made of high-carbon low-alloy steel as usual for rolling bearings. But the planet wheel 2, and axle 7, must not be brittle, so they should be made of low-carbon alloy steel. They should be case-hardened to obtain a hard surface layer while the core remains tough. All the surfaces contacting with the rollers should be case-hardened or carbonitrided, heat-treated to high hardness, and then thoroughly ground and polished.

Planet wheel 12 shown in Figure 6.24b works at lower speeds, and the design of its support is uncommon. Seven rows of rollers 1, without a cage (so-called "full complement design"), are separated by hardened rings 13. Between the planet wheel and planet carrier 14 are placed hardened rings 15 with antifriction coating, which are able to take relatively small axial forces from the spur planet wheel and the rollers. The oil is supplied through hole 16 in the planet carrier, and then through oil ducts in axle 17. It flows out of the bearing through slots 18 made in thrust rings 15.

Thrust rings 15 are fixed against rotation to prevent the wear of the planet carrier. Axle 17 is fixed against axial movement and rotation by strip 19, which is bolted both to the axle and the planet carrier.

6.2.9 Lubrication of RBs

Detailed recommendations for rolling bearing lubrication are given in the manufacturers' catalogs. It is noteworthy that when the speed is not high, the RBs are somewhat sensitive to lubrication grade and mode. The main concern is not to leave the bearing without lubrication at all. Scarcity (or even worse, complete absence) of lubricant may lead to overheating and wear of all bearing elements, and seizure and breaking down of the cage.

It is important to prevent entrance of foreign matter (wear products from inside, dirt, and water) into the lubricant. For this reason, rubbing parts of the mechanism should be made wear resistant by means of hardening and proper lubrication of friction surfaces. The mechanism should be satisfactorily sealed from the surroundings. For trapping of ferromagnetic particles, magnetic plugs are in common use. In spite of this, some amount of foreign particles and products of oil decomposition accumulate in the lubricant bit by bit. When the speed is not too high, these impurities increase the wear of RBs, but not drastically. Therefore, it is common to bear with small amounts of pollution in the lubricant, and not to complicate the mechanism by a circulation lubrication system with filters. At high speed, the lubrication oil should be filtered.

When choosing the lubrication mode, speed and temperature should be taken into consideration. Oil, as compared with grease, provides better heat removal and enables higher rotational speed. Decomposition of the oil proceeds more slowly, and it can be easily replaced. Oil can work at wide-ranging temperatures.

The main advantage of grease lubrication is that the grease doesn't flow out; it makes the sealing problem much easier. In many cases, when the load and speed are moderate, sealed bearings greased for life can be used. Recommendations for the selection of suitable grease and lubrication intervals are given in bearing catalogs. When replenishing grease, don't forget that the new grease must be compatible with the previous formulation.

When adjacent machine parts are lubricated with oil, it is practical to supply the oil to the RBs as well. It should be always checked whether there is a really sufficient supply of oil to the bearing. Sometimes, the bearing may remain dry in the middle of an oil storm, where apparently no dry place could be found. Some information about this issue is given in Chapter 7, Section 7.10.

6.3 SLIDING BEARINGS (SBs)

6.3.1 FRICTION OF LUBRICATED SURFACES

The main problem to be solved when designing an SB is the decrease of friction forces. There may be several modes of friction: liquid friction, semiliquid friction, and boundary friction. The mode of friction depends on many parameters, such as sliding velocity, unit pressure, temperature in the contact area, geometry of the surfaces, and the properties of the lubricant. Among the immense variety of SBs made of metals, rubbers, plastics, or wood and lubricated with oil, grease, water, air, or whatever else, we choose for consideration only the metal bearings lubricated with mineral oil.

The surface of a metal is remarkable for its molecular activity, because the atoms on the surface don't have a full complement of atomic bonds (similar to the atoms inside crystals), and they produce atomic (molecular) attraction. This effect was described in the beginning of Chapter 3. Because of this attraction, the surface of the metal is "sticky," and it is always covered with a very thin film of substances that have come in contact with it. This phenomenon is called *adsorption*.

In particular, because cutting emulsions are used in machining, in the beginning, the film adsorbed on a part consists of molecules of oil, water, and gases. This film may be completely removed only by heating the part in a vacuum, but the composition of the film may be changed if substances with greater surface activity (and, correspondingly, with greater ability of adsorption) are applied to the metal surface.

A distinctive feature of surface-active substances is that in their molecules the centers of the positive and the negative charges don't coincide.[6] In other words, each of their molecules contains two charges, positive and negative, separated in space. Such molecules, when they come in contact with a metal, fasten onto its surface in a fixed manner (perpendicular to the metal surface, Figure 6.25) and create a monolayer 1 of one molecule thick. Lubrication oils containing surface-active additives create a boundary layer 2 over the monolayer 1, in which the molecules are also oriented perpendicular

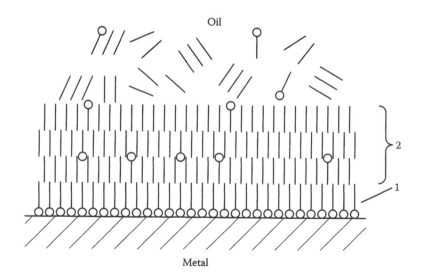

FIGURE 6.25 Adsorbed oil film.

to the metal surface. This layer has a quasi-crystal structure and is able to take very high loads, up to 3000 MPa. But under increased temperatures, the molecular links become weaker and the quasi-crystal layer "melts." These temperatures for different metals lie in the range of 90–150°C.

Two surfaces contact first at microasperities on the top of the macrowaves (Figure 6.26a). As the load increases, the microasperities, and afterward the macrowaves, become deformed elastically and plastically, and thus the contact area grows with increased load. However, this area rarely exceeds 30–40% of the nominal contact area. When there is no relative motion, the surfaces are separated (at points of closest approach) by a quasi-crystal boundary layer of oil, about 0.1 μm thick. When in motion, sliding takes place between the boundary layers. Where the surfaces are close, there may be mechanical interaction and deformation of the surface asperities, with no damage to the boundary layer. Under certain conditions, the heat generated in spots of close contact may "melt" the quasi-crystal boundary layer. Then, in these spots may occur direct contact of metal surfaces, accompanied by local adhesion and seizing (microwelding) (see Figure 6.26b). But in the main, the sliding surfaces are separated by the oil film, which is renewed in the spots where it was destroyed after the local conditions improve (while sliding). This described mode of friction is called *boundary friction*. The coefficient of friction and wear intensity in this mode are small as compared to dry friction (without lubrication).

It is clear that the better the actual fit of the sliding surfaces, the less the unit load in the spots of close contact, and consequently, the risk of boundary layer destruction is less. Better fit is favored by more exact geometry of the surfaces, smoother surface (in the first place, of the harder member), and lesser modulus of elasticity. From this point of view, rubber is a very good material, but in most cases, it is too pliable to ensure the needed exact position of the shaft relative to the housing. (Propeller shaft bearings in boats and ships are a popular application for rubber bearings. They are also used in hydroelectric power plant turbines).

Liquid friction is characterized by a complete separation of the surfaces by an oil film under pressure. In other words, the load, which presses the surfaces against each other, is supported not because of the mechanical strength of the adsorbed boundary layer, but rather, the pressure in the oil film. This pressure may be produced either by a special pump, which forces the oil between the sliding surfaces (such bearings are called *hydrostatic*), or owing to the shape of the sliding surfaces, which forms an oil wedge while sliding (such bearings are called *hydrodynamic*).

Oil layers placed at a distance of more than 0.5 μm from the metal surface (beyond the boundary layer) may move freely relative to each other, and their resistance to movement is minimal. This friction mode is the best with respect to minimizing friction and wear.

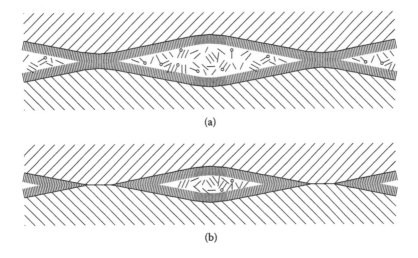

(a)

(b)

FIGURE 6.26 Contact of lubricated surfaces.

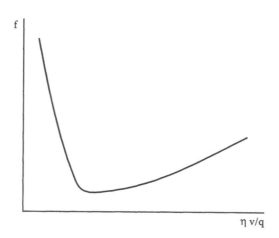

FIGURE 6.27 Hersey–Stribeck diagram.

Semiliquid friction constitutes a mixture of liquid and boundary modes. In hydrodynamic bearings, the percentage of each mode depends on load magnitude, sliding velocity, oil viscosity, and other factors. For example, an ordinary radial SB starts with boundary friction. As the sliding velocity increases, the load capacity of the oil wedge grows, and the percentage of the liquid friction increases correspondingly. When the oil wedge is able to take the entire load, the surfaces are completely separated by an oil film, and the friction losses are minimal. Further increase of the sliding velocity leads to gradual increase in the friction coefficient, because the oil film thickness grows much less than the velocity does, and the velocity gradient within the oil film grows.

The process of transition from boundary friction through the semifluid to the fluid friction mode is reflected in the well-known diagram shown in Figure 6.27, in which is given the coefficient of friction f vs. parameter $\eta v/q$ (η: oil viscosity, v: sliding velocity, q: unit load). Pure liquid friction is achieved when the f magnitude is minimal. On the left side of this point is semiliquid and boundary friction.

In hydrostatic SBs, pressure can always be produced, which is enough to bear the entire load and completely separate the sliding surfaces. In hydrodynamic bearings, this possibility depends, among other parameters, on the sliding velocity, which inevitably begins at zero, when no hydrodynamic effect can be gained. Therefore, in many cases, especially at start or shutdown of the machine, semiliquid and boundary friction modes are a matter of fact. Hydrostatic bearings are used relatively seldom, and so they will not be discussed further.

6.3.2 TYPES OF SBs

Routine design of supports with SBs is depicted in Figure 6.28: types NF (Figure 6.28a), F1 (Figure 6.28b), and F2 (Figure 6.28c). In contrast to RBs, which may take both radial and axial

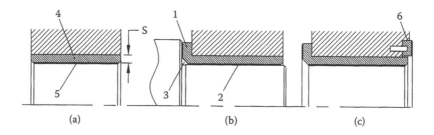

FIGURE 6.28 Types of sliding bearings.

forces, sliding bearings can take either radial or axial load but not both. Physically, both of them may be combined into one, as shown in Figure 6.28c and Figure 6.28d, but their design and calculation are completely independent.

They are designed independently, but often they live in close neighborhood and, as in real life, one flat mate can cause troubles for another. For example, thrust bearing 1 is lubricated with oil flowing out of radial bearing 2 into a circular chamber 3 (Figure 6.28b). If small particles of metal break away from the radial bearing (caused, for example, by wear or local submelting of the antifriction layer), these particles are transferred with the oil to the thrust bearing and may damage it. Being damaged (or wrongly designed), the thrust bearing may close the oil outlet from chamber 3. This leads to decrease of oil flow through the radial bearing, which may result in overheating and failure.

As already mentioned, the necessary condition for continuous duty of SBs is the creation of an oil film between the shaft journal and the bearing. If there is a wedge-shaped clearance between these surfaces, which is opened toward the motion (Figure 6.29), the oil is pulled into this clearance by friction. The oil then produces hydrodynamic forces, which tend to increase this clearance further. These forces are larger when speed or oil viscosity is greater. Because heat generated in a sliding bearing decreases oil viscosity and reduces oil-film thickness, it is important to ensure sufficient heat removal. The most effective method is to increase oil circulation through the bearing, which can be accomplished by increasing the bearing clearance and the input oil pressure. Sometimes, additional oil grooves are made to supply more chilled oil into the bearing.

The wedge-shaped clearance can be created by various means. In radial SBs, it arises automatically when the shaft becomes eccentric about the bearing under radial load (Figure 6.29a). In thrust bearings, this purpose may be served either by machining of the bearing surfaces with needed slope (Figure 6.29b) or by using floating elements (Figure 6.29c). When the load and the speed are low, thrust bearings with flat surfaces are used.

Radial clearance in a radial SB, as well as the wedge angles and clearances in thrust SBs, are shown exaggeratedly in Figure 6.29. In reality, the radial clearance measures 0.1–0.3% of the shaft diameter, and the minimum oil-film thickness h measures microns or, at the most, tens of microns. The inclination angle of the thrust bearing surfaces is also very small. The opening of the wedge $t = (0.8-1.6)h$, so it is a question of tens of microns as well. We will discuss this in the following text.

6.3.3 MATERIALS USED IN SBS

The journal of a shaft is typically made of steel. It may also have a hard metal coating, such as chrome. What is subject to discussion is the material of the bearing or, to put it more precisely, of the working surface, which is directly in contact with the shaft journal. Theoretically, if an oil film separates the bearing and the journal, it doesn't matter what kind of material they are made from,

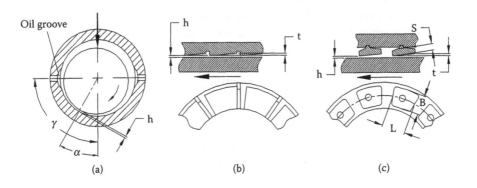

FIGURE 6.29 Formation of hydrodynamic oil film in sliding bearings.

if its thermal conductivity is high enough to ensure good heat dissipation. But the problem is that these surfaces are not always separated. When the mechanism is standing, the adsorbed oil layer separates the surfaces of journal and bearing. When put into operation, the boundary friction may be accompanied by local destruction of the boundary layer and direct contact of the two surfaces in local spots. The oil film sufficient for complete separation of the bearing and journal surfaces doesn't arise immediately. It takes several turns until the oil film becomes thick enough. Heavy machines, to prevent bearing damage during the start, are often provided with special devices enabling filling out the lubrication system and the bearings with heated oil and slow rotation of the shafts before the load start. Afterward, the machine is put in operation and loaded gradually. In smaller machines, the process of putting in operation is certainly not so intricate. But anyway, during the first turns of the shaft, the microasperities of the journal and the bearing may not be separated. This leads to some wear or, under adverse conditions, the damage may be more serious.

Direct contact in SBs may occur in operation as well. Occasional overloading of the bearing, local overheating, insufficient oil supply, and other conditions may cause this. It is clear that in such cases, when two surfaces come into contact at high speed, the nature and extent of the damage depend very much on the properties of the contacting materials.

Failures of SBs are usually caused by some abnormality, such as:

- Lack of, or low pressure of, oil
- Contaminated oil
- Overload caused by work under emergency conditions
- Local overload due to inaccuracy of machining or assembly
- Local overload due to elastic deformations of the shaft and the support under load

Against lack of oil or its contamination, nothing can help. The same can be said about emergency conditions, but these cases are rare. What always exists is the local overload, and it is highly desirable to impart to the SB the ability to somehow repair this problem by itself — some kind of local self-alignment. Therefore, the most important properties of the antifriction layer of the SBs are as follows:

Low hardness, which imparts to the bearing the ability to run in: As was said in the preceding text, the minimal thickness of the oil film in SBs is measured in microns or tens of microns. Therefore, uneven load distribution, caused by corrugation and roughness of the surfaces and by misalignments of all kinds, may lead to local overload, overheating, and destruction of the oil film. Soft material may be easily plastically deformed, worn out, or even melted in this small area, so that a minute amount of material is removed, the local load is decreased, and the proper functioning of the bearing is restored.

Low melting temperature: Quick melting of the overloaded and overheated local spots of the bearing and subsequent better distribution of the load may prevent overheating and damage of the entire bearing. The melted particles of bearing material are carried out with the oil flow.

Lack of tendency to seizing or welding to the shaft material: In the local areas, where the oil film is destroyed, the temperature may rise very significantly. If the material of the antifriction layer is able to adhere or to weld to the shaft, the journal surface becomes rough, and the bearing may fail quickly.

High thermal conductivity: Heat generated in SBs is mainly removed by oil. But in these very areas, where the oil film is damaged, the part of the heat transferred directly into the shaft and (particularly) through the bearing body into the relatively cold housing increases sharply (because there is no oil flow there).

Oil wettability: Those who have had occasion to wash dishes by hand know that the water doesn't stick to a greasy plate, it slides down. But a clean plate, if wetted, remains wet.

That means that a thin film of water covers it. These are two examples of bad and good wettability. If the bearing material has good wetability with oil, its surface is covered by a thin film of adsorbed oil, which prevents dry friction.

Low coefficient of boundary friction: The lower it is, the less the heat generated in the overloaded areas, and the more likely the bearing will run-in owing to local wear or plastic deformation without melting.

Mechanical strength: This requirement contradicts the demand for low hardness. In the following text, we will see how this contradiction may be resolved.

Wear resistance: This natural requirement, at first sight, also contradicts the demand for low hardness. In reality, it is not always so. The wear phenomenon is a multiple factor. It includes mechanical distortion (scratching and microshear of the surface asperities; plastic deformation, embrittlement, and flaking-off of microparticles), fatigue appearances, heat related effects (welding in microvolumes with following abruption), molecular interaction (adhesion and seizure), chemical and electrical processes, and so on. Therefore, there is no direct relation between the hardness and the wear resistance of materials. Specifically, some soft materials such as rubber and tin babbitt are wear resistant in lubricated sliding contact.

The most popular material used for producing of antifriction layer in SBs is called *babbitt*, invented in the 19th century and named for its inventor (I. Babbitt, 1799–1862). This antifriction alloy, which consists mostly of tin with small additions of antimony and copper, complies with the majority of these requirements except mechanical strength. Hardness of babbitt metal at 20°C is HB 20–30, yield strength under compression is 40–60 MPa, and the fatigue limit is about 20 MPa. At 100°C, these parameters decrease by 50%. Therefore, the bearing shell should be made of stronger material (usually steel or hard bronze), and the babbitt is used as a thin antifriction liner. In thin layers, strength of materials is considerably higher. The melting temperature of babbitt is about 240–250°C. The shaft journal that works with the babbitt liner may be of a middle hardness, about HB 250–300. But the higher the hardness, the more is the wear resistance of the journal, so HRC > 50 is preferable.

Hard particles that may enter the bearing with lubrication oil (such as wear debris, pollution particles, and oil aging products) embed into the soft bearing material and scratch the shaft journal. Therefore, its higher hardness is useful because it helps to keep in time the initially high quality of the shaft surface.

For larger unit loads, stronger antifriction materials are needed. One of the best of them is leaded bronze, which consists mainly of 30% lead and 70% copper with small additions of tin, zinc, and nickel. Almost all these characteristics of bronzes are worse than those of tin babbitt. The bronze is twice as hard (HB 40–60) and the ability to run-in is much lower; it is able to seize the journal, and its oil wettability is less. The coefficient of boundary friction is higher by a factor of 1.5–2 than that of tin babbitt. The only advantage of bronze over babbitt is its mechanical strength, which is indispensable for high load applications, such as bearings of diesel engines. To improve the antifriction characteristics of bronzes, they are usually coated with a very thin layer of babbitt (see Subsection 6.3.4). If the antifriction layer is made of bronze, the hardness of the journal should be HRC 50–55 or more.

Leaded bronzes are still not very strong, and they are used as a liner for steel shells. There are many stronger bronzes, which are used, say, for worm wheels. These bronzes (based on copper with addition of about 10% of tin or 10% of tin + 10% of lead) have hardness of HB 70–85. They are also used for solid SBs (without lining, possibly with galvanic babbitt coating of several microns thick). Such bearings are much cheaper, and they are used at relatively small speeds, with boundary and semifluid friction conditions.

Many applications have made use of antifriction alloys based on aluminum with additions of tin, copper, and other elements. Aluminum alloys are strong enough, and they are used both in solid form and as bimetal bearings. Hardness of aluminum antifriction alloys is about HB 40–80.

Running-in capabilities and coefficient of boundary friction are worse than that of the bronzes. Benefits are low cost and low weight. Hardness of the adjoined shaft journal should be HRC 50–55 or more.

6.3.4 Design of Radial SBs

Continuous (nonsplit) SBs usually consist of steel shell 4 and antifriction lining 5 (Figure 6.28a). The shell is made of low-carbon steel (0.1–0.2% of carbon), because with growing carbon content, the adhesion of the liner becomes worse. The sleeve thickness S may be approximately defined from a formula: $S = 0.03d + 4$ mm, where d is the bearing diameter, in millimeters. Split (into two halves) bearings in small-lot production are usually thicker, $S = (0.10–0.15)d$. In large-scale manufacture, thin-walled split bearings have wide application. The halves of such a bearing are machined separately in a jig, which ensures their interchangeability.

An antifriction layer is deposited on the shell by pouring of melted metal into a special form that contains the shell. Surface roughness, which is intended for adhesion of the liner, is usually about $R_a = 3–5$ µm. The final thickness of the liner (after machining) comes to 0.5–1.5 mm, depending on the diameter of the bearing. As mentioned earlier, the material of the lining is usually babbitt or bronze. To ensure the best conditions of friction and running-in, the bronze is usually plated with an "improver": a tin–lead layer of 20–60 µm thickness. To prevent the diffusion of tin from the improver into the bronze, a very thin (of 2–3 µm) layer of nickel should be deposited between them.

Leaded bronze is susceptible to segregation of lead (uneven lead distribution); therefore, the bearings should be checked for segregation by x-raying. The adhesion of the liner to the shell may be checked by the ultrasonic method. At the ends of the bearing, adhesion may be checked using dye penetrant. It is desirable to install SBs in the housing with a tight fit, producing a surface pressure of about 5–10 MPa. At that pressure, the compression stress in the shell should not exceed the yield stress of its material. The tight fit reduces the probability of fretting within the connection and improves the heat dissipation from the bearing into the housing. The last is particularly important when the machine starts and the oil film is not settled yet.

Example 6.8

Shaft journal diameter $d = 100$ mm; bearing shell thickness $S = 0.03 \cdot 100 + 4 = 7$ mm; thickness of the liner is about 1 mm. From this, the outer diameter of the bearing $D = 100 + 2 \cdot 7 + 2 \cdot 1 = 116$ mm. The needed pressure in the connection with the housing $p = 5$ MPa minimum. The housing is made of gray iron, modulus of elasticity $E_h = 1 \cdot 10^5$ MPa, and the outer diameter of the housing is 170 mm. What should be the interference in this connection?

For this calculation, the influence of the liner may be neglected, as it is thin and its modulus of elasticity is low. From Equation 6.7,

$$\xi_1 = \frac{102}{116} = 0.879; \quad \xi_2 = \frac{116}{170} = 0.682;$$

$$\delta = 10 \cdot 116 \cdot 10^3 \left[\frac{1}{2.06 \cdot 10^5}\left(\frac{1+0.879^2}{1-0.879^2} - 0.3\right) + \frac{1}{10^5}\left(\frac{1+0.682^2}{1-0.682^2} + 0.3\right)\right] + 2 \cdot 1.4 \cdot 5^{0.5} = 83.7 \,\mu m$$

The standard fit H7/u6 gives an interference of 109 µm minimum and 166 µm maximum. Because, in this case, the influence of the surface roughness is small, the pressure in the connection may be assumed to be proportional to the interference. Therefore, the minimal and the maximal pressure in the connection are

$$p_{min} = \frac{109}{83.7}5 = 6.5 \, MPa; \quad p_{max} = \frac{166}{83.7}5 = 9.9 \, MPa$$

Let's check the compression stress in the bearing shell at maximum interference:

$$\sigma = \frac{pD}{2S} = \frac{9.9 \cdot 116}{2 \cdot 7} = 82 \, MPa$$

The yield stress of low-carbon steel with 0.2% carbon is about 330 MPa. Therefore, there will not be any plastic deformation of the bearing shell. The next question is, how much will the inner diameter decrease after the shell is installed in the housing with maximal interference:

$$\Delta d = \frac{2pdD^2}{E(D^2 - d^2)} = \frac{2 \cdot 9.9 \cdot 102 \cdot 116^2}{2.06 \cdot 10^5 (116^2 - 102^2)} = 0.043 \, mm$$

The operating diametral clearance c in this bearing (the difference between the bearing and the journal diameters) should be

$$c = 0.001d + 0.05 = 0.001 \cdot 100 + 0.05 = 0.15 \, mm$$

Obviously, the shrinkage of the bearing caused by the tight fit can't be neglected, and the inner diameter of the bearing should be increased (before assembly) by 0.043 mm.

Independently of the tightness of fit, SBs should be fixed against rotation relative to the housing. For this purpose, pins, screws, keys, and other means are used. If the bearing doesn't have shoulders, it should be fixed in the axial direction as well.

Examples of SB fixation are shown in Figure 6.30. A bearing without shoulders (Figure 6.30a) may be fixed in all directions by screw 1. This screw is installed in a threaded hole drilled after the bearing is put in place. When choosing the diameter of the screw, differences in materials should be taken into consideration. For example, when drilling a bearing shell made of steel and a housing

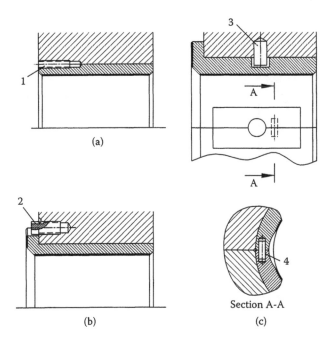

FIGURE 6.30 Fixation of sliding bearings against rotation and axial displacement.

of cast iron or aluminum, there may be some displacement of the hole toward the softer material, and a small screw hole may drift completely out of the bearing shell.

In Figure 6.30b is shown a combined (radial and thrust) bearing. Owing to the flange, the bearing is held in axial direction between the housing and the shaft shoulder. To prevent possible rotation of the bearing, a semicircular groove is machined in the bearing's flange, and screw 2 is screwed through it in the housing. Punching in one of two lengthwise grooves locks the screw.

Figure 6.30c shows a split combined bearing installed in split housing. It is fixed in axial direction by its flange and against rotation by pin 3. But it is not enough in this case. Because there is a thrust bearing, both halves of the bearing must be fixed in axial direction relative to each other to keep the faces of the flange coplanar. Pins 4 are installed with tight fit on both sides before final machining.

If the bearing has failed and gripped the journal, no fixing elements are able to prevent it from rotation. Steel screws and pins are usually bent and squashed damaging the housing. Therefore, the fixing elements should not be too strong. From this point of view, pins and screws of bronze or brass are preferable; should the bearing rotate, the copper alloy elements are sheared neatly and don't damage the housing.

SBs are usually pressure-feed-lubricated. To make the oil enter along the entire length of the bearing, it is usually provided with an oil-distributing groove. Length of the groove is up to 80% of the bearing length. Some shapes of these grooves are shown in Figure 6.31. The shape of the oil distributing grooves is mainly determined by manufacturing limitations. Inside the sleeves, they are usually produced by a mill or a cutter bar and have the shape of a crescent (Figure 6.31a). The width of the groove b (in circumferential direction) should not be too great, so as not to occupy the loaded sector of the bearing, where the oil wedge is supposed to be. If the radial force changes its direction (for example, when reversing the machine), the sector needed to take the load may be much larger. In these cases, the width of the groove may have to be lessened (Figure 6.31b). In any case, the transition from the groove to the bearing surface should be made smooth and tapered to ease the oil entering the small gap between the bearing and the shaft journal. For this reason, the oil should not be too thick, but not too thin either. In split bearings, the inner edges of the parting faces should be beveled by small chamfers to prevent jamming of the shaft by the sharp edges (1 in Figure 6.31c).

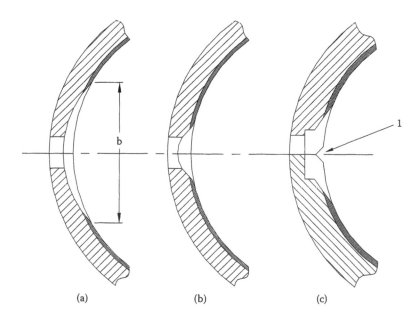

(a)   (b)   (c)

FIGURE 6.31 Oil distribution grooves.

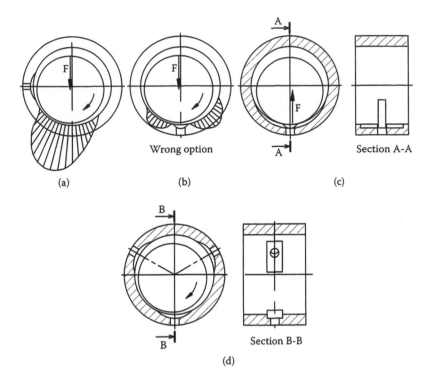

FIGURE 6.32 Location of oil distributing grooves.

The oil distributing grooves should be placed in a nonloaded sector of the bearing, within $\gamma = 90$ to $270°$ from the radial force direction (see Figure 6.29a). Experiments have shown that the minimal temperature in the bearing is achieved when $\gamma = 90°$ or $\gamma = 270°$. At $\gamma = 180°$, the temperature may be higher by 10 to 20%.[7]

It is unacceptable to place the oil distributing grooves in the middle of the loaded sector of the bearing. In this case, not only is the load capacity of the bearing oil wedge gravely reduced (compare Figure 6.32a and Figure 6.32b) but also the oil flow rate is sharply decreased because the oil groove is almost closed up by the shaft journal. This may result very likely in failure of the bearing due to overheating.

If a narrow oil distribution groove is placed in vertical plane below the shaft, as shown in Figure 6.32c, the shaft closes up the groove, when the machine is standing. (Notice that the load vector F is directed upward.) In the case of need to fill up the oil system and the bearings before starting, the hydrostatic pressure may not be able to lift up the shaft, and this bearing may remain dry. Such a situation is unacceptable. There may be a few options to solve this problem, such as the following:

- Changing the location of the oil groove (if possible, in the shown case, it can be displaced by 90° in either direction)
- Increasing the width of the oil groove to increase the lifting capacity of the hydrostatic oil pressure
- A semiannular groove can be made (see Figure 6.32c, section A-A) to ensure oil supply to the bearing in any situation

Sometimes, lifting a heavy shaft by a special device can be considered. One way or another, the bearing must not start dry, if it is required to be filled up before.

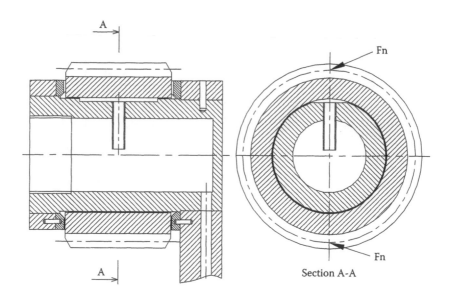

FIGURE 6.33 Sliding bearing of a planet wheel.

If the direction of the radial load changes indefinitely, a system of three smaller oil grooves (Figure 6.32d) may be helpful. The length of these grooves (in axial direction) may be about 25–30% of the bearing length.

If the radial force rotates along with the shaft (or, on the contrary, a fixed shaft serves as a support for a rotating part while the load is unidirectional, as in the case of planet wheels), the oil distributing groove should be made on the shaft as a flat (see Figure 6.33). In such a support, the antifriction layer should be preferably deposited on the shaft, which doesn't rotate relative to the load direction. In this design, the antifriction layer is subjected to a permanent load, and the permissible pressure may be twice as much as in the case of a pulsating load.

In the beginning of design, the dimensions of the SB may be preliminarily estimated reasoning from the relative unit load by the following equation:

$$q = \frac{F_r}{dL} \ MPa \qquad (6.10)$$

where
  F_r = radial load, N
  L = length of the bearing (mm)
  d = diameter of the bearing (mm)

The admissible value of q depends on the strength of the antifriction layer. Because the maximal pressure in the oil film is about 3–3.5 times q, the q value should be correspondingly about 30% of the liner material strength. For babbitt under permanent load and at 100°C, it is about 6 MPa, but in thin layers, it may be threefold or fivefold. For leaded bronze, which is threefold stronger than babbitt and keeps its strength up to 200°C, the admissible cyclic load in thin layers may be as high as 30–50 MPa (in diesel engines). Of course, at that load, the liquid friction in the bearing should be ensured. For stationary machines, when the dimensions of bearings are not highly restricted, the unit load q = 3–5 MPa is usually allowed. The ratio L/d is mostly taken equal to

0.4–0.8. Longer bearings are too sensitive to misalignments. When L/d is greater than 1, self-aligning SBs are mostly used.

Radial clearance c of the bearing may be chosen in a wide range of values. As the viscosity of mineral oils depends on temperature, the sliding bearing is a self-regulating system. With the increase of the radial clearance, the bearing capacity of the oil wedge reduces. But at the same time, the oil flow rate increases because of larger clearance, and this leads to better cooling and lower bearing temperature. This in turn increases the load capacity of the oil wedge, which returns to its initial value, but at lower temperature. Therefore, the radial clearance should not be too small. As a good approximation, its minimal value could be taken as $c = 0.001d + 0.05$ mm.

Now, we have all dimensions of the bearing needed to perform calculation of its hydrodynamic parameters to determine the minimal oil-film thickness h and the maximum temperature of the oil T_{max}. That may be done using a computer program. The condition for reliable operation with liquid friction is

$$h \geq 1.1 h_{crit}$$

where h_{crit} = critical thickness of the oil film, at which there is a risk of boundary friction and of damage to the antifriction layer. Its value depends on the roughness heights of the shaft journal R_{z1} and the bearing R_{z2} and on the misalignment of the shaft relative to the bearing evaluated by angle γ:

$$\eta_{crit} = R_{z1} + R_{z2} + 0.5\gamma L$$

To tell the truth, the minimal oil-film thickness can hardly be calculated with a high accuracy owing to imperfections of the theory and impossibility to take into account all the endless variations in loads and vibrations, in macro- and microgeometry of the surfaces, in deformations of the shaft, housing, and the antifriction layer, and so on. Therefore, it is always worth taking into consideration (in addition to the calculation) the parameters of analogous machines (prototypes), which are in service. Their bearings should be analyzed comparatively by means of Equation 6.10 and using computer programs as well.

Mineral oils at temperatures above 120–140°C fail to form adsorbed layer on the bearing surface and lose their ability to lubricate. Therefore, the admissible T_{max} value is usually limited to 110°C. Decrease of this temperature may be achieved by increase of the radial clearance and pressure of the supplied oil, and by additional cooling of the oil and the bearing (for example, air cooling of the housing).

6.3.5 Design of Thrust SBs

A thrust bearing can be made as a shoulder on a radial bearing (Figure 6.28b, Figure 6.28c, and Figure 6.34), or as a separate flat ring (6 in Figure 6.28c). In a split version, the separate thrust ring can be made of two half-rings centered in a circular recess of the housing (as shown). The separate rings must be also fixed against rotation by pins (Figure 6.28c), screws, or otherwise.

Relation between the outer diameter d_o and the inner diameter d_i of the thrust bearings varies in the range from $d_o/d_i = 1.2$–1.6. The materials are the same as for radial SBs.

Load capacity of the flat thrust bearing shown in Figure 6.34 is relatively small, because there is no hydraulic wedge formed. It is able to work at unit pressure of about 0.4–0.6 MPa and at speed up to 15 m/sec. To get the sliding surfaces lubricated, oil grooves should be made on the working surface of the bearing. These grooves should be deep enough to avoid be plugged up with wear products or particles of slightly melted antifriction layer (say, because of a momentary overload). The oil grooves may take up to 20% of the bearing area.

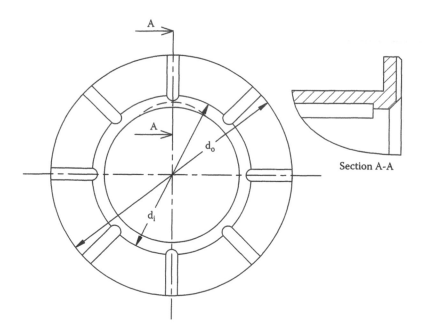

FIGURE 6.34 Sliding thrust bearing with flat working surface.

Clogging of the oil grooves not only impairs oil supply to the thrust bearing, but also hinders the drain of oil from the neighboring radial bearing. The latter may become overheated.

At higher unit loads (up to 4–5 MPa), there may be a need to form angled surfaces to create hydrodynamic oil wedges (Figure 6.29b and Figure 6.35). The radial grooves in this case should have contraction at the output, although the output cross section should be enough to enable oil drain from the neighboring radial bearing. It is significant to emphasize that any wear of such a

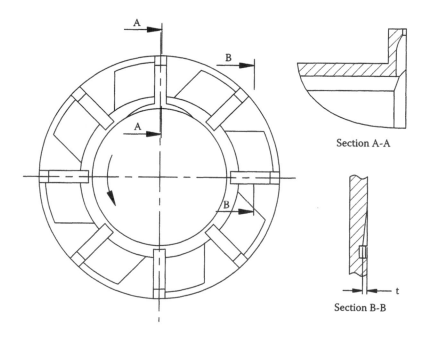

FIGURE 6.35 Sliding thrust bearing with tapered working surface.

bearing is impermissible, because the depth t of the angled surfaces is very small, and any wear wipes them out turning the angled bearing surfaces into flat ones. Therefore, the designer should exclude all the factors promoting wear, such as start under load or without preliminary fill with oil, oil contamination, abnormal working conditions (overload or underspeed, when the load capacity of the oil wedges is insufficient), and so on.

Thrust bearings with self-aligning elements (Figure 6.29c) are much more complicated in design, but because this design provides excellent conditions for creation of a hydrodynamic oil wedge, they have high load capacity and performance reliability. In addition, these thrust bearings are less sensitive to changes in working conditions and to wear. These advantages determine wide application of this type of thrust SBs in heavy engineering.

It is recommended that the self-aligning elements should be designed so that their length in the middle is approximately equal to their width ($L \approx B$, Figure 6.29c).[7] Thickness of the shoe $S = (0.3–0.5)L$. The bearing point of the shoe is usually displaced from the middle of the length by $(0.05–0.10)L$. (In reversible machines, this point should be placed in the middle of the length.) Shoe materials are the same as of the other sliding bearings: body of low carbon steel and antifriction lining (usually babbitt).

6.3.6 SURFACES CONNECTED WITH SBS: FEATURES REQUIRED

Precise geometry, smooth surface, and high hardness — these are the three main features required of a surface adjoined to the sliding bearing, whether it is a shaft neck, shoulder, or disk. It is desirable that geometric errors be small compared to the minimal oil-film thickness in the bearing. With this achieved, the elastic deformations should be taken into account. Shaft deflections may cause increased load on the end of the bearing where the load capacity of the oil wedge is much less because the oil flows freely out of the bearing in this area. Partly, this effect can be compensated by wear and deformation of the antifriction layer of the bearing. But if the shaft deflections are significant, self-aligning SBs should be used (Figure 6.36) or the working surface of the bearing should be shaped in compliance with the deformed shaft.

Figure 6.36a shows a bearing of a camshaft of a large marine diesel engine. These bearings are made of aluminum alloy and mounted with a small clearance into a cylindrical bore of the engine housing. A pin, which is installed with a generous clearance, prevents both rotation and axial movement of the bearing. Clearances allow the bearing to swing on its narrow "leg" and to

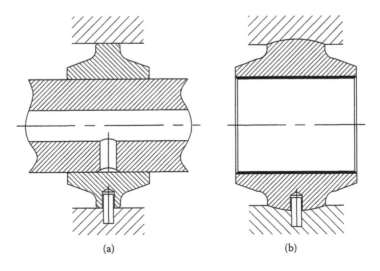

(a)　　　(b)

FIGURE 6.36 Self-aligning bearings.

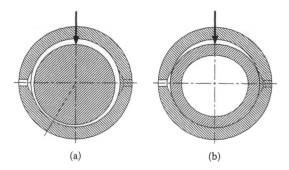

FIGURE 6.37 Deformation of a hollow shaft.

follow the bending deformations of the shaft. Lubricating oil is supplied to the bearings through a common central hole and radial holes in the shaft. Each radial hole is placed in compliance with the neighboring cams, so as to be in the unloaded part of a bearing. The bearing shown in Figure 6.36b has a spherical support in the housing.

Hollow shafts must be checked for roundness. As stated earlier, the hydrodynamic oil wedge develops owing to the clearance between the shaft and the bearing (Figure 6.37a). Because the cross section of a hollow shaft may become slightly oval under load, this clearance may be reduced, or even eliminated, to the point where contact may occur in the plane perpendicular to the load direction (Figure 6.37b). In this case, overheating and failure of the bearing may occur.

A shaft shoulder or disk, which serves as a thrust face for the thrust bearing, should be perfectly perpendicular to the axis of rotation. The part of the shaft adjoined to the bearing may become tilted relative to the housing by a certain angle Θ (due to displacement in the bearings [Figure 6.2] or because of the flexibility of the shaft [Figure 4.15b]). If this angle is less than or equal to the hydrodynamic angle (of about 0.0003–0.001 rad), there may be formed a hydrodynamic oil wedge. But if the angle is larger than stated previously, some measures should be undertaken to ensure uniform load distribution. For example, self-alignment of the bearing may be achieved using a spherical support (Figure 6.38a). Provided that the axial load is not too big, spring support may provide more even load distribution over the bearing circumference (Figure 6.38b).

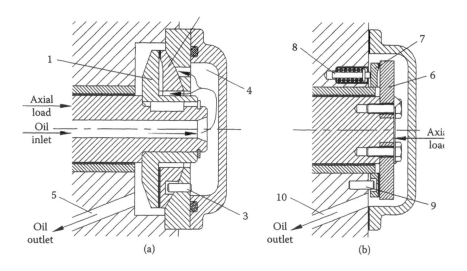

FIGURE 6.38 Self-aligning thrust bearings.

In Figure 6.38a, thrust disk 1 is connected to the shaft by a tight fit, a key, and a snap ring. Thrust bearing 2 (with an antifriction layer) has a spherical support; and its rotation is prevented by two pins 3. Lubrication oil is supplied through a central hole in the shaft, then it enters the oil chamber 4, and from there, to the thrust bearing. Hole 5 is intended for free outflow of the oil after it has worked in the radial and thrust bearings.

In Figure 6.38b, thrust disk 6 is bolted to the shaft, and thrust bearing 7 is supported by several spring-loaded pushers 8 distributed evenly. The thrust bearing is stopped against rotation by several pins 9, and it is lubricated by oil, which flows out of the radial bearing. Used oil can flow out freely through bore 10.

Low roughness of the sliding surfaces is of primary importance. The minimal oil-film thickness in the bearing measures microns or tens of microns; therefore depending on the roughness height, either the surfaces will be completely separated by an oil film or they will be in direct contact with each other. In the latter case, increased friction, heat generation, and wear may result in damage of the bearing. While starting, when the oil wedge has not been generated yet, low roughness contributes to decreasing the wear of the antifriction layer. Usually, the roughness height should not exceed $R_a = 1.25$ μm for the antifriction layer and $R_a = 0.63$ μm for the shaft journal. The less the calculated minimal oil film thickness h, the less should be the roughness height of the sliding surfaces.

The smoothness of antifriction layer is less important because it is soft and may change its surface conditions appreciably during the running-in process. In the beginning, the journal takes off mechanically a thin surface layer from the bearing. Soon the journal's asperities become blunt, and then they deform plastically the surface layer of the bearing in the sliding direction. Now, the hard particles of the antifriction compound move deeper into the soft matrix,[8] and the soft component, being easily deformed, fills all the dimples and wears off the elevations. This process leads to increase in the real area of contact and improvement in the friction conditions. The plastic deformation, along with the oil and temperature influences, form a special amorphous structure impregnated with carbon from the oil and oxygen from the air. The thin layer of this material prevents the sliding surfaces from direct contact and imparts the wear resisting property of the bearing. The roughness of the bedded-in surfaces may be smaller or bigger than before running-in. It is determined by the parameters of the running-in process, such as materials and initial roughness of the journal and bearing, sliding velocity, unit load, kind of lubricant, and rate of change of the velocity and load during the running-in period.

6.3.7 OIL SUPPLY TO SBS

SBs are usually fed with oil through oil ducts made directly in the housing. Such ducts, which don't have threaded and pipe connections, are remarkable for their simplicity and reliability. The indispensable requirement for the oil ducts formed in the housing is the easy access for their cleaning and control of cleanliness. There should not be blind holes that can't be cleaned and visually checked through and through.

Figure 6.39a shows oil ducts drilled in the lugs of a cast housing. The lugs also serve as stiffeners, which increase the strength and rigidity of the walls. The roughness of the holes should be made reasonably low (sometimes they are reamed) to provide better cleaning conditions.

Figure 6.39b shows a part of a welded housing, in which the oil feeding pipes are welded in as an element of the structure. Notice that the pipes are welded throughout the contour inside the bearing housing as well (welds 1). If the gap between the pipe and the hole is not fully welded, there may remain chips and scale or dirt, which may afterwards fall out, get to the bearing, and damage it. Welding inside the bearing housing is not easy, especially if the bore diameter is not big enough. It may be replaced by tight fit of the pipe into the hole or by beading the pipe, but somehow the gap, which can't be cleaned and checked, must be avoided.

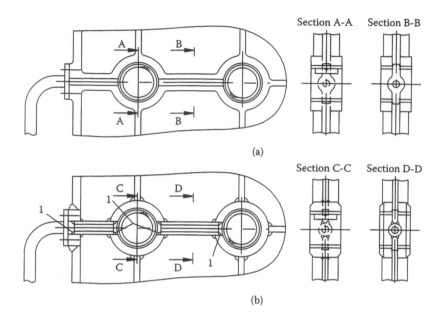

FIGURE 6.39 Oil ducts in a housing.

It is not always practical to supply lubrication oil through the housing ducts; sometimes, it is more convenient to do that through a hollow shaft. Figure 6.40 shows a typical arrangement using a hollow shaft as an oil duct. The oil is fed from pipe 1 into bearing 2 through circular groove 3 in the housing, slot 4, and circular groove 5 in the bearing shell.

Groove 3 is made to avoid failure of oil supply to both bearings because of accidental displacement of slot 4 from the opening of pipe 1. This displacement may be caused either by wrong machining, or assembly, or by shearing the fixing elements during a momentary seizure of the bearing.

From groove 5, the oil flows to the oil-distributing groove 6 (to lubricate bearing 2) and, through radial bore 7, to chamber 8. From this chamber, the oil is supplied through radial bore 9 and circumferential groove 10 to oil-distributing groove 11 of bearing 12.

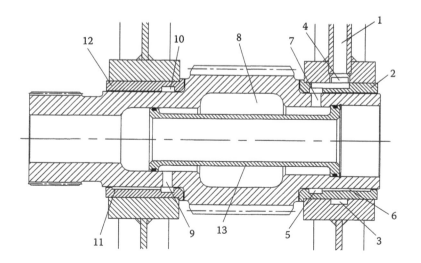

FIGURE 6.40 Oil supply through a hollow shaft.

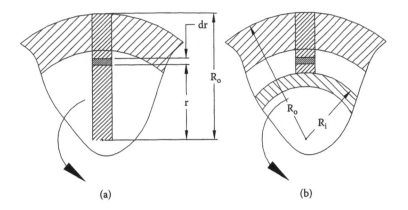

(a) (b)

FIGURE 6.41 Pressure in the rotating oil.

The designer should consider that the oil, on its way to chamber 8, has to overcome the hydrodynamic pressure created by centrifugal forces. To calculate this pressure, let's examine the rotating column of oil presented in Figure 6.41a. Its sectional area is A. Let's take in this column an elementary volume dr in height, placed at a distance of r from the center of rotation. The mass of this volume is

$$dm = A\frac{\gamma}{g}dr$$

where
  γ = unit weight of the oil
  g = acceleration of gravity

Elementary centrifugal force developed by this mass

$$dF = \omega^2 r \cdot dm = A\frac{\gamma}{g}\omega^2 r \cdot dr$$

The entire centrifugal force of the oil column shown in Figure 6.41a is

$$F = \int_0^{R_0} dF = \int_0^{R_0} A\frac{\gamma}{g}\omega^2 r \cdot dr = A\frac{\gamma}{g}\omega^2\frac{R_0^2}{2}$$

To draw the pressure p, the force F should be divided by the area A:

$$p = \frac{F}{A} = \frac{\gamma}{g}\omega^2\frac{R_0^2}{2}$$

EXAMPLE 6.9

How much may be the influence of the centrifugal forces? Let's assume that the shaft diameter is 200 mm, and the rotational speed $n = 2000$ r/min. The unit weight of the oil $\gamma = 0.9 \cdot 10^{-3}$ kg/cm³, and the acceleration of gravity $g = 981$ cm/sec². The angular speed is

$$\omega = \frac{\pi n}{30} = \frac{\pi \cdot 2000}{30} = 209.4 \; s^{-1}$$

(The physical meaning of the angular speed is number of radians per second, but the radians have no dimension.)

Because centimeters are used in both the unit weight and the acceleration of gravity, the shaft radius should be taken in centimeters as well: $R_0 = 10$ cm. Now we are ready to calculate the hydrodynamic pressure:

$$p = \frac{0.9 \cdot 10^{-3}}{981} 209.4^2 \frac{10^2}{2} = 2 \; kg/cm^2 = 0.2 \; MPa$$

The oil pressure must be more than 0.2 MPa; otherwise, oil can't enter inside the shaft.

As the rotational speed increases, the hydrodynamic pressure increases as well. For example, if $n = 3000$ r/min, $p = 0.45$ MPa. It is a lot and must be taken into consideration when defining the minimal oil pressure needed in the lubrication system. It is clear that the pressure losses in the ducts caused by flow resistances must be calculated as well. During the trials, the oil pressure should be checked as close to the bearings as possible.

If there is an insert in the shaft (part 13 in Figure 6.40), the hydrodynamic pressure will be less. It may be obtained from a formula (see also Figure 6.41b):

$$p = \frac{\gamma}{g} \omega^2 \frac{R_0^2 - R_i^2}{2}$$

Often, the oil is supplied to the bearings through pipes connected to the housing with fittings. (These pipes should be rigid enough to avoid the possible resonance with the frequencies of the mechanism.) But there may be bearings that can be lubricated only through ducts made in the parts of the mechanism. As a typical example, the crank mechanism of a diesel engine may be considered. The way of oil supply to these bearings is shown in Figure 6.42. The oil is delivered from a pressure pipeline to the main bearing 1 of the crankshaft, which has a semiannular groove 2. This groove is always connected with one of holes 3 drilled in the crankshaft journal; thus, chamber 4 permanently communicates with the pressure pipeline (through holes 3 and groove 2).

It is clear that bearing 1 has oil-distributing grooves as well, but here we are considering the way of oil supply to those bearings that move relative to the housing.

From chamber 4, the oil flows through holes 5 to chamber 6, which is connected by radial holes 7 with the rod journal bearing 8. The latter has two semiannular grooves: on the inner surface (9) and on the outer surface (10). These grooves are communicating, and the oil flows through holes 7 and grooves 9 and 10, and then through the central bore 11 in the connecting rod to the piston pin bearing 12.

Figure 6.43 shows another way to deliver oil through rotating shafts. Gears 1 and 2 rotate in sliding bearings, and they are connected with a spline coupling 3. The oil is supplied to steel cover 4 from pressure pipeline 5. Oil transition from cover 4 to gear 1 is carried out owing to aluminum pipe 6 installed between them. The pipe has spherical ends slide-fitted to accurately machined holes

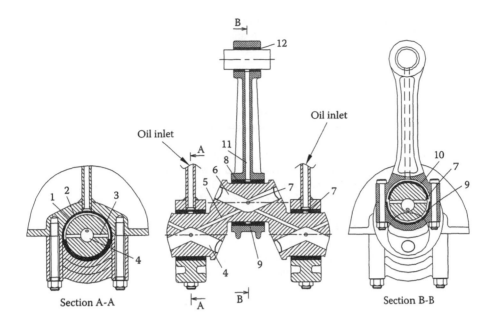

FIGURE 6.42 Oil ducts in diesel engine crank mechanism.

in both the cover and the gear. When the gear rotates, pipe 6 may rotate with any speed from zero to the speed of the gear.

From the central hole of gear 1, the oil flows to the bearings through radial holes 7 and 8. The bearings have semiannular grooves 9 and 11 and also oil-distributing grooves 10 and 12 in the nonloaded areas of the bearings. When the gear rotates, radial holes 7 and 8 always communicate with one of the ends of the semiannular grooves. In such a way, permanent oil supply is ensured to the bearings.

Pipe 13 is similar to pipe 6, and serves as an oil duct between gears 1 and 2. In pipe 13, there is a small hole 14, through which oil flows out to lubricate the spline connections. Lubrication of the bearings of gear 2 is carried out in the same way as the bearings of gear 1, but the grooves in the bearings may have different angular positions, in accordance with the radial load direction.

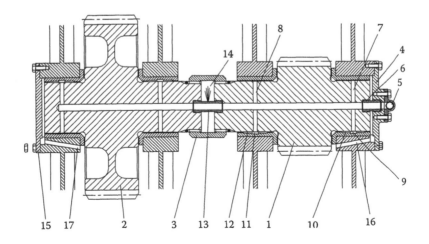

FIGURE 6.43 Oil supply to plain bearings and spline connections.

Not only the oil supply, but also the free oil outflow should be ensured. In this case, covers 4 and 15 close the bearing chamber, and holes 16 and 17 serve to drain them.

REFERENCES

1. Dudley, D.W., *Gear Handbook*, McGraw-Hill Book. Co., New York, Toronto, London, Sydney, 1962.
2. Nikitin, Y.M., *Design of Elements of Arts and Units of Aircraft Engines*, Mashinostroenie, Moscow, 1968 (in Russian).
3. Edunov, V.V., Gankin, B.M., Klebanov, B.M., and Shumiliver, R.G., Resistance of a shaft in cylindrical roller bearings to axial displacement, *Transactions of the Rostov-na-Donu University of Agriculture Engineering: Problems of Mechanics in Agriculture Engineering*, Rostov-na-Donu, 1985 (in Russian).
4. Levina, Z.M. and Reshetov, D.N., *Contact Stiffness in Machines*, Mashinostroenie, Moscow, 1971 (in Russian).
5. *Rolling Bearings Handbook*, Spizin, N.A. and Sprishevsky, A.I., Eds., *Mashgiz*, Moscow, 1961 (in Russian).
6. Garkunov, D.N., *Tribo-Engineering*, Mashinostroenie, Moscow, 1989 (in Russian).
7. *Reduction Gears of Power Plants* Handbook, Derzhavets, Y.A., Ed., Machinostroenie, Moscow, 1985 (in Russian).
8. Karasik, I.I., Running-in capability of materials for sliding bearings, *Nauka*, Moscow, 1978 (in Russian).

7 Gears

Gear engineering is not an exact science. It is both an art and a science.

Darle W. Dudley

This chapter seems, at first glance, to be more difficult than the previous ones, and we all like to avoid difficulties. Keeping this in mind, the authors would like to begin this chapter with a story of a man whose doctor gravely prescribed him a gymnastics regimen. Not wanting to do the exercises, the lazy patient made an agreement with himself: "OK, I'll exercise once a week, and then we'll see." So he got up in the morning, begrudgingly gathered all his will, and started doing gymnastics, comforted by the thought that once it is done, he will live a free man for a week. Upon finishing the thought and the exercise, the man was pleased and at ease.

The contented feelings remained with the man through the following morning. "How wonderful! I don't need to do gymnastics today — and tomorrow." But then a cloud covered the man's high spirits. "Today and tomorrow are fine, indeed, but in a week, this nasty day will come again! What if I exercise today, so I won't have to think about it for two whole weeks?" The contemplation of an extended period of liberty agreed well with the man, and so he exercised. The following day he decided to prolong his prospective period of bliss for a month and exercised again. He continued until he finally earned himself a whole gymnastics-free year. Reflecting upon his hard work, the man was filled with pride and satisfaction. "That's it! Enough!" he declared to himself, looking forward to the break he so richly deserved. When the man woke up the next morning, however, he realized that he simply could not start his morning right without gymnastics anymore.

Very often the difference between what is difficult and what is easy, what is unpleasant and what brings gratification is a matter of habit. The authors solemnly swear that your patience will not be abused, dear reader. Ready? Let's go!

Each machine consists in principle of three basic parts: a motor or another prime mover (that sets the machine in motion), an actuating mechanism (which is directly doing the needed work: digs a ditch, weaves cloth, peels potatoes, propels a ship, and so on), and a transmission that connects the first two parts.

Sometimes, the transmission may be lacking, for example, in a fan the impeller is mounted directly on the shaft of the electric motor. But in most cases, the transmission is indispensable, to match the speed of the mover with that of the actuator, to set in motion several actuators from one mover, or for another reason. For instance, on a big ship a steam turbine is usually used as a mover. Its rotational speed is several thousands r/min, whereas the speed of the propeller (for better efficiency) should be about 100 r/min. Hence, between the turbine and the propeller shall be a speed reducer.

On a small motor boat the propeller, with respect to the speed, may be mounted directly on the shaft of the engine. But the engine is desired to be placed above the water (just to be maintained easily), and the propeller to work should be submerged; so in this case the transmission is needed as well.

A gear is the most common type of transmission; therefore, it is one of the mandatory subjects when machine elements are discussed. The geometry of gears is well known, and it is not the subject of investigation here. But the analysis of strength and deformations in gears is impractical without some minimal presentation of tooth geometry. This is the theme of the following section.

7.1 GEOMETRY AND KINEMATICS OF GEARING

Among the great number of gear types, we will take for our consideration the simplest, involute spur gear with a standard mesh. *Standard mesh* means that, when cutting the gear, the cutter is not "held out" and not "fed in," so the pitch line of the cutting rack is tangent to the pitch circle of the gear. This point will be clarified in detail in the following text.

Involute gearing was invented by Leonard Euler (1781), and till now, it has been mostly used for power transmission between parallel shafts. Some basic elements of gear geometry are shown in Figure 7.1. The working profiles of involute teeth have the shape of an involute of a base circle d_b. The diameter of the base circle is equal to

$$d_b = d \cos\varphi,$$

where
  d = pitch diameter

$$d = m \cdot N \ mm$$

$$m = \frac{25.4}{P} \ mm$$

  N = number of teeth
  P = diametral pitch (in the American system)
  φ = pressure angle (see Figure 7.3a, its routine value $\varphi_0 = 20°$)

All the involutes of one circle, if developed in the same direction, are equidistant. In other words, the distance between two involutes is the same in each point and equals the length of the arc of the base circle between the beginnings of these involutes. For example, the distance between involutes 1 and 2 equals the length of arc $a–b$. Because the teeth are equally spaced, the distance between the working surfaces of two neighboring teeth p_b, called *base pitch*, is equal to

$$p_b = \frac{\pi \cdot d_b}{N}$$

where N = number of teeth.

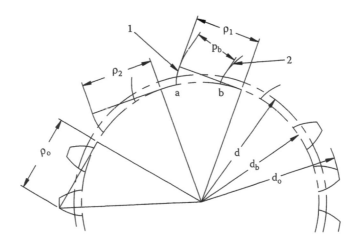

FIGURE 7.1 Involute teeth geometry.

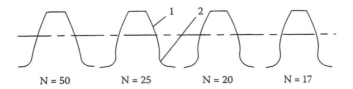

N = 50   N = 25   N = 20   N = 17

FIGURE 7.2 Tooth shape dependence on the number of teeth.

The radius of curvature of the involute ρ in each point equals the length of the tangent to the base circle from this point (for example, ρ_1 and ρ_2 in Figure 7.1). Hence, the working profile of a spur gear tooth is a cylinder with a variable radius of curvature that may change from zero (if the real tooth profile begins at the base circle) to its maximal value ρ_0 on the outside diameter d_0 (Figure 7.1):

$$\rho_0 = 0.5\sqrt{\left(d_0^2 - d_b^2\right)}$$

The shape of the teeth depends on the number of teeth. Figure 7.2 shows teeth of the same pitch. It is evident that the less the number of teeth, the thinner the tooth root, and these teeth are less strong in bending. Besides, the gear with a smaller number of teeth (pinion) is exposed to more stress cycles. That is why pinion teeth fail more often than gear teeth. Balanced life may be achieved by making the pinion from stronger material than the gear. Another way to increase the bending strength of the pinion is to "hold out" the cutter, meaning to make the teeth with a "long addendum" (see Section 7.5 and Figure 7.12).

The tooth profile has two main parts: involute profile 1 and fillet 2. As a rule, not the whole of the involute profile participates in the engagement; usually, there is a small part adjacent to the fillet that doesn't contact the mating gear. It is important to know the location of the boundary point between the active profile and the "idling" part of it, especially when a nonstandard profile with an undercut in the root is to be designed.

Figure 7.3 shows elements of the engagement geometry. The profiles of the contacting teeth must have a common normal in the contact point. Because the profiles are of involute form, this normal is tangent to the basic circles of both gears. This line is called the *line of action* and designated a_1a_2. All possible points of contact are placed on the line of action. It is clear that contact can't be realized outside the outer diameters of the gears (d_{01} and d_{02}). Segment b_1b_2 of the line of action cut by the outer diameters is called the *length of action* (b_1 and b_2 are the points of intersection of the diameters d_{01} and d_{02} with the line of action a_1a_2). At the indicated direction of rotation, each pair of teeth begins contacting in point b_2, and then the point of contact moves toward point b_1, where the teeth disengage.

Figure 7.4 shows an enlarged view of the engagement zone taken from Figure 7.3. In Figure 7.4a, the pair of teeth B_1-B_2 contacts in the beginning point. The preceding pair, A_1-A_2, moves ahead of pair B_1-B_2 at a distance of a base pitch p_b. From this moment on, there are two pairs of teeth engaged, until pair A_1-A_2 reaches point b_1 and gets disengaged. Figure 7.4b shows this very moment, when pair A_1-A_2 contacts in point b_1, and then begins the zone of one-pair engagement until pair C_1-C_2 reaches point b_2 and comes into contact.

The quotient of the length of segment b_1b_2 and base pitch p_b is called *contact ratio*:

$$\varepsilon = \frac{b_1b_2}{p_b}$$

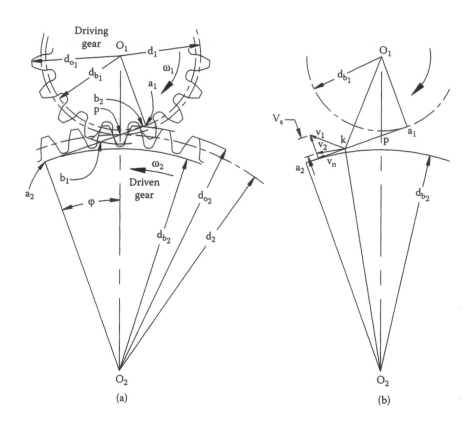

FIGURE 7.3 Spur gear geometry.

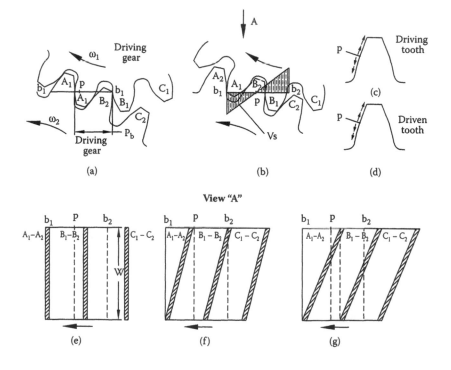

FIGURE 7.4 Tooth engagement.

If, for instance, $\varepsilon = 1.3$, then 30% of the time, two pair of teeth are in contact and 70%, one pair. If $\varepsilon = 2.15$, 15% of the time three pairs of teeth are engaged and 85% of the time, two pairs.

From Figure 7.3a, we can see that it is very easy to calculate the contact ratio: the center distance O_1O_2, base diameters d_{b1} and d_{b2}, and the outside diameters d_{o1} and d_{o2} are known. But it should be kept in mind that the outside diameters must be taken with their tolerances minus tip round radii. It is usually recommended to have $\varepsilon \geq 1.1$.

This recommendation seems to be reasonable. But, sometimes, we are forced to decrease the outer diameters of the gears (for otherwise they don't enter the available space), or we have to use gears with very small numbers of teeth. What on earth if the contact ratio is a bit less than 1? Is this completely impracticable? Let's consider.

Engagement of gears is a cyclic process of alternation of teeth. If $\varepsilon < 1$, in each cycle, there is a small period of time when the gears are disengaged, and the force applied to the teeth is zero. At the same time, the torques applied to the gears from the shafts remain practically unchanged, because there remains the elastic torsion deformation of the shafts. Under influence of these torques (not balanced with the teeth forces), the gears change their speed of rotation: the driving gear accelerates, and the driven gear slows down. Until the moment when the teeth contact resumes, the relative angular position of the gears is unmatched, and this leads to an impact between the teeth. If this error in the relative angular position of the teeth multiplied by the base radius doesn't exceed the admissible error of the base pitch, the gear will work normally.

It is clear that the error in the angular position depends on duration of the disengagement, inertia of the gears, and the torque applied to the gears. The lesser the load on the gear, the higher the inertia of the gears; and the higher their speed of rotation, the lesser the contact ratio that may be allowed. The thinking reader can say that, apparently, light-loaded high-speed gearing may work satisfactorily with only one tooth on each gear. That is right! If the gearing is somehow spinned up in the beginning to high speed, it will continue working. But how can this be done?

One day a new diesel engine was inspected after successful trials. In the pump drive, lack of three neighboring teeth of a pinion was suddenly found. The drive worked, and in the severe noise of the engine, it was impossible to hear any additional noise. But it would be impossible to start the engine with these broken-down gears, because the teeth may come to rest against each other tip to tip, and in this case, all the mechanism may be completely destroyed.

Here is one more interesting case from our practice. A technician who operates a diesel generator set came from the back lands to a plant manufacturing diesel engines. He said that he needs urgently a new gearwheel, which drives an oil pump of an old diesel generator. "This gear lacks seven teeth already, and four of them are neighboring. We are worried about the further service," said the unpretentious man. "But how has it worked till now?" asked the amazed engineers. "We made wooden teeth and attached them with screws," answered the countryman, "But now, I think, is the right time to replace the gear with a new one." Certainly, he was given a new gearwheel, but the case needs some comment. The wooden teeth are not able to take the working load, but they enabled starting the engine, and then the drive worked using inertia of the rotating parts.

The authors have investigated spur gears with contact ratio $\varepsilon = 0.98$, 0.95, and 0.9.[1] There was no problem with them, but the load capacity was found to change in direct proportion to the contact ratio, as well as it was when $\varepsilon > 1$.

Let's get back to Figure 7.3 and turn our attention to point p. In this point, the line of action intersects with the center line O_1O_2. This point is called *pitch point*. From the similarity of triangles O_1a_1p and O_2a_2p, it follows that point p divides each of the lines O_1O_2 and a_1a_2 into two parts that are directly proportional to the base circles radii, and consequently, to the numbers of teeth:

$$\frac{O_1p}{O_2p} = \frac{a_1p}{a_2p} = \frac{N_1}{N_2} = \frac{1}{i}$$

where i = gear ratio.

Motion in the contact of teeth is mixed; there is both rolling and sliding. Angular speed of the rolling motion is ω_1 for the driving gear and ω_2 for the driven gear. The sliding velocity may be obtained from Figure 7.3b. When the teeth contact in point k, the vectors of linear speed are V_1 and V_2. They meet the following conditions:

- Vector V_1 must be perpendicular to radius O_1k.
- Vector V_2 must be perpendicular to radius O_2k.
- Projections of vectors V_1 and V_2 on the line of action a_1a_2 (V_n) must be equal because the point of contact moves along this line and the teeth are in continuous contact.

Because

$$V_n = \omega_1 \frac{d_{b1}}{2} = \omega_2 \frac{d_{b2}}{2} = const$$

the vectors V_1 and V_2 may be easily determined in any point of contact. The vector of sliding velocity

$$\vec{V}_S = \vec{V}_1 - \vec{V}_2$$

When the teeth contact in point p, vectors V_1 and V_2 are unidirectional (perpendicular to line O_1O_2), and their difference $V_S = 0$. That means that at this point there is no sliding.

The direction of the friction forces is different before and after the pitch point. From Figure 7.4a, we can see that when pair B_1-B_2 comes into contact (before the pitch point), the teeth enter the tooth spaces. The friction forces on both contacting teeth (B_1 and B_2) are directed from the contact point to the tooth root. On the contrary, when the pair of teeth contacts after the pitch point (pair A_1-A_2, Figure 7.4b), the teeth go out of the tooth spaces, and the friction forces are directed from the point of contact toward the tip of each tooth. Now, we see that the driving tooth contacts with its dedendum before the pitch point and with the addendum after it. In the pitch point there is no sliding; hence, the friction forces on the driving tooth are always directed from the pitch line toward the root and the tip of the tooth, Figure 7.4c. On the driven teeth, the friction forces are directed from the tip and the root of the tooth toward the pitch line, Figure 7.4d.

When the teeth are heavily loaded and working for a long time, the material of their working surfaces becomes plastically deformed by the friction forces. Figure 7.9h and Figure 7.9i show a driving tooth with a kind of depression along the pitch line and a driven tooth with elevation in this place.

A sliding velocity chart is given in Figure 7.4b. It is maximal at points b_1 and b_2. At pitch point p, the sliding motion and the friction forces change their direction. Cyclic change of the friction force direction is one of the reasons spur gears generate greater vibration and noise than helical gears.

The terms *addendum* and *dedendum* used in the preceding text are not just a matter of geometrical abstraction. These terms are physically determined, and the pitch line represents a boundary between them. Thus, the addendum is the area from the pitch line to the tip of the tooth, and the dedendum is the rest of the working surface, from the pitch line to the border of the active profile. These two areas differ in load conditions and, consequently, in their strength. The point is that the addendum of one tooth always contacts the dedendum of its mate, and in this pair, the addendum has the greater speed. In other words, the surface of the addendum leads in the sliding motion, and the dedendum lags. For example, in Figure 7.3b the point of contact k is located in the after-the-pitch-point zone, in which the driving tooth is engaged with the addendum (its speed is V_1) and the driven tooth with its dedendum (its speed is V_2). As we can see, V_1 is greater than V_2, so the addendum is the leading member of the pair.

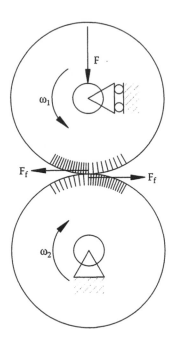

FIGURE 7.5 Leading and lagging surfaces in contact.

The differences in load conditions can be represented more clearly on rollers. Figure 7.5 shows two identical rollers that are pressed against each other by a force F and rotated with different angular speeds (ω_1 is a bit greater than ω_2). Because the upper roller is leading, it imitates the addendum, and the lower roller (lagging) the dedendum. Friction forces F_f make the surface layers near the point of contact compressed on one side (this is symbolized by dense lines) or tensioned on the other side (sparse lines). As we can see, the addendum surface comes into contact compressed, and the dedendum surface comes tensioned. The effect of this condition on the stress of the contacting surfaces is discussed in Section 7.3. But, meantime, you can keep in mind that sometimes (not always) it is preferable to be leading.

Figure 7.4e shows the contact zone (view A) of a spur gearing. At this moment, two pairs of teeth are engaged: A_1-A_2 and B_1-B_2. Next moment, pair A_1-A_2 leaves the contact zone, and B_1-B_2 will be the only engaged pair until pair C_1-C_2 enters the zone. When pair A_1-A_2 disengages, it becomes a one-pair contact, and the total length of the contact lines changes stepwise from $2W$ to W (here W is the face width).

One-pair contact is less rigid than two-pair contact. The cyclic change of the rigidity excites vibrations of gears (called *parametric excitation*) and may be responsible for high noise levels. This defect may be eliminated using gears with contact ratio $\varepsilon = 2$ where two pairs of teeth are always engaged. Otherwise, the number of teeth or the rotational speed should be changed so as to avoid the resonance.

Figure 7.4f shows the same contact zone but for helical gears. At this moment, two pairs of teeth, A_1-A_2 and B_1-B_2, are engaged with all their length. Next moment, pair A_1-A_2 begins to disengage, but gradually, not at once. Before this pair disengages completely, pair C_1-C_2 comes into contact, so at least two pairs of teeth are always engaged. But in this case, the total length of the contact lines is cyclically changing; when pair A_1-A_2 begins to leave gradually the contact zone, the length of the contact lines decreases until pair C_1-C_2 comes into contact. If the helix angle is increased so that the end of engagement of the preceding pair coincides with the beginning of engagement of the following pair (Figure 7.4g), the total length of the contact lines remains constant.

Indeed, from the depicted position on, the pair A_1-A_2 gradually leaves the contact zone, and the following pair B_1-B_2 enters this zone by the same amount. When pair B_1-B_2 completely enters the contact zone, pair A_1-A_2 continues leaving it, but pair C_1-C_2 enters the contact zone by the same amount, so that the total length of the contact lines remains constant. The constancy in time of the total length of the contact line is important to decreasing gear noise and vibrations. It is achieved in helical gears when the value

$$\frac{W \cdot \sin \psi}{\pi \cdot m} = whole\ number$$

Here
W = gear face width
ψ = helix angle
m = normal module

Cyclic change of the total length of the contact lines means cyclic change in the load capacity of the gears. It is known that in spur gears the maximal wear of the working surfaces occurs in the one-pair zone, in which the load is greater. To maintain contact in this zone, the teeth shall approach each other by the amount of the wear. This approach occurs partly because of following factors:

- Decreased bending and contact deformations (as the load becomes lesser because of the wear)
- Change in the rotational speed of the gears

Because the load in the worn-out zone of the teeth decreases, the torque produced by the tooth force becomes less than that applied to the gear by the shaft. In this situation, the driving gear accelerates, and the driven gear becomes slower. (Please compare this with the operation of gears described in the preceding text with contact ratio less than 1.) According to the principle of dynamic equilibrium (suggested by D'Alembert [1743] based on Newton's second law), the sum of all forces applied to a body, including inertial forces, must be zero. In our case, the originated inertial torque supplements the decreased torque from the teeth forces to the amount of the shaft torque, so that the sum of all torques is zero.

In the two-pair zone, in which the wear is less, the process of operation is opposite, and the tooth load rises. Thus, owing to inertial forces (as the gears rotate nonuniformly because of uneven wear of the working surfaces), redistribution of the load takes place between the one-pair and two-pair zones. Outwardly, it is expressed as increasing vibrations and noise of a spur gear as it wears. (To a lesser degree, the effect of increasing vibrations may take place also in helical gears, if the total length of the contact lines changes too much.) Considerable decrease in the amplitude of vibrations may be achieved when the contact ratio $\varepsilon = 2$. For this purpose, the height of the teeth should be increased by 25–30%.

It is seen from Figure 7.4f and Figure 7.4g that the contact lines of helical gearing lie always on both sides of the pitch point. Therefore, the friction forces on the teeth are always applied in opposite directions and counterpoise each other. This decreases the noise and vibrations.

When gears rotate, the point of contact moves along the profile of the teeth: on the driving tooth, from the root to the tip and, on the driven tooth, from the tip to the root. In a spur gear, the engaged tooth is positioned identically across its width. Therefore, the contact lines pass parallel to the tip (Figure 7.6a). A helical gear may be represented as a large number of thin spur gear slices, each turned by a small angle relative to its neighbor (Figure 7.6b and Figure 7.6c). The phase of contact is different for each slice, so the contact line, which constitutes the locus of the contact points, is positioned at an angle α to the gear axis. The larger the helix angle ψ, the larger is the angle α of the contact line (Figure 7.6d). When the contact line moves over the tooth, it also takes up a position near the upper corner of a tooth. This may cause higher bending stress in the root, and so, usually, the corner is chamfered (see Chapter 8, Section 8.6 and Figure 8.26).

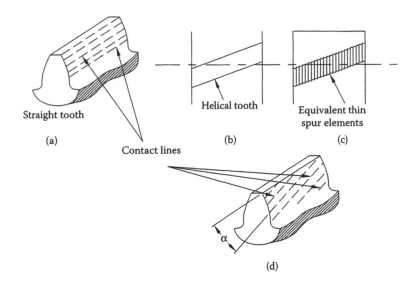

FIGURE 7.6 Contact lines on spur gear teeth and helical teeth.

7.2 FORCES IN SPUR GEARING AND STRESSES IN TEETH

As the torques are applied to the shafts of a gear, there appears force F_n in the contact of the teeth (Figure 7.7). This force creates the balancing torque. As long as the friction force is neglected in the strength calculations, force F_n is directed at a tangent to the base cylinder d_b and is given by

$$F_n = \frac{2T}{d_b}$$

where T = torque applied to the gear with base diameter d_b.

Sometimes, it is convenient to resolve vector F_n into two components: tangential to the pitch diameter F_t and radial F_r. But for our analysis it is not essential.

Coming to the stress of the tooth loaded by force F_n, we should take into consideration that this load is distributed along the contact line as shown in Figure 7.8a. For spur gears, the minimal

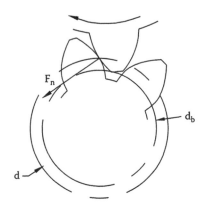

FIGURE 7.7 Tooth load.

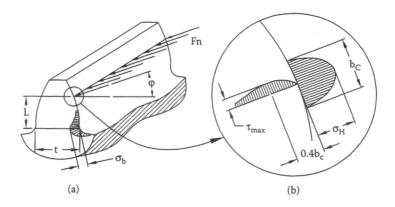

FIGURE 7.8 Contact (Hertzian) stress σ_H and bending stress σ_b in the loaded tooth.

length of the contact line equals the face width W. In helical gears, it is greater because more than one pair of teeth are engaged concurrently.

When the contact stress is calculated, the teeth are considered as two cylinders of radii ρ_1 and ρ_2 pressed against each other with force F_n. Elastic deformation of the teeth under load forms a contact area along the line of contact. The width of this area (b_C) measures usually fractions of a millimeter. The surface compression stress is distributed across the contact area in a semielliptic form as shown in Figure 7.8b. The maximal value of the contact stress σ_H is determined from Hertz's formula (1895):

$$\sigma_H = 0.418\sqrt{\frac{F_n E}{\rho_e L}} \qquad (7.1)$$

where
   E = modulus of elasticity (MPa)
   L = length of the contact line (mm)
   ρ_e = equivalent radius of curvature (mm), that is

$$\rho_e = \frac{\rho_1 \rho_2}{\rho_1 \pm \rho_2}$$

In the last equation "plus" is used for two convex surfaces, and "minus" for contact of a convex surface with a concave surface (internal toothing). The ρ_e value must be positive.

The width of the contact area is given by

$$b_C = 3.04\sqrt{\frac{F_n \rho_e}{LE}} \qquad (7.2)$$

Example 7.1

Two steel spur gears with $m = 5$ mm, $N_1 = 20$, $N_2 = 40$, and $W = 50$ mm are loaded with a torque $T_1 = 1000$ N·m applied to the smaller gear. Pressure angle $\varphi_0 = 20°$. What are the maximal Hertzian stress σ_H and the width of contact area b_C?

We begin with calculation of the base diameter of the smaller gear:

$$d_{b1} = mN_1 \cos 20° = 5 \cdot 20 \cdot 0.9397 = 93.97 \ mm$$

The tooth contact force is

$$F_n = \frac{2T_1}{d_{b1}} = \frac{2 \cdot 1000 \cdot 10^3}{93.97} = 2.13 \cdot 10^4 \ N$$

The maximal contact stress is not constant, because the radii of curvature ρ and the total length of the contact lines L vary as the gears rotate. Usually, the maximal surface stress is determined when the teeth contact at the pitch point, because there is a one-pair contact zone. From triangle O_1a_1p (Figure 7.3), we can find the radius of curvature of the pinion and gear teeth:

$$\rho_1 = a_1 p = 0.5 \, d_{b1} \tan 20° = 0.5 \cdot 93.97 \cdot 0.3638 = 17.1 \ mm$$

$$\rho_2 = a_2 p = \rho_1 \frac{N_2}{N_1} = 17.1 \frac{40}{20} = 34.2 \ mm$$

The equivalent radius of curvature is

$$\rho_e = \frac{17.1 \cdot 34.2}{17.1 + 34.2} = 11.4 \ mm$$

The modulus of elasticity of steel $E = 2.06 \cdot 10^5$ N/mm²; length of the contact line $L = 50$ mm. Substitution of these data into Equation 7.1 gives:

$$\sigma_H = 0.418 \sqrt{\frac{2.13 \cdot 10^4 \cdot 2.06 \cdot 10^5}{11.4 \cdot 50}} = 1160 \ MPa$$

The width of the contact area is obtained from Equation 7.2:

$$b_C = 3.04 \sqrt{\frac{2.13 \cdot 10^4 \cdot 11.4}{50 \cdot 2.06 \cdot 10^5}} = 0.467 \ mm$$

The load also initiates bending stress σ_b, which peaks in the root fillet (Figure 7.8a). The teeth of spur gears are calculated for bending strength as a short beam with a root fillet as a stress raiser. Two potentially dangerous cases are considered:

1. When part of force F_n is applied to the highest point (tip) of a tooth (this part to be found by calculating the load distribution between two pairs of engaged teeth, taking into account their rigidity and worst combination of errors in base pitch).
2. When the entire force F_n is applied to the upper point of the one-pair zone. (For a more detailed description see Section 7.5.)

Bending strength of helical teeth is much more complicated to calculate, because the contact line is placed at an angle on the tooth-working surface. This problem may be solved using finite element method (FEM) or approved recommendations that are based on many experimental data.

7.3 KINDS OF TOOTH FAILURE

Gears fail as a rule because of tooth breakage or degradation of their working surfaces. Very seldom, the failure is caused by breakage of the gear rim. Some of the commonly encountered kinds of gear failure are discussed in the following text.

Tooth breakage may be caused by a momentary overload, for example, when foreign matter like a cotter-pin or bolt enters the engaged teeth, or when a heavy impact loads the gear. In this case, the fracture surface is convex as shown in Figure 7.9a.

But mostly, tooth breakage is of fatigue nature, and the fracture surface is concave, (see Figure 7.9b). The initial fatigue crack is oriented at right angles to the fillet surface and then gradually changes its direction toward the opposite fillet of the same tooth. Load capacity of non-hardened teeth (say, HB ≤ 350) is limited usually by contact strength of their working surfaces. At normal service conditions, breakage of these teeth may result from uneven load distribution along the teeth (Figure 7.9c) or because severe wear has reduced the tooth's cross section (Figure 7.9g).

When teeth are surface-hardened (case-hardened or nitrided), the load capacity of their working surfaces increases three to five times, while the bending strength rises by 1.5 to 2 times only. Therefore, in surface-hardened gears, the margin of safety against tooth breakage is relatively small (as compared with that in nonhardened gears), and this kind of failure occurs more often.

Nonhardened teeth are less sensitive to uneven load distribution. On the one hand, they are able to run-in because of local wear, pitting, plastic deformation, and other kinds of surface modification. And on the other hand, the increased safety margin against tooth breakage gives them the time needed for the running-in process. (It is important to point out that after the teeth have become aligned, the process of surface deterioration in nonhardened gears ceases in most cases.) Surface-hardened teeth are much more wear resistant, and their ability to run-in is very low. Their relatively small safety margin against tooth breakage makes them very sensitive to misalignment. In some cases where high accuracy can't be achieved, it may be preferable to make at least one of the gearwheels without surface hardening to keep the ability to run in.

Chipping or *flaking* (Figure 7.9d) occurs on surface-hardened teeth near the top. This defect may be associated with the fragility of the hardened layers, the impact load applied to the top area (in the beginning of engagement, point b_2 in Figure 7.4a), or possibly with burnings or even microcracks created while grinding. Chipping may be prevented by profile modification (see Section 7.7 and Figure 7.20), rounding the sharp edges, and preventing grinding defects.

Deterioration of the working surfaces of the teeth may be of different kinds, and quite often several of them, like pitting, scoring, plastic deformation, and others, can be observed concurrently.

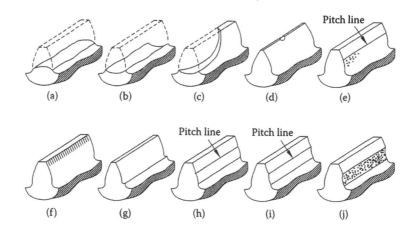

FIGURE 7.9 Gear tooth failures.

Pitting (Figure 7.9e) is the most widespread kind of working-surface failure when there is no abrasive wear. Pitting is a specific form of fatigue destruction of the surface layer under cyclically changing contact load. It shows up as rounded pits with dark bottoms, without metallic luster, created by separated flakes of metal. Usually, the pits appear on the dedendum (below the pitch line). In spur gears, the one-pair zone of the dedendum is the weakest place.

Pitting appears mostly on the dedendum because the actual stress on the dedendum is greater than on the addendum, although the Hertzian stress is the same. This is because the Hertzian formula is derived for the static case and doesn't take into account the friction forces F_f. These forces substantially change the stress in the surface layer.

In Figure 7.5, it was shown that the addendum surface comes into contact compressed and the dedendum surface comes tensioned. In the contact area, the surface layer is stressed in two perpendicular planes, it is compressed (or tensioned) by the friction forces in the tangential direction and compressed by the Hertzian stress in the normal direction. It is known that in the case of tension or compression in two perpendicular planes the maximal stress is

$$\tau_{max} = \frac{\sigma_y - \sigma_x}{2}$$

For the leading surface (addendum), stresses σ_x and σ_y have the same sign (minus, because they are compressive), so in the numerator there will be the difference of their absolute values. For the lagging surface (dedendum), these two stresses have different signs (compressive and tensile), so their absolute values sum up, and the resulting maximal stress is higher.

This is the simplest explanation of the fact that the load capacity of the addendum is considerably greater than that of the dedendum. It was experimentally found,[2] that at higher speed, when the oil film between the teeth is thicker and the friction coefficient is less, the load capacities of the addendum and dedendum approach.

The pitting shown in Figure 7.9e is caused by uneven load distribution across the face width. After certain local wear and redistribution of the load, the farther expansion of pitting is most likely to stop. But if the gear is underdesigned, the pitting finally covers all the dedendum surfaces, and then it may appear on the addendums of the teeth.

Moderate local pitting doesn't impair the functioning of the gear. But the crumbled-out particles of metal dirty the oil, and it should be filtered or cleaned by any means (for instance, by a magnetic plug).

The size of the pits on nonhardened surfaces may be of 0.5 mm to 5 mm in diameter, and even more. On the surface-hardened teeth, the initial pitting may appear as very small pores visible only under magnification. Later on in the course of work, they grow up and become visible to the naked eye.

Pitting may be prevented by choosing the dimensions and materials of the gear in conformity with the given loads, and providing for correct load distribution across the face width and between the teeth. An important role is played by the correct lubrication oil, which should not be too thin.

The main function of the lubricant is to create an oil film between the contacting teeth. The greater the oil viscosity, the harder it is to squeeze it out of the contact area and, consequently, the thicker is the oil film and lesser the damage to the teeth surfaces. The thicker oil better damps the impacts in the gearing and the vibrations of the gears and housing, reducing thereby the noise level of the entire gear set. Therefore, for low- and middle-speed gears, high-viscosity oils are usually used. As the speed increases, the energy waste by oil agitation grows, and the oil viscosity should be decreased. Approximately, the needed viscosity of petroleum oil at 50°C (v_{50}) may be determined as follows[3]:

$$v_{50} = \frac{(100 \div 200)}{V^{0.4}} cSt$$

where V = rim speed (m/sec).

On surface-hardened teeth (case-hardened and nitrided) may occur *spalling* caused by origina-tion and propagation of fatigue cracks under the hardened layer. This kind of surface destruction is more dangerous than pitting because big areas of the working surface suddenly fall off the teeth. The usual reasons for this defect are insufficient thickness of the hardened layer and inadequate hardness of the core. They also may be the cause of *waviness* of the working surfaces as a result of plastic deformation of the core. The maximal shear stress acts at a depth of $0.4b_c$ (see Figure 7.8b) and is given by

$$\tau_{max} = 0.3\sigma_H$$

But the stress curve is sloping slightly. At the depth of $0.8b_c$, $\tau = 0.24\sigma_H$.

The possibility of underlayer destruction may be excluded by choosing a thicker hardened layer and more durable core material. In particular for case-hardened gear teeth, the recommended case depth Δ may be determined from the following equation:[4]

$$\Delta = (0.28 - 0.007m)m \pm 0.2 \ mm$$

A thicker case is undesirable for two reasons:

1. The tip of the tooth becomes through-carburized and brittle.
2. The compressive stress in the hardened layer decreases (the smaller core is not able to effectively withstand the expansion of the hardened layer). Core hardness can be as high as 35–45 HRC. Higher hardness can make the tooth fragile.

The processes of heating, carburization, quenching, and low-temperature tempering involve change in the crystal lattice of the metal and, consequently, warpage (form distortion) of the part. Therefore, case-hardened teeth should be finished (ground or lapped). In the course of finishing, part of the hardened layer is taken away. To decrease the depth of finishing, the shape of the teeth should be made exact enough before the heat treatment. This includes exact gear cutting (possibly grinding or lapping) before carburization. Warping can be diminished by a reasonable sequence of heat treatment operations and devices that restrict the possible form distortion of the part.

Nitrided gears fail from spalling more often because the nitrided layer is rather thin (usually not thicker than 0.5–0.6 mm). The core can't be very hard because the process of nitriding is utterly time consuming and carried out at a high temperature (about 600°C). Because the thin nitrided layer is not practical to machine, no shape distortion is allowed during the nitriding process and afterward. For this purpose, the tempering of the part before nitriding should be performed at higher temperature than nitriding by at least 10°C. The moderate hardness achieved from tempering remains unchanged after nitriding; therefore, nitrided gears have lower load capacity than case-hardened ones.

One of the typical kinds of surface damage is *scoring*. Scoring occurs when, under high load, the microasperities break through the oil film, contact metal to metal, weld to each other, and then tear off. This damage shows up as slight scratches (Figure 7.9f) that usually appear in the area of higher sliding velocity, near the tip or the root of the tooth. The direction of the scratches coincides with the direction of sliding, across the tooth. Severe scoring may cover the entire working surfaces of the teeth and even result in welding and tearing off larger fragments of the surfaces. The gears may become completely unusable.

Increasing the hardness and smoothness of the contacting surfaces improves their resistance to scoring. If the gear is inclined to scoring, the tip edges of the teeth should be rounded and polished thoroughly. Sharp edges may cause stress concentrations and break through the oil film in the beginning of tooth engagement (point b_2, Figure 7.4a). Profile modification (Figure 7.20) is also helpful against scoring because it decreases the tooth load in the beginning of engagement.

It is important to reduce sliding velocity in the beginning and the end of tooth engagement. Therefore, it is desirable to put the pitch point p in the middle of the length of action b_1b_2 (Figure 7.4). It may be helpful to move point p toward the beginning of engagement to decrease the sliding speed in this place. For this purpose, the driving tooth addendum shall be made longer and the driven tooth addendum shorter.

Creation of an additional intermediate layer proved to be an effective method to prevent scoring. This layer may be produced by copper coating, phosphatizing, or antiscoring additives in the lubrication oil. The additives react with the tooth material and produce the molecular separating layer that renews after each cycle of engagement.

Useful means for prevention of scoring are those that increase the oil-film thickness, such as

- Using thicker oil
- Reducing temperature in the contact region (by cooling the oil and intensive sprinkling of the oil on the gears to cool them)
- Reducing the sliding speed and increasing the speed of rolling (by increasing the pressure angle φ to 25–28°)

Accidental scoring may occur due to oil contamination. For example, if a cast housing is not carefully cleaned of all molding sand, the paint may separate from the interior walls and let the sand damage everything moving inside the mechanism. Another cause for accidental scoring may be malfunction of the lubrication system: disabling of the oil pump, clogging of filter or nozzle, destruction of oil pipe, etc.

Tooth *wear* occurs more often in gears that are poorly lubricated or not protected from contamination. Figure 7.9g shows a severely worn tooth. Such excessive wear sometimes happens in slow-running, poorly lubricated gears made of mild steel. These gears have a large margin of safety against breakage, so they usually are allowed to work until the teeth become as sharp as a knife.

Considerable wear is usually observed in gears of machines used in agriculture, construction, and road building. The wear intensity may be as great as 0.1 mm per 1000 working hours. Because the teeth of these gears are surface-hardened, their service life depends on the thickness of the hardened layer and the quality of the seals.

Very intensive wear may occur when a tempered gear contacts a surface-hardened pinion, if the hardened teeth are not finished. Rough and hard surfaces of the hardened teeth work as a file and chip the mating teeth.

Plastic *deformation* of the working surfaces is inevitable because the contact stress is very high, but visible changes need a certain amount of time to become apparent. This time decreases with increasing load and decreasing hardness of the teeth. The typical kind of plastic deformation is displacement of the surface layers relative the pitch line and is induced by the friction forces. The direction of these forces is shown in Figure 7.4c and Figure 7.4d. When the working surfaces of the teeth are of low or middle hardness, one can observe a depression along the pitch line on the driving teeth (Figure 7.9h) and a prominence on the driven teeth in this area (Figure 7.9i).

Damage of the working surfaces of teeth is usually less dangerous than tooth breakage and results in failure only when the damage covers most of the working surfaces. In other words, this damage is time dependent; its development can be inspected and appropriate measures taken to prevent sudden failure of the gear. In practice, quite often, local pitting or moderate scoring caused by uneven load distribution in the beginning of service stops developing, and later on may be smoothed down. Yet progressive pitting that covers most of the dedendum surfaces may lead to severe change in the tooth shape. The dedendum may become concave (Figure 7.9j), and the bending strength of the tooth would be impaired.

7.4 CONTACT STRENGTH (PITTING RESISTANCE) OF TEETH

As was said in Section 7.2, the contact stress value is obtained using Hertz's formula derived for two round cylinders pressed against each other with no motion. Equation 7.1 and Equation 7.2 are valid for perfectly elastic and smooth cylinders. In reality, the teeth are involute cylinders (not round), and they have deviations from their theoretical shape, such as surface roughness and waviness. Therefore, force F_n is never distributed evenly along the line of contact.

Perpendicular to the line of contact, the stress σ_H is not distributed across the contact area in a semielliptic form as shown in Figure 7.8b. This theoretical form is distorted because of form deviations, friction forces, hydrodynamic effects, and the effect of time delay in deformation. As the gears rotate, the new surfaces come into contact, and the pattern of the surface stress permanently changes, quickly and unpredictably.

Nobody knows definitely what the contact stress pattern is because there is no possibility to measure it. But there is one more problem. For the strength evaluation not only the stress, but also the metal condition is important. In the course of rolling and sliding under load, the surface layers are plastically deformed and the structure of metal is changed. In addition, chemical reactions and electric effects take place in the contact area, so the processes that affect the contact strength are very complex.

Because of this complexity, calculation of the contact strength of gears is based on experimental data and operating experience with gears of similar purpose and duty. As a rule, the Hertzian stress σ_H is used as a crucial factor for evaluation of the stress level and service life. That means the admissible value of σ_H may be safely chosen relying on the operating experience with gears of the same purpose and service conditions. If these data are not available, general recommendations may be used for design, but thereafter experimental testing should be carried out.

To derive the relationship between the gear dimensions, torque transmitted, and stress σ_H, we shall return to the geometry of a spur gear (Figure 7.3 and Figure 7.4). A pair of teeth come into contact at point b_2 and disengage at point b_1. In each point on this path, the radii of curvature and sliding velocities are different. For contact strength calculations, the pitch point p is usually considered because it is convenient. In addition, in spur gears, the pitch point lies in the one-pair zone, which is the weak place. The radius of curvature of the pinion tooth in the pitch point

$$\rho_1 = a_1 p = \frac{d_1}{2} \sin \varphi$$

The radius of curvature for the gear,

$$\rho_2 = a_2 p = \frac{d_2}{2} \sin \varphi = \rho_1 i$$

The equivalent radius of curvature is

$$\rho_e = \frac{\rho_1 \rho_2}{\rho_1 \pm \rho_2} = \rho_1 \frac{i}{i \pm 1} = \frac{d_1 \sin \varphi}{2} \frac{i}{i \pm 1}$$

Remember that the "plus" in the denominator is valid for external gearing and the "minus" for internal gearing.

If $i = 1$ in internal gearing, the equivalent radius $\rho = \infty$. But don't be delighted with this fact: it is not a gear; it is just a gear coupling.

The length of the contact line $L = W$, where W is the face width.

Force F_n is given by

$$F_n = \frac{2T_1}{d_{b1}} K_{WH}K_d = \frac{2T_1 K_{WH}K_d}{d_1 \cos\varphi}$$

where
T_1 = nominal torque applied to the pinion (N·mm)
K_{WH} = factor of load distribution across the face width (for contact strength calculation)
K_d = factor of dynamic load

Now we have all the needed factors to insert into Equation 7.1. Raising this equation to the second power gives

$$\sigma_H^2 = 0.418^2 \frac{F_n E}{\rho_e L}$$

or, after substitution of F_n, ρ_e, and L,

$$\sigma_H^2 = \frac{2T_1(i\pm 1)}{Wd_1^2 i} K_W K_d \frac{0.7E}{\sin 2\varphi} \tag{7.3}$$

In Equation 7.3 parameter E (modulus of elasticity) is nearly constant for all kinds of steel, and angle φ doesn't change much (usually 20–22°). Therefore, several variables in Equation 7.3 may be substituted by one variable C_H, which conforms to the square of Hertzian stress:

$$C_H = \sigma_H^2 \frac{\sin 2\varphi}{0.7E}$$

There is a linear dependence between the load and the stress factor C_H. The following formula derived from Equation 7.3 yields an admissible value of C_H by calculation of tested and field-proven gears made of similar materials:

$$C_H = \frac{2T_1(i\pm 1)}{Wd_1^2 i} K_{WH}K_d \; MPa$$

The factors K_{WH} and K_d should be determined taking into account the design features, speed, and load of the prototype gear. If the gear that is to be designed is similar to the field-proven prototype, factors K_{WH} and K_d can also be eliminated because they are nearly the same. In this case, the formula for comparative calculation becomes simpler:

$$K_H = \frac{2T_1(i\pm 1)}{Wd_1^2 i} \; MPa \tag{7.4}$$

It follows that two gears of similar function and design, and having the same K_H value, will have nearly the same Hertzian stress σ_H and supposedly the same service life.

Approximately, the admissible value of K_H for general-purpose gears may be defined as

$$K_H = K_0 K_{HL}$$

where

K_0 = contact stress factor admissible for infinite service life (MPa)

K_{HL} = service life factor for the contact strength calculations

For the preliminary calculation of gears, the following formulas for K_0 can be used:

1. For helical gears with case-hardened teeth: $K_0 = 0.126HRC - 1.78$ MPa
2. For spur gears with case-hardened teeth: $K_0 = 0.099HRC - 1.22$ MPa
3. For helical gears with surface-hardened pinion and tempered gear: $K_0 = 0.0127HB - 1.1$ MPa
4. For helical gears with both pinion and gear tempered: $K_0 = 0.0107HB - 1.1$ MPa
5. For spur gears with both pinion and gear tempered: $K_0 = 0.008HB - 0.87$ MPa

Here, HRC and HB are the surface hardness of the larger gear (wheel) in Rockwell-C and Brinell hardness units, respectively.

In cases 4 and 5 it is recommended to make the pinion harder than the gear by approximately 40 HB. The increased hardness of the pinion shall compensate for its greater number of cycles and lower bending strength of its teeth as compared to the gear (compare teeth $N = 17$ and $N = 50$ in Figure 7.2).

Factor K_{HL} is made to express the dependence of admissible load on the run time that is given as a number of loading cycles. It is clear that the less the specified number of cycles, the greater the load can be applied to the part (within certain limits) with the same probability of failure during that number of cycles. Nowadays, this idea may seem trivial, but it was first thoroughly investigated by A. Wöhler only 150 years ago (see Chapter 12). The experimental points look as shown in Figure 7.10a. Each point represents one experiment and is characterized by a certain stress amplitude, σ_a, and number of cycles to failure N. (Points with arrows belong to experiments in which the part did not fail.) As is seen from the chart, the parts tested at the same stress amplitude have scattering number of cycles to failure, though they have been made of the same material and by the same drawing. Fatigue life line 1 is usually curved so that 90 or 95% of the experimental points remain above it. This is to decrease the probability of failure when the part is calculated for durability using the fatigue life curve. Please have a sense of humor and don't count the number of points above and below line 1 in Figure 7.10a!

The reasons for that scattering are numerous, but in experiments for bending or torsional stress, we at least know exactly the number of cycles to failure. In experiments for contact strength, the point of failure is a matter of opinion. In fact, a gear with initial pitting continues working as if nothing has happened. You can't detect any negative changes in its behavior; the temperature is the same, and the noise even decreases (because of running-in). With time, the pitting may become smaller and even disappear, but it may also grow and finally cover most of the working surfaces of the teeth. It depends on the hardness of these surfaces, evenness of load distribution along the

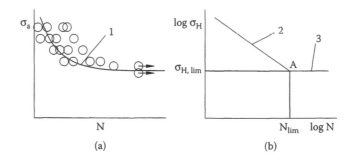

(a) (b)

FIGURE 7.10 The dependence between stress amplitude and the number of cycles to failure N.

contact lines, and the rim speed. Therefore, the fatigue lines for contact strength of the same gear may differ, depending on the experience and criteria of the investigator.

When the dependence σ vs. N is plotted in double-log or semilog coordinates, the fatigue life curve looks nearly as a broken line, (Figure 7.10b). (Because we are speaking here about the contact strength, the stress σ_a is replaced by Hertzian stress σ_H.) Its breaking point A is characterized by fatigue limit $\sigma_{H,lim}$ and number of cycles N_{lim}.

Inclined part 2 of this broken line is approximated by this equation:

$$N \cdot \sigma_H^m = const$$

For pitting deterioration of the teeth surfaces, the value of exponent m is approximately equal to 6 (to be more accurate, $m = 6 \pm 2$). That is,

$$N \cdot \sigma_H^6 = const \tag{7.5}$$

Horizontal part 3 of the fatigue life curve has two meanings. First, it means that if the stress is equal to or less than fatigue limit $\sigma_{H,lim}$, the fatigue failure is not likely to occur. The second meaning is that if, during the number of cycles N_{lim}, the fatigue failure didn't occur, the part can be considered as passing the test without failure. As mentioned in the preceding text, the number of cycles in this kind of experiment can't be estimated exactly. One of the most reliable sources[5] recommends the following formula for N_{lim}:

$$N_{lim} = 30 \cdot HB^{2.4} \tag{7.6}$$

where HB is the Brinell hardness of the working surfaces.

On the basis of Equation 7.5 we can write the following:

$$N_i \cdot \sigma_{H,i}^6 = N_{lim} \cdot \sigma_{H,lim}^6;$$

$$\frac{\sigma_{H,i}}{\sigma_{H.lim}} = \sqrt[6]{\frac{N_{lim}}{N_i}} \tag{7.7}$$

Here

N_i = specified number of cycles that is less than N_{lim}

$\sigma_{H,i}$ = admissible Hertzian stress for N_i cycles of loading

Inasmuch as the admissible torque, as well as the admissible K_H value, is proportional to the square of the Hertzian stress (see Equation 7.3 and Equation 7.4), the expression for K_{HL} derived from Equation 7.7 looks as follows:

$$K_{HL} = \frac{\sigma_{H,i}^2}{\sigma_{H.lim}^2} = \sqrt[3]{\frac{N_{lim}}{N_i}} \tag{7.8}$$

When determining the service life factor K_{HL} using Equation 7.8, the following remarks should be taken into account:

- The maximal working load should not exceed the magnitude characterized by $K_{HL} = 6–7$ for teeth hardness of 350 HB or less and $K_{HL} = 3–3.5$ for harder teeth (40 HRC and more).[7]
- Infrequent momentary overloads should not be considered, but their amount is limited by the maximum admissible K_{HL} value: for hardness 350 HB and less $K_{HL,max} = 10–12$, and for harder teeth surfaces, $K_{HL,max} = 4–5$.[7]

- When the hardness of the pinion and gear is nearly the same, determination of the factor K_{HL} should be based on the rotational speed of the pinion. When the pinion teeth are much harder (for example, surface-hardened pinion teeth and high-tempered gear wheel), the rotational speed of the gear counts.
- If a gear is engaged with two or more others (such as sun wheel of a planetary gear), its rotational speed should be multiplied by the number of mating gears.

If a gear works at a variable load and speed and it is known that the maximal torque T_1 is applied during n_1 hours in all, the lesser torque T_2, during n_2, T_3, during n_3, etc., the equivalent number of cycles n_e at the maximal torque

$$n_e = n_1 + n_2 \left(\frac{T_2}{T_1}\right)^3 + n_3 \left(\frac{T_3}{T_1}\right)^3 + \dots \qquad (7.9)$$

This equation is based on the hypothesis about the linear summation of damaging effects (see Chapter 12, Subsection 12.1.3).

Remark:

The working surfaces of non-surface-hardened teeth continue to deteriorate slowly even if the stress is less than the fatigue limit. This is reflected in the fatigue life line for contact strength of such teeth. The line has no horizontal part and should look as shown in Figure 7.11. It is recommended[5] that the K_{HL} value, should be obtained at $N_i > N_{lim}$ by the formula

$$K_{HL} = \sqrt[12]{\frac{N_{lim}}{n_i}}$$

but not less than $K_{HL} = 0.8$.

Example 7.2

A helical gear works 1 hour per day, 200 days a year, and the needed service life is 20 years. The pinion teeth are induction-hardened to 55 HRC, and the gear is tempered to 300 HB. The rotational frequencies of the pinion and gear are $n_1 = 1000$ r/min, $n_2 = 150$ r/min. The gear works 10% of the time at full load, 30% at 70% of the full load, and 60% at half-load.

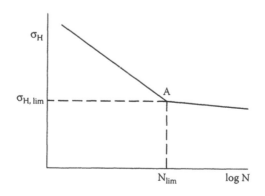

FIGURE 7.11 The fatigue life line for non–surface-hardened teeth.

From the equation for helical gears with surface-hardened pinion and tempered gear (see the preceding text),

$$K_0 = 0.0127 \cdot 300 - 1.1 = 2.71\, MPa$$

The total working time is 20 (years) × 200 (days) × 1 (hour) = 4000 h. The gear set works 400 h at full load, 1200 h at 70% load, and 2400 h at half-load. Now, we can determine the needed equivalent service life at full load:

$$L_{SL} = 400 + 1200 \cdot 0.7^3 + 2400 \cdot 0.5^3 = 400 + 411.6 + 300 = 1112\, h$$

The number of cycles of the gear wheel equals $1112 \cdot 150 = 1.67 \cdot 10^5$.
From Equation 7.6,

$$N_{\lim} = 30 \cdot 300^{2.4} = 26.4 \cdot 10^6$$

From Equation 7.8,

$$K_{HL} = \sqrt[3]{\frac{26.4 \cdot 10^6}{1.67 \cdot 10^5}} = 5.4$$

Thus, the admissible contact stress factor $K_H = 2.71 \cdot 5.4 = 14.6$ MPa. In this case, the bending strength of the teeth may limit the load capacity of the gear; it must be checked.
Using the K_H value, the permissible pinion torque T_{1H} limited by its contact strength is given by

$$T_{1H} = \frac{K_H W d_1^2 i}{2(i \pm 1)} \quad N \cdot mm$$

For bevel gears,

$$T_{1H} = \frac{K_H \psi_b (1 - \psi_b) d_1^3 i}{4.5 \sin \Sigma} \quad N \cdot mm$$

where
$\psi_b = W/A_0$
W = face width of the bevel gear
A_0 = outer cone distance
d_1 = pitch diameter of the pinion (mm)
i = gear ratio
Σ = shaft angle (in most cases $\Sigma = 90°$ but can be less)

If a gear should be designed to transmit the given torque T_1, the diameter of the pinion of spur or helical gear can be found from the following equation:

$$d_1 = \sqrt[3]{\frac{2T_1(i \pm 1)}{\psi i K_H}} \tag{7.10}$$

Here $\psi = W/d_1$, the ratio of face width to pinion diameter, usually, 0.8–1.2.

For bevel gears,

$$d_1 = \sqrt[3]{\frac{4.5T_1 \sin \Sigma}{\psi_b(1-\psi_b)iK_H}}$$

The ψ_b value usually equals 0.25–0.30.

As soon as the pinion diameter is determined, all other dimensions can be easily calculated using the gear ratio and factor ψ (ψ_k for a bevel gear).

When using these equations it is necessary to be careful with the dimensions of the variables. Sometimes, the gear dimensions and load data are given in different systems, and they must be converted to be compatible. If the K_0 values are given in MPa (that is, in N/mm²), it is convenient (in this case) to express all the linear dimensions in millimeters (mm), and the torque in newton millimeters (N·mm).

EXAMPLE 7.3

Design a speed reducer to run at speeds from 2000 to 285 r/min. The transmitted power is 1000 kW. The assumed service life is indefinite, meaning that $K_{HL} = 1$ for surface-hardened teeth or $K_{HL} = 0.8$ for tempered teeth.

First calculate the pinion torque T_1:

$$T_1 = 9550\frac{1000}{2000} = 4775 \, N \cdot m = 4.78 \cdot 10^6 \, N \cdot mm$$

The gear ratio is

$$i = \frac{2000}{285} = 7.018$$

Let's take $\psi = W/d_1 = 1$. Now, we have to choose the hardness of the teeth. To decrease the dimensions and weight of the gear, we choose a helical gear with case-hardened teeth, surface hardness ≥ 56 HRC. For this gear, $K_0 = 0.126 \cdot 56 - 1.78 = 5.28$ MPa. Because $K_{HL} = 1$, $K_H = K_0 = 5.28$ MPa. Now we have all the needed variables for Equation 7.10:

$$d_1 = \sqrt[3]{\frac{2 \cdot 4.78 \cdot 10^6(7.018+1)}{1 \cdot 7.018 \cdot 5.28}} = 128 \, mm$$

The other dimensions of the gears are as follows:

Face width $W = \psi \cdot d_1 = 128$ mm
Gear wheel diameter $d_2 = i \cdot d_1 = 7.018 \cdot 128 = 898$ mm
Center distance of the gear

$$C = \frac{d_1(i+1)}{2} = \frac{128(7.018+1)}{2} = 513 \, mm$$

This calculation is approximate, and it is not obligatory. The strength of the predesigned gear should be checked using methods recommended in the respective branch of industry. To fulfill some requirements of standards, the center distance C can be expressed in round numbers: 500 mm or 20 in. (depending on accepted system of units). The other dimensions can be also changed to meet the designer's requirements.

7.5 BENDING STRENGTH (BREAKAGE RESISTANCE) OF GEAR TEETH

Possibly, you have already noticed that in Example 7.3 the dimensions of the gear were defined irrespective of tooth bending strength, as if the teeth can't break. The reason is that the needed bending strength of the teeth can be achieved usually by increasing the module m (or, what is the same, decreasing the diametral pitch P) without increasing the diameters and face width of the gears.

Usually, $m = (0.015–0.025)C$, where C = center distance. But in high-speed gears, the module is often made less to reduce the sliding speed and lessen the chances of scoring.

The tooth is considered for bending strength as a cantilever beam loaded with a distributed force F_n (Figure 7.8). The maximal bending stress σ_b takes place on the fillet surface. It can be found using FEM or the hypothesis of broken-line cross sections (BCS) suggested by A. V. Verhovsky (for details see Chapter 3, Section 3.2).

For spur gears, the calculation of force is performed separately for one-pair and two-pair engagement. Figure 7.4a shows the position in which pair B_1-B_2 comes into contact in point b_1, but pair A_1-A_2 is already engaged. In this position, the dangerous situation is for tooth B_2 (because some part of F_n is applied to the tip of the tooth) and for tooth A_1 (because the full force F_n is applied at the upper point of the one-pair zone). The load on tooth A_1 in this position (border point of the one-pair zone) is

$$F_n = \frac{2T}{d_b}$$

The load on tooth B_2 in this position depends on the base pitch error. Let's denote as p_{b1} the actual value of the base pitch between teeth A_1 and B_1 and as p_{b2} the pitch between teeth A_2 and B_2. Now, we can calculate the static load distribution between pairs A_1-A_2 and B_1-B_2. The worst case is when $p_{b1} < p_{b2}$ because then tooth B_2 is loaded with a greater force and comes to contact with an impact. Because the calculation of an impact load is too complicated, the load distribution between two pairs of teeth is usually determined in a static statement. Let's suppose that $p_{b2} - p_{b1} = \delta_1$. Because $p_{b1} < p_{b2}$, in unloaded state, pair A_1-A_2 must disconnect while pair B_1-B_2 is in contact. As the torque is gradually applied to the gears, pair B_1-B_2 deforms first (elastically) by δ_1. Then teeth A_1 and A_2 come into contact and begin to take load. From this point on, both pairs, A_1-A_2 and B_1-B_2, deform together by the same magnitude; we denote it as δ_2. Now, to calculate the load distribution between the two pairs, we need the stiffness of the teeth. For a start, we symbolize them as C_A and C_B for pairs A_1-A_2 and B_1-B_2, respectively.

The dimension of the stiffness is MPa (that is, N/mm²). The physical meaning of this factor is the force in newton per millimeter of contact length needed to elastically deform the pair of teeth by 1 mm. This deformation includes both the contact and bending deformations of the teeth.

The equilibrium equation is

$$C_B\delta_1 + (C_A + C_B)\delta_2 = \frac{F_n}{W}$$

where W = face width of the gear. From this equation, we obtain

$$\delta_2 = \frac{\dfrac{F_n}{W} - C_B \delta_1}{C_A + C_B}$$

The unit load F_B/W in pair B_1-B_2 is

$$\frac{F_B}{W} = C_B(\delta_1 + \delta_2) = \frac{C_B}{C_A + C_B}\left(C_A \cdot \delta_1 + \frac{F_n}{W}\right) \tag{7.11}$$

EXAMPLE 7.4

Unit load in a gear $F_n/W = 500$ N/mm, the admissible error of the base pitch is 5 μm. What is the load applied to the pair of teeth that contact at point b_1?

The stiffness of a pair of teeth can be approximately taken as $C_A = 14000$ MPa (contact in the upper point of the one-pair zone) and $C_B = 10000$ MPa (contact of a tooth tip with the dedendum of its mate).[6]

The maximal difference in the base pitches $\delta_1 = 5 + 5 = 10$ μm $= 0.01$ mm. So the maximal load obtained from Equation 7.11 is given by

$$\frac{F_B}{W} = \frac{10000}{14000 + 10000}(14000 \cdot 0.01 + 500) = 267 \; N/mm$$

This calculation should be supplemented with some remarks.

1. The difference in the base pitches is caused not only by manufacturing errors but by the deformation of the teeth as well. A moment before pair B_1-B_2 comes into contact, pair A_1-A_2 is loaded and elastically bent. Tooth A_1 is bent to the right, and the pitch between teeth A_1 and B_1 (p_{b1}) decreases. On the contrary, tooth A_2 is bent to the left, and the pitch between teeth A_2 and B_2 (p_{b2}) increases. These deformations increase by far the difference $p_{b2} - p_{b1}$. In practice, this situation is eliminated by tip relief of the teeth (see Section 7.7 and Figure 7.20); therefore, it has not been taken into account here.
2. This load is obtained from the equation of static equilibrium. In reality, it should be multiplied by a dynamic load factor K_d. For more details see Section 7.7.

As mentioned earlier, whereas tooth B_2 is loaded on its tip, tooth A_1 is loaded (a moment before that) at the upper point of the one-pair zone by full load F_n. Factor K_d is seemingly irrelevant in this position because pair A_1-A_2 has been already engaged. But the torsional oscillation is a continuous process, and some dynamic factor should be taken for this point as well.

To take into account the influence of the uneven load distribution across the face width on the bending stress of the teeth, factor K_{Wb} should be used.

The values K_{WH} and K_{Wb} are different for two reasons:

1. The uneven load distribution, which is characterized by factor K_w, influences differently the permissible load of a gear, depending on what kind of possible failure is considered. For example, teeth of low and middle hardness are able to run in, and the initial pitting is not dangerous. Therefore, the K_{WH} value may be much less than K_w. But the "initial

breakage" that may occur during the run-in period cannot be agreed to. So the K_{Wb} value is expected to be closer to K_W.

2. The K_{Wb} value depends also on the length of the teeth; shorter teeth are less sensitive to the unevenness of load distribution.

These issues are discussed in detail in Section 7.6.

As soon as the force applied to the tooth is known, the maximal bending stress σ_b can be determined. The teeth have a complex form and for their strength calculation FEM should be used. Teeth that differ only by their module (or diametral pitch) are geometrically similar. Therefore, for "standard" teeth there are recommendations in technical literature that enable simpler calculation for bending strength. But in many cases the tooth shape differs from the standard.

In Figure 7.2, we could see how a "standard" tooth shape depends on the number of teeth. Figure 7.12 shows how the tooth shape changes depending on displacement Xm of the cutting rack (hob) from the "standard" position, in which the pitch line of the rack is tangent to the pitch diameter d of the cut gear. In this picture, only positive displacements $+X$ are shown ("long addendum"). Any negative displacement for $N = 17$ leads to undercut of the tooth root. But when the number of teeth is large enough, some negative displacement $(-X)$ may also be practical.

As is evident from Figure 7.12, the greater the displacement coefficient X, the thicker is the tooth in the root and thinner its tip. The radii of curvature of the working surfaces grow as well; therefore, the "long addendum" option has been often used both to decrease the contact stress, and to increase the bending strength of the teeth. The thickness of the tip should not be less than the admissible value (see Chapter 8, Section 8.6).

When both the pinion and the gear are made with positive displacement X, the pitch point may be kept in the middle of the length of action (segment $b_1 b_2$), so that the maximal sliding velocity is the same before the pitch point and after it. This enables keeping the sliding velocity as small as possible. If the pinion is made with "long addendum" and the gear with "short addendum," i.e., X_1 is positive and X_2 is negative, the segment $b_1 b_2$ becomes displaced relative to the pitch point, and the maximal sliding velocity increases. This may lead to scoring.

Not only the rack (hob) displacement, but also the change in the shape of the rack (for example, addition of protuberances intended to form tooth undercut as shown in Figure 7.13d) makes the shape of the tooth nonstandard. Teeth cut with a shaping cutter also have a shape different from hobbed teeth. Overall grinding of the tooth surfaces, including the fillets, may essentially change the shape obtained when cutting (depending on the grinding machine kinematics and the shape of the grinding wheel). In these cases, FEM should be used to calculate the maximal bending stress.

The gear designer should take into consideration the tooth generation process. If the gear cutting or finishing processes must be changed, the possible influence on the teeth shape and strength

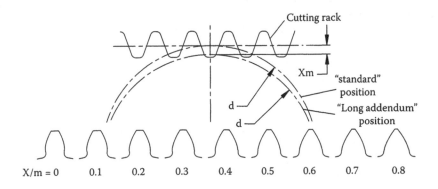

FIGURE 7.12 Tooth shape dependence on generating profile displacement X ($N = 17$).

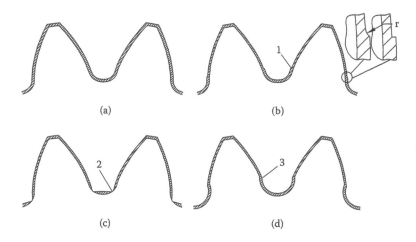

FIGURE 7.13 Influence of grinding on tooth root shape.

should be checked. Let's consider a gear that was hobbed and then case-hardened (in Figure 7.13a, the case-hardened layer is hatched). After heat treatment, the teeth should be ground to achieve high accuracy. In the process of grinding, part of the case-hardened layer is removed, and there is created a step between the ground and nonground surfaces. If only that part of the involute profile that is in contact with the mating tooth (active profile) is ground, step 1 may be sufficiently far from the tooth root (Figure 7.13b) not to impair the bending strength. Fillet radius r in the step depends on the method of grinding. A conical grinding wheel (shown in Chapter 8, Figure 8.23b) may have a rounded tip and provide a generous fillet radius. Saucer-shaped grinding wheels have sharp edges. When installed in the 15° position (Chapter 8, Figure 8.23c), they provide a certain fillet radius because of generating motion. But when they are installed in the 0° position (Chapter 8, Figure 8.23d), the fillet radius of the step is about zero. The stress concentration factor in this case is rather large. (Enlarged sketches of the two kinds of steps are shown in Figure 7.13b.)

The position of the boundary point of the active profile can be found from Figure 7.3a; this point is located on radius O_1b_2 for the pinion and O_2b_1 for the gear. Deeper grinding of the tooth profile brings the step nearer to the tooth root, and this may decrease its bending strength. If you want to grind all the tooth space to make it shining and eliminate the step, you should check carefully the geometry of hobbed and ground teeth. Figure 7.13a shows hobbed and case-hardened teeth. Figure 7.13c shows the same teeth after the tooth space was fully ground. The cutting teeth of the hob have a profile rounded off with a radius of $(0.3–0.4)m$, and the saucer-shaped grinding wheels have sharp edges. So the grinding wheels, which remove a constant-thickness layer from the involute part of the hobbed profile, take off significantly more material in the root area. Therefore, as is seen in Figure 7.13c, the fillet radius may become rather small, and the case-hardened layer may be completely cut through in this area (point 2). The bending strength of such teeth is very low in comparison with that expected for case-hardened teeth.

Irrespective of the foregoing, grinding of the tooth root is undesirable because it removes the hardest part of the case-hardened layer with greatest compressive stress. In addition, local high temperature caused by grinding may result in local stress relief and even in formation of tensile stress in the surface layer. Undercut of teeth before case hardening (Figure 7.13d) enables grinding of the involute profile only, without any steps. The undercut decreases slightly the thickness of the tooth root; but instead, the highest quality of the surface layer and highest reliability may be guaranteed.

Point 3 should be placed on lesser radius than the radius of the lower point of the active profile (b_2 for the pinion and b_1 for the gear, Figure 7.3a).

In most cases, the bending strength (as well as contact strength) of standard gears is calculated by easy methods comparing the gear to be designed with experimentally tested and service-proven gears of the respective kind. For this purpose, simplified formulas are derived.

As was said, with respect to bending-stress calculation, the tooth is a cantilever beam loaded with a distributed force F_n. The bending stress can be calculated using Equation 3.3:

$$\sigma_{max} = M \frac{6}{bh^2}$$

But some substitutions and supplements should be made in this formula:

- $b = W$ (face width).
- $h = t$, thickness of the tooth root (see Figure 7.8a).
- $M = F_n \cdot \cos \varphi \cdot L$, where L = arm of the bending force (see Figure 7.8a).
- From the tension stress caused by bending moment M should be subtracted compressive stress caused by the radial component of the tooth load.

$$\sigma_{compr} = \frac{F_n \sin \varphi}{W \cdot t}$$

- The tooth force F_n should be multiplied by factors K_d (dynamic load) and K_{Wb} (uneven load distribution across the face width).
- The net stress should be multiplied by effective stress concentration factor K_e.

In elaborate calculations, there may be some additional factors, such as temperature factor, roughness factor, size factor, and others; but here we only explain the basics of the gear strength calculations, so the factors that are secondary in usual working conditions are not mentioned.

All in all, the maximal bending stress from the tension side

$$\sigma_b = \left(F_n \cos \varphi \cdot L \frac{6}{W \cdot t^2} - \frac{F_n \sin \varphi}{W \cdot t} \right) K_d K_{Wb} K_e$$

Substituting

$$F_n = \frac{2T}{d_b} = \frac{2T}{d \cos \varphi}$$

we obtain the following equation for bending stress:

$$\sigma_b = \frac{2T}{dW} K_d K_{Wb} K_e \left(\frac{6L}{t^2} - \frac{tg\varphi}{t} \right) \tag{7.12}$$

In this beautiful equation, we have a lot of unknown components. First of all, we don't know the place of the weak point, and therefore, we don't know the values t, L, and K_e. We can find these values using FEM, but once we use it, we don't need Equation 7.12. So what do we need it for at all? Wait a moment, and you will see.

As mentioned, the teeth that differ only in their module (or pitch) are geometrically similar. It means that we can write $L = C_1 \cdot m$, $t = C_2 \cdot m$, where C_1 and C_2 are constants for a certain number of teeth. Because the pressure angle φ is nearly constant (about 20°) and $K_e = C_3$ (constant for the certain number of teeth), we can make the following substitution:

$$\left(\frac{6L}{t^2} + \frac{tg\varphi}{t}\right)K_e = \left(\frac{6C_1 m}{C_2^2 m^2} + \frac{tg\varphi}{C_2 m}\right)C_3 = \frac{1}{m}\left(\frac{6C_1}{C_2^2} + \frac{tg\varphi}{C_2}\right)C_3 = \frac{1}{m}Y$$

where Y is some new constant value for a certain number of teeth.

The factor Y is called the *geometry factor* because it includes such parameters as the shape of the tooth (thickness t and stress concentration factor K_e) and the position of load application (L value). The shape of the teeth is determined by the number of teeth (see Figure 7.2). So the Y value may be constant for a certain number of teeth and for a certain point of the force application, for example, at the tip of a tooth or at the upper point of the one-pair zone. (In the last case, the teeth numbers of both the gear and the pinion should be known.) For standard gears, these Y values can be found graphed or tabulated. If the teeth are nonstandard (they have "long addendum" or "short addendum," the hob teeth have protuberances, etc.), new Y values should be calculated for them.

As a result, Equation 7.12 looks simpler:

$$\sigma_b = \frac{2TY}{dWm}K_d K_{Wb} \tag{7.13}$$

The Y value for "standard" teeth can be found in reference books. Usually, $Y = 3$–3.5 (greater value for smaller numbers of teeth).

In correctly designed gears, the product $K_d K_{Wb}$ doesn't usually exceed 1.3–1.5. Assuming some average value of the geometry factor, the torque limited by bending strength of the teeth may be approximately evaluated for the spur gears as

$$T_{1b} = 0.1 d_1 Wm\sigma_{b,\lim} \tag{7.14}$$

where $\sigma_{b,\lim}$ = admissible bending stress given in figure 7.14. In this figure, the lines are intended for:

1. Tempered gears, load reversible (e.g., planetary gears and other idlers)
2. Tempered gears, load irreversible
3. Case-hardened gears, load reversible
4. Case-hardened gears, load irreversible

For helical gears the admissible torque is greater by about 50%. For straight bevel gears

$$T_{1b} = 0.07 d_1 Wm(1 - \psi_b)\sigma_{b\lim}$$

where m = module at the pitch diameter.

EXAMPLE 7.5

Determine the permissible torque limited by bending strength of the teeth for a case-hardened gear from Example 7.3. In this gear, $d_1 = W = 128$ mm. The admissible bending stress we find from Figure 7.14: for the core hardness of HB 360 and nonreversing load $\sigma_{b\lim} = 430$ MPa.

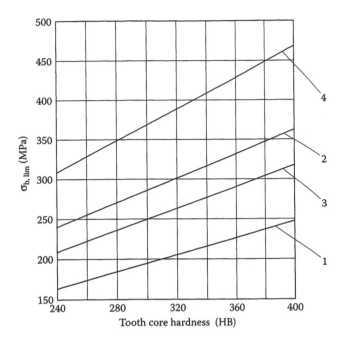

FIGURE 7.14 Permissible bending stress.

Taking $m = 7$ mm ($N_1 = 18$), we obtain using Equation 7.14:

$$T_{1b} = 0.1 \cdot 128 \cdot 128 \cdot 7 \cdot 430 = 4.93 \cdot 10^6 \text{ N} \cdot \text{mm}$$

As we see, the bending strength is sufficient (the torque to be transmitted $T_1 = 4.78 \cdot 10^6$ N·mm, see Example 7.3). Anyway, a more exact strength calculation should be performed. If needed, the module can be increased, say, to 8 mm ($N_1 = 16$).

When the proven gears taken for comparative calculation and the gear to be designed are similar with respect to their design, dimensions, rotational speed, and service conditions, the factors K_d and K_{Wb} can be omitted, because they are nearly the same. When the numbers of teeth don't differ a lot, the geometry factor Y may be also omitted. In this case, the bending stress obtained is comparative (let's denote it as $\sigma_{bcompar}$), and Equation 7.13 looks as follows:

$$\sigma_{bcompar} = \frac{2T}{dWm} \, N/mm^2 \tag{7.15}$$

7.6 UNEVENNESS OF LOAD DISTRIBUTION ACROSS THE FACE WIDTH (FACTOR K_W)

Manufacturing errors and elastic deformations may be the cause of misalignment of teeth under load. Figure 7.15a shows two nonparallel teeth touching each other at angle γ. Under load (Figure 7.15b), the teeth become deformed, and their contact extends for some length L. The load distributes approximately as shown.

The deformation consists of two parts: contact deformation and bending compliance. Both are linear with respect to load, so the load distribution along the line of contact is also linear. But the

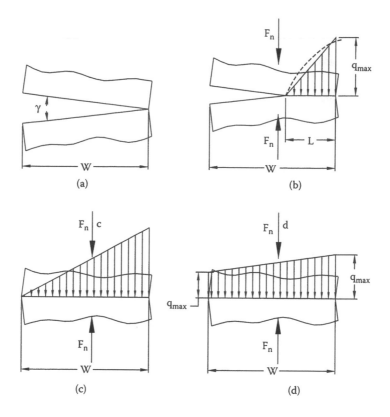

FIGURE 7.15 Tooth misalignment and load distribution along the contact line.

end of the tooth is more pliable, and this gives some decrease in the unit load magnitude in the end area (see the dashed line in Figure 7.15b). Because the load is not as sharply concentrated as is shown in the picture, the influence of the ends is secondary and has not been taken into consideration here. Besides, this neglect increases the safety against failure (i.e., it is a conservative assumption).

In this case, the K_W value is determined from the following considerations:

The maximal unit load (N/mm) is obviously

$$q_{max} = \frac{2F_n}{L} \tag{a}$$

The maximal deformation of the teeth (on the right end) is

$$\Delta = \gamma L$$

Hence, the maximal unit load

$$q_{max} = \gamma L C \tag{b}$$

where C = unit stiffness of the teeth, N/mm² (see Section 7.5).

Equating the previous two equations (Equation a and Equation b), we obtain the contact length:

$$L = \sqrt{\frac{2F_n}{\gamma C}} \qquad\qquad \text{(c)}$$

The average unit load (when the load is equally distributed)

$$q_a = \frac{F_n}{W} \qquad\qquad \text{(d)}$$

The load distribution factor, taking into account Equation a, Equation c, and Equation d:

$$K_W = \frac{q_{max}}{q_a} = \frac{2W}{L} = \sqrt{\frac{2\gamma WC}{q_a}} = \sqrt{\frac{2\gamma W^2 C}{F_n}} \qquad\qquad \text{(7.16)}$$

When the whole length of the teeth is in contact ($L = W$), but the unit load on the left end is still zero (Figure 7.15c), the maximal unit load equals, according to Equation a,

$$q_{max} = \frac{2F_n}{W}$$

It is twice as much as the average unit load (see Equation d), so in this case, $K_W = 2$. Consequently, when L is less than W, K_W is greater than 2.

Equation 7.16 is valid when only part of the tooth length is in contact (Figure 7.15b). This condition must not exist in satisfactorily designed and manufactured gears. Usually, the bearing pattern covers the whole working surfaces of the teeth. And when the teeth contact along the whole length (Figure 7.15d), the following equations are relevant:

The difference between the maximal and minimal unit load

$$q_{max} - q_{min} = \gamma WC \qquad\qquad \text{(e)}$$

The average unit load

$$q_a = \frac{q_{max} + q_{min}}{2} = \frac{F_n}{W} \qquad\qquad \text{(f)}$$

From Equation e and Equation f, we obtain

$$K_W = \frac{q_{max}}{q_a} = 1 + \frac{\gamma WC}{2q_a} = 1 + \frac{\gamma W^2 C}{2F_n} \qquad\qquad \text{(7.17)}$$

The K_W value obtained using this formula must be less than 2. If it proves to be greater, a repeated calculation should be done using Equation 7.16.

In operation, the working surfaces of the teeth run in because of wear, surface plastic deformation, and pitting. This leads to a more even load distribution along the teeth. The intensity of the running-in process depends above all on the hardness of the surfaces but also on the load, the rotational speed, and the purity and viscosity of the oil. To consider the influence of the running-in ability, the effective factor K_{WH} for the contact strength calculation may be taken as

$$K_{WH} = 1 + (K_W - 1)q_H$$

Here q_H = factor of sensitivity of the gear to the unevenness of load distribution. Values of this factor proved in the course of many decades are given in Figure 7.16.[7]

In some cases, the teeth wear rate obtained from field experience can be used to determine the parameters of the running-in process.[8] For example, wear-prone gears lose a certain part (V) of their thickness during a certain operating period, and this value is statistically known. It is known also that this amount of wear (V, mm) depends linearly on the hardness of the metal (H), unit load (q, N/mm), and the number of cycles (N):

$$V = \psi \frac{q}{H} N \quad mm \tag{7.18}$$

Here, ψ depends on the wear-resisting property of the material under given working conditions. Because the parameters V, q, H, and N are known from experience, factor ψ can be easily calculated from Equation 7.18.

Figure 7.17a shows the case where $K_W < 2$. If the initial angle between the teeth was γ_o, after x cycles of engagement it would be γ_x ($\gamma_x < \gamma_o$). For this moment, the following equations can be written:

$$q_{max,x} = q_a + \frac{\gamma_x WC}{2} \tag{g}$$

$$q_{min,x} = q_a - \frac{\gamma_x WC}{2} \tag{h}$$

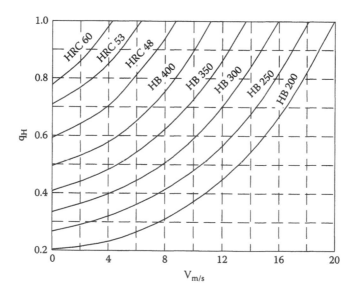

FIGURE 7.16 Sensitivity of gears to uneven load distribution across the face width.

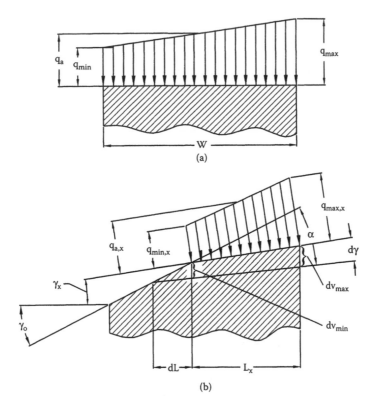

FIGURE 7.17 Run-in process in a wear-prone gear.

After a very small number of cycles (dN), the greatest additional wear ($dV_{max,x}$) will occur in the zone of greatest unit load (q_{max}); and the smallest additional wear ($dV_{min,x}$) will take place where the load is minimal (q_{min}). In conformity with Equation 7.18,

$$dV_{max,x} = \psi \frac{q_{max,x}}{H} dN$$

$$dV_{min,x} = \psi \frac{q_{min,x}}{H} dN$$

This uneven wear leads to a decrease in the angle between the teeth:

$$-d\gamma = \frac{dV_{max,x} - dV_{min,x}}{W} = \psi \frac{q_{max,x} - q_{min,x}}{W \cdot H} dN$$

Substitution of q_{max} and q_{min} from Equation g and Equation h gives

$$d\gamma = -\psi \frac{\gamma C}{H} dN; \quad \int_{\gamma_0}^{\gamma_x} \frac{d\gamma}{\gamma} = \int_0^{N_x} \psi \frac{C}{H} dN$$

$$\ln \frac{\gamma_x}{\gamma_0} = -\frac{\psi C}{H} N_x$$

$$\gamma_x = \gamma_0 e^{-\psi \frac{C}{H} N_x}$$

(7.19)

Here, γ_x is the angle between the teeth after N_x cycles of engagement. So at every moment, we can know the angle of misalignment and, using Equation 7.19, the K_W value.

Figure 7.17b shows a tooth that works with a part of its length. On the idle part of the tooth, the angle of misalignment remains unchanged (γ_0), whereas on the working part it gradually decreases. Thus the length of the working part increases. The following equations may be written on the basis of Figure 7.17b:

$$\frac{dV_{min}}{dL} = tg\alpha \approx \alpha; \quad \alpha = \gamma_0 - \gamma_x + d\gamma \approx \gamma_0 - \gamma_x$$

$$\frac{dV_{min,x}}{dL} = \gamma_0 - \gamma_x; \quad dV_{min,x} = \psi \frac{q_{min,x}}{H} dN \qquad (7.20)$$

$$dL = \frac{\psi \cdot q_{min,x}}{H(\gamma_0 - \gamma_x)} dN$$

We have signed the average unit load q_a when the entire length of the tooth is in contact. For a partly loaded tooth, we have to add the average unit load on the loaded portion ($q_{a,x}$) that is larger:

$$q_{a,x} = q_a \frac{W}{L_x}$$

$$q_{min,x} = q_{a,x} - \frac{\gamma_x L_x C}{2} = q_a \frac{W}{L_x} - \frac{\gamma_x L_x C}{2} \qquad (7.21)$$

Substitution of Equation 7.21 and Equation 7.19 into Equation 7.20 gives the differential equation of the run-in process under abrasive wear:

$$dL = \frac{\psi}{\gamma_0 H \left(1 - e^{-\psi\frac{C}{H}N_x}\right)} \left(q_a \frac{W}{L_x} - \gamma_0 e^{-\psi\frac{C}{H}N_x} \cdot \frac{L_x C}{2}\right) dN$$

Changing of variables turns this equation to type

$$Z' + P(u) \cdot Z = Q(u)$$

The result is

$$L_x = \sqrt{\frac{2q_a W \psi N_x}{\gamma_0 H \left(1 - e^{-\psi\frac{C}{H}N_x}\right)}} \qquad (7.22)$$

The angle of misalignment γ_x can be calculated using Equation 7.19.

Now that we have described the run-in process, we are able to perform more exact calculation of the teeth for bending strength using the sum of damaging cycles.

Among the factors involved in Equation 7.17, the misalignment angle γ, the face width W, and the average unit load q_a are those that may be influenced by us with the goal of decreasing the K_W value. (The stiffness of the teeth C can't be changed considerably unless we use material with a different modulus of elasticity.) Decreasing the misalignment under load is achieved by higher manufacturing accuracy and more rigid design of the shafts, bearings, and housing. The other two factors can be changed profitably by using stronger materials. For example, case-hardening enables a significant decrease in the gear dimensions (including, certainly, W) and a corresponding increase in q_a (as compared with tempered steel). Thus, using high-strength materials, in addition to other benefits, makes easier the problem of reaching a more even load distribution across the face width of a gear.

In very large gears in which case-hardening can't be used for technological reasons, nitriding is successfully used. But the dimensions of this kind of gear remain large, and the problem of proper load distribution is one of the hardest to resolve.

EXAMPLE 7.6

Let's analyze Equation 7.17:

$$K_W = 1 + \frac{\gamma WC}{2q_a}$$

In this equation, $\gamma \cdot W$ is the misalignment of teeth in millimeters. What does it do if, for instance, $\gamma \cdot W = 0.01$ mm? Let's take the case-hardened gear from Example 7.3 and Example 7.4. The average unit load in this gear

$$q_a = \frac{2T_1}{d_1 W \cos \varphi} = \frac{2 \cdot 4.78 \cdot 10^6}{128 \cdot 128 \cdot \cos 20^0} = 621 \, N/mm$$

Stiffness of "standard" helical teeth may be determined from the following equation[4]:

$$\frac{1}{C} = 10^{-3} \left(0.0514 + \frac{0.1425}{N_{1,e}} + \frac{0.1860}{N_{2,e}} \right)$$

where

$$N_{1,e} = \frac{N_1}{\cos^3 \psi}; \quad N_{2,e} = \frac{N_2}{\cos^3 \psi}$$

Here ψ = helix angle.

In this gear, $N_1 = 18$, $N_2 = 126$, $\psi = 10°$; so $N_{1,e} = 18.85$, $N_{2,e} = 131.9$, and $C = 17380$ N/mm². Now we can calculate the K_W value:

$$K_w = 1 + \frac{0.01 \cdot 17380}{2 \cdot 621} = 1.14$$

Notice that misalignment is only 10 μm.

Uneven load distribution leads to increase of both the contact stress and the bending stress. But the latter increases less. Let's consider two cantilever beams loaded on the corner by a single force (Figure 7.18a). These beams are of the same width but different in length. It is quite obvious that the short beam must have a sharp maximum of bending stress in the vicinity of the load whereas

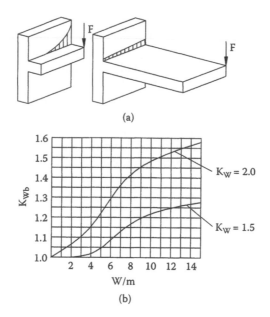

(a)

(b)

FIGURE 7.18 Influence of load distribution on bending stress in the tooth root.

the parts distant from the load don't even feel it. In the long beam, the bending stress changes moderately across the width. As applied to the gear tooth, the greater the height relative to its length, the less is the influence of the load distribution on the bending stress. Factor K_{Wb} introduced previously (see Section 7.5) is given by

$$K_{Wb} = \frac{\sigma_{b\,max}}{\sigma_{b,a}}$$

where
σ_{bmax} = maximal bending stress
$\sigma_{b,a}$ = average bending stress (in the case of even load distribution)

Factor K_{Wb} is given in Figure 7.18b vs. K_W and W/m (the latter defines the ratio between the length and the height of the tooth). This chart is calculated using FEM for a straight tooth of $2.25m$ height, with the load linearly distributed along the tip edge. (The nearer the contact line to the tooth root, the closer is the K_{Wb} value to K_W.) We can see that the bigger the module m at the same face width, the less is the influence of the uneven load distribution on the bending stress pattern.

There are certain limitations for the module increase:

• The diameters of gears are calculated from contact strength considerations. When the module increases, the number of teeth decreases automatically. This is associated with decreased contact ratio and increased sliding velocity, which in turn leads to increased chances of pitting and scoring. Nevertheless, in low-speed, heavily loaded gears are used pinions with 10 teeth, and even fewer. (To avoid undercut, these gears must have a positive generating contour displacement X.)

• The larger the module, the thicker should be the rim of the gear; thus its weight increases.

Ratio W/m may be decreased also by decreasing the face width of the gear. In this case, the diameters should be increased to meet the strength requirements. The increase in diameter and decrease in width enable the pinion shaft to be made more rigid and the teeth misalignment, somewhat less.

7.7 DYNAMIC LOAD IN THE GEAR MESH AND FACTOR K_d

Estimation of load applied to the teeth begins by calculating the nominal force F_n: the nominal torque is divided by the radius of the base circle. But there are some additional forces that should be taken into account.

It is known that in rotating systems, torsional vibrations are unavoidable. In important applications, a dynamic analysis is usually performed, and if the dynamic torque exceeds 5–10% of the nominal torque, the needed changes in mass and torsional stiffness of certain elements should be made to reduce the amplitude of the alternating torque.

Apart from what has been stated earlier, the dynamic load in the gear mesh is practically independent of the other members of the mechanism. Two engaged gears make a torsionally oscillating system of two rotating masses connected by an elastic element (the gear mesh). The stiffness of this element changes in spur gears when the one-pair and two-pair modes take turns. This cyclic change in the stiffness parameter excites vibrations and may lead to what is called *parametric resonance*. The resonance frequency may be calculated; if it is close to the tooth engagement frequency, either the moments of inertia of the gears or the number of teeth should be changed, or the contact ratio should be increased to 2 so as to eliminate the parametric excitation. And finally, there are dynamic loads in the gear mesh caused by errors in the geometry and base pitch of the teeth under load. This is what the factor K_d is intended for.

Early in the 20th century, when large steam turbines were developed, high-speed, large-capacity reduction gears came into demand. The technology of gear manufacturing was not able then to provide very high accuracy of the teeth, and there were cases of teeth breakage during bench runs *without* any load. Nowadays, when the rim speed may exceed 200 m/sec (40,000 ft/min), the industry is equipped with super-precision gear-cutting and finishing machines that allow manufacturing to very close tolerances, within 1–3 μm for gears of several meters in diameter. So the problem of accuracy can be solved now quite satisfactorily. But some moderate dynamic load remains.

Figure 7.19 shows two cases of impact in spur gearing. In the first case (Figure 7.19a), the base pitch across teeth A_1-B_1 is less than that across teeth A_2-B_2. (In the picture, it is shown as interference between teeth B_2 and B_1.) As a result, pair B_1-B_2 comes into contact with an impact. This case is called *edge impact* because one of the teeth gets the impact at its tip.

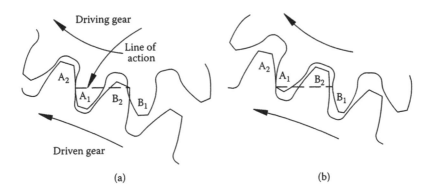

FIGURE 7.19 Tooth impact caused by pitch deviations.

In the other case shown in Figure 7.19b, the base pitch across teeth A_1-B_1 is bigger than that across teeth A_2-B_2. In the position shown, pair A_1-A_2 disengages and pair B_1-B_2 remains, kind of hanging in the air. Because there is no contact between the teeth (for a moment), the driving gear accelerates, and the driven gear becomes slower. (The same process was described in Section 7.1 with reference to a spur gear with contact ratio less than 1.) Therefore, the teeth come into contact with a "middle impact."

The edge impact is more dangerous because the force is applied to the tip of a tooth. It results in increased noise and, possibly, pitting, scoring, and teeth breakage. Increasing tooth accuracy is not enough to prevent the edge impact. Because of elastic deformation of the teeth under load (shown dashed in Figure 7.20a), the base pitch decreases on the driving gear (teeth A_1-B_1) and increases on the driven gear (teeth A_2-B_2). It is exactly what leads to the edge impact. To prevent it, the involute profile of the teeth should be modified by making the tip thinner (Figure 7.20b). The profile relief begins in the upper point of the one-pair zone (about $0.4m$ from the tooth tip). It is of involute shape with a larger pressure angle. The maximal depth of the profile modifications usually equals the elastic deformation of the pair engaged in the end of the one-pair zone (A_1-A_2, in this case).

Because the addendum of one tooth is in contact with the dedendum of its mate, the profile relief may be done on the addendum, or on both the addendum and the dedendum (in the latter case, with lesser deviation from the initial involute).

Perhaps a reader who has followed the process this far might still wonder how to obtain the K_d value? After all, for the strength calculation, the number only is needed, not the understanding as to how it was reached? At this point, the authors might start speaking in a roundabout way. "Look ... " (This is the usual beginning, when we don't know exactly what to say, and we try to involve the listener in our process of thinking.) "You know, ... " (Possibly, you don't know, but never mind.) The only answer the authors can give in the context of this book is as follows.

Dynamic load is a result of a continuous oscillatory process in the vibrating system of two gears. The dynamic load on a certain pair of teeth depends not only on the geometry of this very pair but also on geometry of the preceding pairs. That is why the exact calculation of the dynamic load should be based on exact knowledge of the real geometry of all teeth. It is rather unreal, particularly, taking into consideration that the teeth geometry changes in operation. This process is also influenced by the oil film between the contacting teeth. Nevertheless, there are some acknowledged methods of gear analysis that include the dynamic load calculation as well. These methods are approximate, but experimentally proved and for that reason quite reliable. And the most comforting thing is that when the gears are manufactured with an appropriate accuracy (conforming to the rim speed) and profile modification, the K_d value lies in the range of 1.15–1.35.

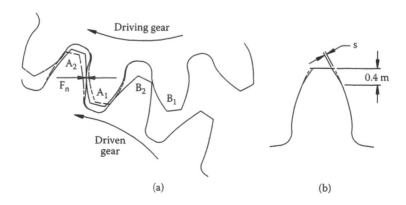

(a)   (b)

FIGURE 7.20 Pitch change caused by deflection under load.

(Certainly, it is not valid for the resonant mode mentioned in the preceding text.) So the possible error in estimation of the K_d value looks quite acceptable as compared to other uncertainties of the gear strength calculations. In important cases, the real bending stress should be measured during trials. Sometimes, these measurements are performed periodically in service to control the influence of the changes in gear geometry and in the alignment of teeth.

7.8 LOAD DISTRIBUTION IN DOUBLE-HELICAL GEARS (FACTOR K_{Wh})

Double-helical ("herringbone") gears are an example of a mechanism with split power transmission (Figure 7.21). Uniform distribution of the load between two halves of the gear is supposed to be achieved by axial self-alignment of the pinion. In the ideal case, when only the meshing forces are applied to the pinion, the static equilibrium of the pinion implies the equality of axial forces F_{a1} and F_{a2}. The pinion moves axially until this equality is reached. Because the helix angles on the right and on the left are equal, the load should be distributed equally.

It doesn't rule out possible unevenness of load distribution across each of the half-gears caused by bending and torsion deformations of the pinion under load.

In fact, there are friction forces in bearings, couplings, and between the teeth that hinder the pinion from axial motion. Inertia of the pinion plays the same part. As a result, there is always a certain difference between forces F_{a1} and F_{a2} and unevenness in load distribution between the respective halves of the gear.

The ability of the pinion for self-alignment depends on several design factors. First of all, the bearings and couplings must enable the needed axial motion. This may be a radial sliding bearing or cylindrical roller bearing without lips on one of the rings (Figure 7.21b). Bearings with one lip may be used on the understanding that the rings are somewhat displaced from their nominal position, so as to produce axial play needed for the pinion to take the equilibrium position (Figure 7.21c).

The bearings shown in Figure 7.21d seemingly enable the axial movement of the pinion. But the friction between the housing and the outer ring of the bearing is too high, and this results in a large difference between forces F_{a1} and F_{a2}. In the extreme case, at a small helix angle, even the full-load axial force may be insufficient to move the pinion under load, and then, only one half of the gear will transmit the entire load.

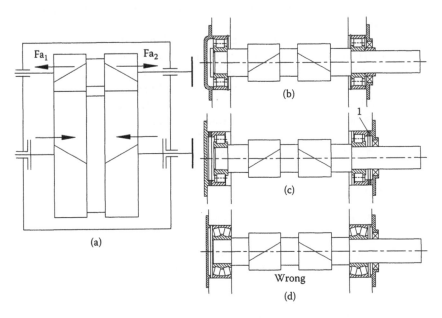

FIGURE 7.21 Bearings for double-helical gears.

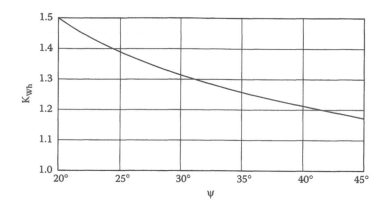

FIGURE 7.22 Overload factor for a double-helical gear.

Use of appropriate bearings is only a part of a proper design. If the bearing is fitted too tightly to the shaft, so that there is no internal clearance between the rings and rollers, the shaft may be squeezed in the bearing. To cut out that possibility, the shaft should be checked by hand for free axial movement (after the gear is completely assembled).

Any devices that produce additional load on the bearings, such as heavy flywheels, belt-drive pulleys, and so on, should be avoided on self-aligning shafts. Extra accuracy of the gear mesh helps to decrease the amplitude of the pinion's oscillating motion. Owing to this, the speed and acceleration of this motion are reduced, and the unevenness of load distribution, which is caused by friction forces and inertia, decreases.

Information about the resistance of cylindrical roller bearings to axial displacement under load is given in Chapter 6, Subsection 6.2.3, Equation 6.5.

When a double-helical gear is properly designed, the overload factor for a half-gear K_{Wh} may be approximately taken from Figure 7.22 depending on helix angle ψ. From the chart, we see that the helix angle should be big enough to keep the overload factor within reasonable limits.

The reader may ask, why on earth do they make herringbone gears? It is much easier to make helical gears, is it not? That is right, and small- and medium-sized gears are usually helical. But the axial forces in these gears cause asymmetric deformations of the gears and housing. For big gears, symmetry of all deformations is a very important principle that helps to keep the teeth aligned and decrease the K_W value. In double-helical gears, the axial forces are balanced and don't load the bodies of gears and housing, so they provide the needed feature.

Another merit of the double-helical gear is that the total face width of the two halves is greater than can be made in a helical gear (because of lesser influence of the pinion twisting deformation on the load distribution across the face width). Owing to that, the diameters of gears may be reduced, and this is very important for large reduction gears, in which the diameters come up to 5–6 m (20 ft).

7.9 BACKLASH IN THE GEAR MESH

The less the backlash the better — this is undoubtedly true when related to kinematical gears, meaning that there must be an exact relationship between the angular positions of the engaged gears. (In these cases, a zero backlash or even prestress is usually adjusted.) But as a rule in power transmissions, the backlash is chosen to range from $0.05m$ to $0.1m$. This should prevent jamming of the gears at the worst combination of tolerances and the greatest temperature differential of the gears as compared to the housing. The increase in backlash doesn't usually have any detrimental influence on the gear, with the exception of mechanisms with sharp changes in their rotational speed, when the clearances are taken up in both directions with an impact.

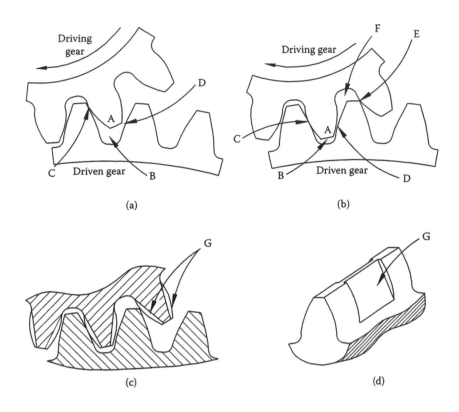

FIGURE 7.23 Oil drain from the tooth spaces.

In spur gears that work in plenty of oil, and in gear pumps, problems may develop because of the displacement of oil from the tooth space by the mating tooth. Figure 7.23a shows the situation in which tooth A enters the tooth space B; at point C, it is in contact with the mating tooth, and the oil can flow out of the volume B through the opening D (and from the ends, of course). The oil pressure developed in volume B may be so high that the bearings may be damaged. Increasing the backlash (say, to $0.1m$) will increase the opening D and may sufficiently reduce the oil pressure in this volume and will also reduce power required to drive the pump.

Figure 7.23b shows the beginning of the two-pair zone. The teeth are in contact at points C and E. At point D, there is a backlash that joins together volumes B and F. As the gears rotate, this closed volume (B + F) contracts, and the oil can flow out of it only through the ends. The ends must be open, and the covers of the pump housing must be provided with grooves that connect the outlet from volumes B and F with the pumping chamber. But if the pump gears are long, it is not enough, and very high pressure may develop in volumes B and F, which may damage the bearings or cause shaft breakage. Recesses G in the teeth shown in Figure 7.23c and Figure 7.23d reduce the pressure in the tooth spaces.

Certainly, such recesses can't be used in power transmissions because they decrease the active face width and reduce the contact and bending strength of the teeth. But the quantity of oil in transmissions is much less, and the problem of oil pressure may come into being only at high speed. In these gears, the oil is sprayed onto the teeth after they leave the contact zone or fairly before it, so that when they come to contact again, they are covered with a thin oil film only. The tooth spaces before entering the contact zone are filled with compressible air and oil mist.

In helical gears, tooth contact begins on one end and moves to another, so the oil is freely displaced in the direction of the moving contact. In a double-helical gear, the oil may be displaced from the center to the sides or from the sides to the center, depending on the direction of teeth and direction of rotation. When the oil is displaced to the center, it may cause an additional heating of the gears in the middle, increase in their diameters, and uneven load distribution across the face width.

7.10 LUBRICATION OF GEARS

The system of lubrication is intended to provide the rubbing surfaces with an oil film and take away the heat. The kind of lubricant and type of gear lubrication depend in the first place on the pitch line velocity of a gear. Low-speed gears, particularly those that are open (that means, they don't have a hermetically sealed housing), are often greased. The grease is applied to the teeth by brush once a day or once a week depending on the need. Because the grease is not able to take off the heat from the teeth, this type of lubrication is used only for light-duty applications, where the small amount of heat may be successfully taken off by convective heat transfer. The recommended speed limit for the grease-lubricated open gears is up to 4 m/sec. At a higher speed or continuous duty, the periodic manual greasing becomes too laborious and is replaced by dipping one of the gears into an oil-filled pan.

Gears which have a hermetically sealed housing are usually lubricated with oil. As compared with grease, oil is more heat resistant, lasting, and easily replaceable. It washes away the wear debris from the wearing surfaces and, by permanent filtering of the oil, the mechanism can be kept clean during long-term usage.

It is particularly important for bearings to be lubricated with clean oil. But the other wearing parts also have sufficiently lesser wear rate when the wear debris is permanently removed from the friction surfaces. Oil transfers the heat from the teeth to the housing walls, which can be provided, if needed, with fins and fans for more effective cooling.

The simplest way to use liquid oil is by oil bath lubrication (Figure 7.24). To diminish hydraulic losses, the gears should not be immersed too deeply, but not less than the teeth height. In bevel gears, it is also recommended that the teeth should be entirely immersed in oil.

The oil level should be checked while the gear is running because the oil splashed on the walls of the housing flows down slowly, and the oil level may become considerably lower than it was in the nonoperated gear. The authors have observed recession of oil level in a reduction gear by as much as 25 mm.

If the oil gauge is placed far from the oil-immersed gear wheel, even a small inclination of the gear housing may result in lack of lubrication. Therefore, the gear should be mounted level on its foundation, and the housing should be provided with special horizontal surfaces for mounting the level.

To guarantee reliable bath lubrication in a multistage gear set, one gear of each stage should dip into the oil. For this purpose, in low-speed gears the oil level can be increased as shown in Figure 7.24a. At higher speed, the smaller gear wheel can be lubricated with an additional idler gear 1 (Figure 7.24b) made of gray cast iron or some kind of plastic. It rotates freely on an axle and supplies oil to the power train. The face width of the idler should be equal to 0.7–0.8 that of the lubricated gear. If the width of the idler is relatively small, the load concentrates in the lubricated

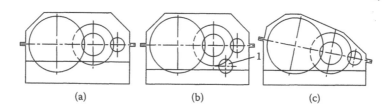

(a)   (b)   (c)

FIGURE 7.24 Oil bath lubrication of gears.

area, because this part of a tooth "protrudes" because of thicker oil film. This may result in severe pitting in the area of flooded lubrication. Sometimes the gear is designed with an inclined casing joint that provides equal dipping for the gears of all stages (Figure 7.24c).

Rolling bearings are usually splash oil fed. It is believed that at oil viscosity less than 250 cSt, reliable splash-oil-feed is achievable when the centrifugal force of the oil drops is greater than their weight. That means

$$\frac{mV^2}{R} \geq m\,g; \quad V \geq \sqrt{gR}$$

where

    V = linear speed of the mesh (m/sec)
    R = radius of the gear (m)
    g = 9.81 m/sec^2 = acceleration of gravity

EXAMPLE 7.7

When R = 0.2 m, the linear speed, for satisfactory oil sprinkling, should be not less than

$$V = \sqrt{9.81 \cdot 0.2} = 1.4\,m/s$$

$$n = \frac{60V}{\pi D} = \frac{60 \cdot 1.4}{\pi \cdot 2 \cdot 0.2} = 67\,rpm$$

If the oil is sprinkled on the walls of the housing, it doesn't mean that it gets into the bearings. Investigation of splash oil feed for bearings has been performed on a two-stage reduction gear with a transparent housing made of acrylic plastic. The scheme of this gear and the results of the investigation are given in Table 7.1. No special grooves were made to lead the oil to the bearings, but they were completely open and accessible for the air and oil splashes from within. Before each experiment, all the bearings were dismounted from the shafts and degreased. Then the gear was assembled and filled up with oil of needed viscosity, so that the intermediate gear wheel was immersed into the oil as deep as two heights of the teeth (as shown in Figure 7.24a). Then the gear was rotated for 5 min with a certain rotational speed in a certain direction. After that, the gear was dismantled, the bearings dismounted from the shafts, and the presence of oil on the races and rolling elements of the bearings was checked by white filter paper. We have seen that inside the gear there is an oil storm, particularly at higher speeds. It was impossible to imagine that inside the housing there can be any dry places. But there were, and, in fact, the dry locations were the races and rolling elements of the bearings!

In Table 7.1, the hatched cases are those in which the bearings were oiled. In the nonhatched cases, the bearings remained dry. The experiments have shown that without appropriate facilities, splash lubrication of the bearings can't be guaranteed. There were cases in which (at a high speed) the oil streamed abundantly down the walls, but tore off the edge of the bearing seat and didn't enter the bearing. When the gear was stopped, the oil flow was getting less intensive, and some small amount of oil was entering the bearing. But such lubrication can't be considered as satisfactory.

Reliable oil supply to the bearings is achieved by oil-catching grooves 1 (Figure 7.25) made on the joint surfaces of the housing. In middle-sized gears the grooves are 5–10 mm in depth and 10–20 mm in width. The grooves catch the oil that flows down the walls. Dripping-pans serve the same purpose (Figure 7.26); they collect the splashed oil and lead it to the bearings.

TABLE 7.1
Bearing Lubrication

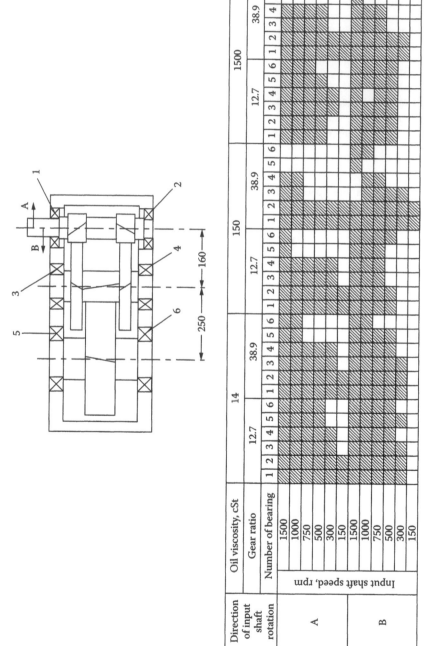

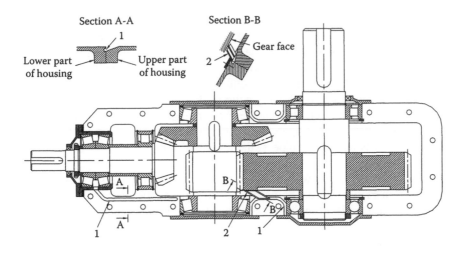

FIGURE 7.25 Lubrication grooves and scraper.

If the gear runs slowly and the sprinkling is insufficient, the oil can be taken off the gear face by a scraper (2 in Figure 7.25). The tip of the scraper can be installed at a distance of 0.5–1 mm from the gear face. If the tip is made of rubber or plastic, it may touch the gear. The scraper shown in Figure 7.25 is designed so that it can take oil off the gear face and lead it to the oil groove independent of the gear's rotational direction.

The oil grooves and ducts are configured so that the oil can flow to the oil reservoir only through the bearings. Therefore, bearing lubrication is assured. Gears designed to operate at a wide range of working conditions (such as rotational speed, direction of rotation, load, temperature, etc.) should have a lubrication system that assures the proper functioning at any specified conditions. For small gears, it may be increasing the oil level until all the gear wheels and bearings are partly or completely immersed. In these cases, the seals should be of good quality and reliability. For larger gears, a force-feed lubrication system (with an oil pump) is usually used.

The capacity of the oil reservoir should be big enough to minimize the rate of oil decomposition and give the wear debris a place to settle. Depending on the power transmitted, the volume of oil

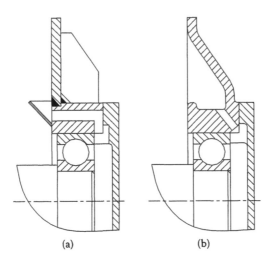

(a)   (b)

FIGURE 7.26 Oil collector and lubrication holes.

is recommended in the range of 3.5–10 l/kw of lost power. For instance, if a two-stage gear transmits power of 100 kW, and its efficiency is 97%, the recommended oil content is 0.03 · 100(3.5 to 10) = 10.5–30 l. In practice, it is not always achievable, particularly in lightweight machines, such as vehicles, planes, and the like, where the entire mechanism is rather small, and there is no place available for this volume of oil. In these cases, field experience is the best adviser.

The oil level is usually monitored using a dip stick. Figure 7.27a shows a design of dip stick that allows measuring the oil level while the gear is running. Pipe 1 protects the stick from the oil splashes. Take note of hole 2 in the upper part of the pipe; the rubber knob of the stick serves also as a plug, and if there is no hole in pipe 1, the knob will force the oil out of the pipe and make the stick indication false.

Widely used level indicators with a transparent face ("sight glasses") from glass or acrylic plastic are shown in Figure 7.27b. (Some designs use glass tubes rather than flat glass plates.) It is preferable to place the level indicators and dip sticks closer to the gear that is less immersed in oil.

The circulation lubrication system is unavoidable if design or service conditions don't allow satisfactory lubrication by immersion and sprinkling; for example, when the gear shafts are vertical, or when the speed is too high or too slow, or when the lubrication system contains oil filters, heat exchanger, and other devices that need forced circulation. Sliding bearings, usually, also need force-feed lubrication.

Oil circulation is commonly provided by a gear-type pump or plunger pump, attached to the gear housing and driven by one of the gear shafts. The gear-type pump is preferable in a reversible version (with appropriate valving) so that the lubrication system and the gear will not be damaged if the gear is accidentally reversed.

In middle-sized gears, the pump usually takes the oil from the gear housing and supplies it through oil ducts to the gear mesh and the bearings. The capacity of the oil reservoir is usually chosen as large as the pump can supply in 0.5–2 min, depending on the design features.

In bigger gears, which transmit great power, the lubrication system may include, besides the associated pump, the following:

• An electric pump for oil circulation before setting the machine in operation
• An oil tank with a capacity that equals the pump supply during 3–4 min or more

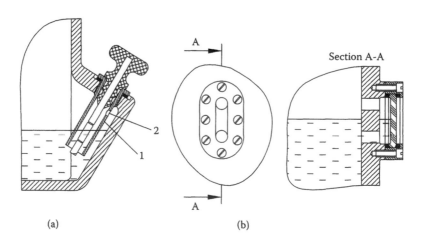

(a)   (b)

FIGURE 7.27 Oil level indicators.

- An oil cooler and, possibly, also a heater (to heat the oil before starting, lest the oil pressure in the system be too high in the beginning because of increased flow resistance)
- Oil filters
- Indicator of metal particles in the oil
- Safety valves
- Manometers, thermometers, and alarm devices that alert the personnel when the pressure and the temperature of the oil deviate from the permissible values, and other devices

The recommended quantity of oil to be sprinkled on the gear teeth is about 5–8 l/min per kilowatt of power wasted in the mesh (about 1.5% of the power transmitted). At this quantity, the oil warms up by 8–5°C, correspondingly.

With force-feed lubrication, uniform distribution of oil across the face width of the gears is achieved using pipelike sprayers with several holes (Figure 7.28a). To avoid their clogging, the diameter of these holes should not be less than 0.8 mm (usually, 1.2–3.0 mm). The oil flow through one hole depends on its diameter, oil pressure, and oil viscosity. At the viscosity of 110 cSt, the oil flow is approximately

$$Q = 1.6d^2\sqrt{p} \quad l/\text{min}$$

where
d = diameter of the hole (mm)
p = oil pressure (MPa)

For example: $d = 2$ mm, $p = 0.3$ MPa, $Q = 3.5$ l/min.

Narrow gears may be lubricated by a sprayer as shown in Figure 7.28b and Figure 7.28c. Dimensions of the jet and oil flow for the option as shown in Figure 7.28c are given in the charts and table.[9]

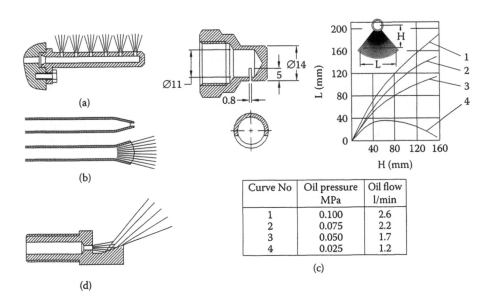

Curve No	Oil pressure MPa	Oil flow l/min
1	0.100	2.6
2	0.075	2.2
3	0.050	1.7
4	0.025	1.2

FIGURE 7.28 Oil sprayers.

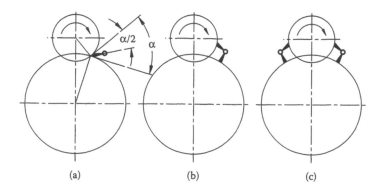

FIGURE 7.29 Direction of oil spray on gears.

In high-speed gears, there may occur erosion of metal in places where the teeth hit the oil jet. In these cases, a sprayer as shown in Figure 7.28d may be useful. It fragments the oil jet and creates some mixture of air and small particles of oil.

The most effective cooling of teeth is achieved when the oil is directed at the beginning of the teeth engagement, as shown in Figure 7.29a. This method is recommended up to a speed of 20 m/sec for spur gears and up to 50 m/sec for helical gears. For high-speed gears, particularly spur gears, this method is unsuitable, because when the oil is forced out of the tooth spaces in a very short time, there may be created a very high hydrodynamic oil pressure. This pressure occurs cyclically with the teeth engagement frequency and may lead to severe vibrations and failure of gear elements (bearings, shafts, housing, etc.). In these cases, the oil should be sprinkled as shown in Figure 7.29b. If the oil picks up insufficient heat, additional sprayers can be installed (Figure 7.29c).

The sprayers and all elements of the piping should be rigid and strong lest they be broken from vibrations. The natural frequencies of the elements of lubrication system should not be close to the teeth engagement frequency. It is desirable to locate the bolt connections so that an undone nut, a piece of a broken cotter pin, or lock wire can't fall and get into the gear mesh. Among the suitable locking element types, tab washers and locking plates are the best.

When the prototype gear is assembled, the lubrication system should be pumped with oil to check whether the jets get exactly to the needed places. This is particularly important when the sprayers are distant from the lubricated surfaces. If the rotational speed of the gear changes in the course of operation, the pump delivery changes as well. In this case, the pump should be designed for excessive oil supply at high speed and provided with an overflow valve to keep constant oil pressure in the system. In addition, the shape, position, and range of the oil jets must be designed to deliver adequate lubrication at all pressures and speeds.

Example of force-feed lubrication for a reduction gear with vertical shafts is shown in Figure 7.30. Pump plunger 1 is set in motion by cam 2 mounted on the end of the intermediate shaft of the gear. The oil comes through ducts 3 and 4 into the upper housing, and there it is distributed among the bearings and gears. The low-speed stage is lubricated by sprayer 5. A small part of the oil flows through oil indicator 6, so that the oil flow can be seen inside a glass pipe.

Figure 7.31 shows a two-stage reduction gear with a bevel input stage and planetary output stage. The gear is lubricated by a gear-type pump 1, which sucks the oil from the lower housing through inlet pipe 2 and forces it into outlet pipe 3. From there the oil flows through a filter to the bearings and gears. Bearing 4 is greased through pipe 5. Labyrinth seal 6 protects the bearing chamber from ingress of dirt and retains the grease. Rotating collar 7 prevents the entrance of liquid oil into that chamber.

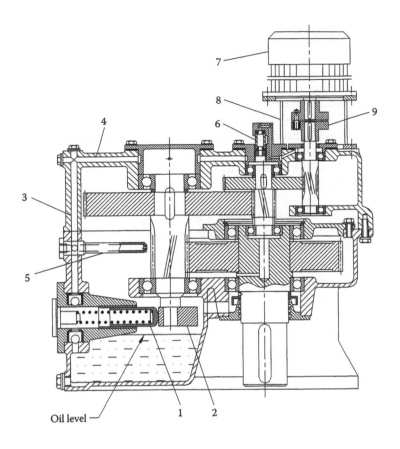

FIGURE 7.30 Three-stage reduction gear with vertical shafts.

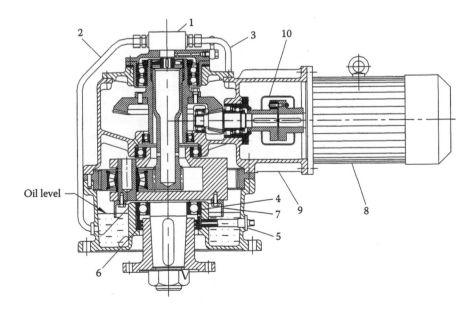

FIGURE 7.31 Two-stage bevel-planetary gear with vertical output shaft.

7.11 COOLING OF GEARS

Almost the entire power wasted in mechanisms turns into heat. The efficiency of gears depends on their kind. Ordinary spur, helical, and bevel gears are quite efficient. From 1.5% to 3% of the power transmitted is spent in one gear stage partly on friction in the mesh and bearings, and the rest on oil agitation (called *hydraulic loss*). Planetary gears have the same or higher efficiency. In small gears of this kind, the cooling problem usually doesn't exist, because the convective heat exchange satisfies the requirements. But the power transmitted by a gear is nearly proportional to the third power of its dimension, whereas the area of the heat-exchange surfaces is proportional to the second power only. So in larger gears that transmit several tens of kilowatts and more, cooling may present difficulties and needs thorough calculation and organization.

In gears with a high ratio in one stage (such as worm gears, harmonic drives, and some special kinds of planetary gears called *fixed differentials*[10]), the power loss may be as great as 50% and more at a ratio of 1:80. Though these gears usually transmit relatively small power, their cooling may present difficulties and restrict the allowable working load at a constant duty. So, cooling problems can be dealt with in small gears as well as in the big ones.

The calculation of cooling begins with the determination of heat input. The efficiency η of a gear is mostly taken relying on recommendations; otherwise, it should be determined experimentally. Virtually all the power loss turns into heat, so the heat input P_{in} is given by

$$P_{in} = P(1-\eta)10^3 \ W \qquad (7.23)$$

Here P = power on the input shaft of a gear (kW).

The output energy P_{out} (that is, the heat removal from the gear) depends on the area A (m²) of the heat-exchange surfaces, the temperature difference Δt (°C) between the gear (t_g) and the surrounding medium (t_s), and also on factor k that shows how much energy (W) can be removed from 1 m² of a surface if the temperature difference $\Delta t = 1°C$:

$$P_{out} = A(t_g - t_s)k = A \cdot \Delta t \cdot k \ W \qquad (7.24)$$

Factor k is called the *heat transfer factor* and is dimensioned in $W/(m^2 \cdot °C)$. It has a perfectly clear physical meaning, but very vague numerical value that depends on the material of the gear housing, on how it is painted if at all, on how it is cleaned from dust and oil, on the kind of surrounding fluid medium and its speed relative to the heat-exchange surface, and others. The rotational speed of the shafts affects the heat exchange very much: the higher the speed, the more intensive and uniform the heat transfer is from the oil to the housing walls and the greater the fanning effect of the rotating couplings. The practical recommendations are as follows:[11]

- For housings without fan cooling,

$$k = 10\text{--}25 \ W/(m^2 \cdot °C)$$

- The smaller values are recommended for painted housings of steel and gray iron, low rotational speed, and dusty air; the greater values, for aluminum nonpainted housings, rotational speed of 1500 r/min and more, and clear air.
- Bottom surfaces of the housing are less efficient in convection, but part of the heat energy is transferred in that area through metallic contact of the mount feet into the frame or

base the housing is attached to. Therefore, with regard to inaccuracy of all this calculation, the bottom surface can be calculated together with the other walls, if there is enough space for the air to pass under it.

- Ribs made to increase the heat-exchange surface are less effective than the wall because their temperature gradually decreases with distance from the wall. Therefore, it is recommended to take into account 50% of their surface (say, only one side of a rib).
- For blower cooling,

$$k = 7 + 12\sqrt{v}\ W/(m^2 \cdot {}^{\circ}C)$$

where v = air speed (m/sec).

- For example, if $v = 10$ m/sec, $k = 45$ $W/(m^2 \cdot {}^{\circ}C)$. As one can see, fanning is a very effective means to intensify the cooling process.

Further reduction of the oil temperature can be achieved by insertion of a water pipe into the oil reservoir. The value of factor k for the water pipe depends on the circumferential speed of the oil-dipped gear wheel and the speed of the water in the pipe. For copper alloy tubes, it ranges from about 150 $W/(m^2 \cdot {}^{\circ}C)$ for a gear speed of 4 m/sec and a water flow speed of 0.1 m/sec to about 200 $W/(m^2 \cdot {}^{\circ}C)$ for a gear speed of 12 m/sec and a water speed of 0.4 m/sec. Steel tubes provide a factor k smaller by 5–10%.

As shown, the participants in the heat extraction process may have different capabilities. For each of them the P_{out} amount should be calculated separately using Equation 7.24 and then summarized:

$$P_{out} = \Sigma P_{out,i} = \Sigma (A_i \cdot \Delta t_i \cdot k_i) \tag{7.25}$$

Before the gear is set in motion, its temperature equals mostly that of the surroundings: $t_g = t_s$ (unless it had been previously warmed up or chilled). When the gear begins to work, it immediately begins receiving the P_{in} amount of heat energy and starts warming up. As the temperature increases, the P_{out} value rises in direct proportion to the temperature difference, but the rate of temperature increase gradually approaches zero (Figure 7.32a) as P_{out} approaches P_{in}. From this moment on, the gear doesn't change its temperature, because the input and output amounts of heat have become equal:

$$P_{in} = P_{out}$$

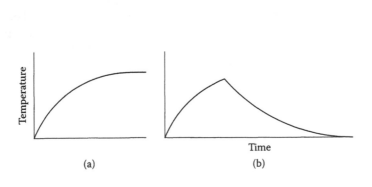

FIGURE 7.32 Process of heating and cooling.

Equating of Equation 7.23 and Equation 7.25 gives directly the steady-state temperature difference between the gear and the surroundings for the case when Δt is nearly the same for all the participants of the heat extraction process:

$$\Delta t = \frac{P(1-\eta)10^3}{\Sigma A_i \cdot k_i}$$

If Δt differs (for example, the air temperature is 30°C, and the water in the pipe is 15°C), calculation of Δt value can be done by iterations.

As soon as the Δt value has been determined, we can know the gear temperature: $t_g = t_s + \Delta t$. The gear temperature should not exceed the value permissible for the oil, which equals approximately 70°C (rarely 80–90°C). If the cooling calculation shows that the gear temperature is higher than admissible, appropriate measures must be taken to ensure an acceptable level of cooling. Besides those mentioned previously, an additional heat exchanger of needed capacity with oil circulation system can be used. For large gears that transmit hundreds and thousands of kilowatts, this is the only means to keep the oil temperature within admissible limits.

In fact, t_g is the temperature of the outer surface of the housing. The temperature of the oil is a bit higher. The gear wheel dipped into the oil has nearly the same temperature, but the pinion is usually distinctly warmer. What is mainly of interest to us is the oil temperature, because the oil, if overheated, is not able to form a protective layer on the metal surface (see Chapter 6, Subsection 6.3.1). It should be kept in mind that the oil undergoes an additional heating in the gear mesh, so the recommended temperature of the oil should not be exceeded.

The preceding formulas are valid for constant duty. When the load is changing cyclically, the P_{in} value in Equation 7.23 can be determined as follows:

$$P_{in} = \frac{\Sigma P_{in,i} t_i}{t_c},$$

where

$P_{in,i}$ = power loss in the cycle while transmitting power P_i
t_i = time of transmission of power P_i within one cycle
t_c = cycle time

The less the load, the less is the efficiency of the gear. The difference can be significant, and it should be taken into consideration when calculating the $P_{in,i}$ values using Equation 7.23.

For intermittent duty, when the idle time is long enough to get the gear cooled off, the temperature curves at warming up and chilling as shown in Figure 7.32a and Figure 7.32b. For this process, the calculation should take into account that part of the energy loss is spent to warm up the elements of the gear and the rest is dissipated into the surroundings. The temperature of the gear at the end of the work cycle (when Δt is nearly the same for all the participants of the heat extraction process) can be obtained from the following formula;[6]

$$t_g = t_s + \frac{N(1-\eta)10^3}{\Sigma A_i k_i}\left(1 - \frac{1}{e^\alpha}\right) \tag{7.26}$$

where

$$\alpha = \frac{t_c \Sigma A_i k_i}{m_m c_m + m_o c_o}$$

m_m = mass of metallic parts of the gear (kg)

m_o = mass of oil (kg)

$c_m \approx 0.13\ W \cdot h/(kg \cdot °C)$, specific heat of metallic parts

$c_o \approx 0.56\ W \cdot h/(kg \cdot °C)$, specific heat of oil

EXAMPLE 7.8

The reduction gear shown in Figure 7.25 transmits a continuous load of 30 kW. The efficiency of the gear $\eta = 0.96$. The gear is indoors, protected from solar heating, and the temperature of the air in summer $t_s = 30°C$. What can be the oil temperature without any additional cooling means?

From Equation 7.23, the input heat energy is

$$P_{in} = 30(1 - 0.96)10^3 = 1200\ W$$

The outer area of the housing A = 0.8 m². Taking $k = 18\ W/(m^2 \cdot °C)$, we determine the output energy:

$$P_{out} = 0.8 \cdot 18 \cdot \Delta t = 14.4 \cdot \Delta t\ W$$

Equating P_{in} and P_{out}, we obtain the temperature difference between the housing and the air:

$$\Delta t = \frac{1200}{14.4} = 83.3°C$$

Because the air temperature $t_s = 30°C$, the housing temperature $t_g = t_s + \Delta t = 113.3°C$. Such a high temperature is unacceptable, and the gear must be provided with some additional cooling devices. Let's try to apply blower cooling to one side and the top walls of the gear. At an air speed of 10 m/sec, $k = 45\ W/m^2 \cdot °C$. The fanned area is 0.3 m², and the rest is 0.5 m². Therefore, the output energy is given by

$$N_{out} = (0.5 \cdot 18 + 0.3 \cdot 45)\Delta t = 22.5 \cdot \Delta t\ W$$

$$\Delta t = \frac{1200}{22.5} = 53.3°C$$

$$t_g = 30 + 53.3 = 83.3°C$$

We can see that blowing is helpful, but the temperature is still too high, we don't want it to exceed 70°C. Let's cancel the blowing and build a water pipe into the oil reservoir. A pipe of 30 mm in diameter bent in U form measures about 1 m in length. The area of its outer surface equals 0.094 m². The average value of $k = 170\ W/m^2 \cdot °C$. The water temperature is 20°C. So, for the water pipe the temperature difference $t_w = 70 - 20 = 50°C$, while for the outer surface of the housing that contacts with the air $t_a = 70 - 30 = 40°C$. Now let's calculate the output energy at the desired 70°C of the oil:

$$N_{out} = 0.094 \cdot 170 \cdot 50 + 0.8 \cdot 18 \cdot 40 = 1375\ W$$

That amount is greater than the input energy ($N_{in} = 1200\ W$), so the real temperature of the oil is believed to be less than 70°C. We can see that the water pipe is very effective.

Let's calculate now how much time that gear with no additional cooling devices can transmit power of 30 kW until the oil temperature reaches 70°C. The weight of the gear is about 100 kg, and the weight of oil is 4.5 kg. For this purpose we can use Equation 7.26:

$$t_g = 30 + \frac{30(1-0.96)10^3}{0.8 \cdot 18}\left(1 - \frac{1}{e^\alpha}\right) = 70\,°C;$$

$$\left(1 - \frac{1}{e^\alpha}\right) = 0.48; \quad e^\alpha = 1.923; \quad \alpha = 0.6539;$$

$$\alpha = \frac{t_c \cdot 0.8 \cdot 18}{4.5 \cdot 0.56 + 100 \cdot 0.13} = 0.6539;$$

$$t_c = \frac{0.6539 \cdot (4.5 \cdot 0.56 + 100 \cdot 0.13)}{0.8 \cdot 18} = 0.705\,h$$

That means that the gear may work 42 min before it warms up to 70°C. Do you believe in that? We don't! Well, we will check that result using the simplified approach and different units of energy. As we know from school, 1 W · h = 0.86 kcal. From here the specific heat of the metal $c_m \approx 0.11$ kcal/(kg · °C), and that of the oil $c_o \approx 0.48$ kcal/(kg · °C). The amount of heat needed to warm up 4.5 kg of oil and 100 kg of metal by 40°C (from 30°C to 70°C) is given by

$$Q_{needed} = (0.48 \cdot 4.5 + 0.11 \cdot 100)40 = 526\,kcal$$

The heat energy supplied for this purpose Q_{sup} is given by

$$Q_{sup} = Q_{in} - Q_{out}$$

The former calculation using Equation 7.26 had shown that the Q_{needed} is supplied during 0.705 h. Let's check this result. The input energy during this time we know exactly (provided that the efficiency of the gear we know for certain):

$$Q_{in} = 860 \cdot N(1 - \eta)0.705 = 860 \cdot 30(1 - 0.96)0.705 = 728\,kcal$$

We can't calculate the output energy exactly because the temperature is changing during this hour. The temperature difference Δt changes from zero to 40°C. Let's calculate approximately the Q_{out} assuming that the temperature difference doesn't change and equals its average value: $\Delta t = 20$°C. From Equation 7.24 (with changed units) and taking into account that factor $k = 18$ $W/(m^2 \cdot °C) = 15.5$ kcal/(m² · h · °C), the output energy during this time is

$$Q_{out} = 0.8 \cdot 15.5 \cdot 20 \cdot 0.705 = 175\,kcal$$

From here, the supplied heat energy is given by

$$Q_{sup} = 728 - 175 = 553\,kcal$$

This result differs only 5% from 526 kcal, so the calculation using Equation 7.26 was correct.

REFERENCES

1. Klebanov, B.M., *Contact Strength of Spur Gears*, in collection of articles entitled, *Toothed and Worm Gears*, Kolchin, N.I., Ed., Mashinostroenie, Leningrad, 1974 (in Russian).
2. *Reduction Gears of Power Plants Handbook*, Derzhavets, Y.A., Ed., Mashinostroenie, Leningrad, 1985 (in Russian).
3. Niemann, G., *Maschinenelemente*, Bd. II., Springer-Verlag, Berlin, Heidelberg, New York, 1965 (in German).
4. Kudriavtsev, V.N., Kuzmin, I.S., and Philipenkov, A.L., Calculation and design of gears, *Politechnika*, Sanct-Petersburg, 1993 (in Russian).
5. *Planetary Gears Handbook*, Kudriavtsev, V.N. and Kirdiashev, V.N., Eds., Mashinostroenie, Leningrad, 1977 (in Russian).
6. Kudriavtsev, V.N., *Machine Elements*, Mashinostroenie, Leningrad, 1980 (in Russian).
7. Kudriavtsev, V.N., *Simplified Calculations of Gears*, Mashinostrienie, Leningrad, 1967 (in Russian).
8. Klebanov, B.M., About the permissible unevenness of load distribution across the face width of a gear, *Transactions of the Leningrad Mechanical Institute*, No. 61, Leningrad, 1967 (in Russian).
9. Mazirin, I.V., Lubrication devices of machines, *Mashgiz*, Moscow, 1963 (in Russian).
10. Dudley, D.W., *Gear Handbook*, McGraw-Hill, New York, Toronto, London, Sydney, 1962.
11. Levitan, Y.V., Obmornov, V.P., and Vasiliev, V.I., *Worm Gears Handbook*, Mashinostroenie, Leningrad, 1985 (in Russian).

8 Gear Design

All gears have one common element: their teeth. In all other respects, the shapes and structures of gears may differ very much. Figure 8.1 shows several typical shapes of gears: pinion shafts (Figure 8.1a and Figure 8.1b), gears mounted on shafts (Figure 8.1c to Figure 8.1f), sun wheels of planetary gears (Figure 8.1g and Figure 8.1h), planet wheels (Figure 8.1i and Figure 8.1j), and ring gears (Figure 8.1k and Figure 8.1l). In this chapter, we will mainly discuss simple gearings.

8.1 GEAR AND SHAFT: INTEGRATE OR SEPARATE?

When the gear must be rigidly connected to the shaft, the designer is often unsure about what is preferable: to make the gear as an integral part of the shaft or to make it apart and connect it to the shaft by some means. Sometimes, the split design is impracticable because the diameters of the shaft and the pinion are very similar (Figure 8.2a).

The diameter of the pinion can even be less than that of the shaft (Figure 8.2b). It may be associated with the need to increase the bending rigidity of the shaft or with unification of its bearings (when gears with different ratios are produced in the same housing).

Sometimes the split design is dictated by assembly needs or by manufacturing limitations (for example, when the shaft is too long to be installed on an available gear-cutting machine). Perhaps, the gear teeth must be case-hardened but this can't be done along with the shaft on the existing equipment or it requires some additional equipment to decrease the warpage of the shaft. When there are no such constraining conditions, economic considerations prevail in making design decisions.

Yet the split design may result in increased weight, lesser accuracy because of tolerance summation (also called *stack-up*) and decreased reliability. Figure 8.3 shows some design options for forged gears: those made together with the shaft (Figure 8.3a) and separate from the shaft (Figure 8.3b and Figure 8.3c). Figure 8.4 shows the same for welded gears. Not only is the dismountable connection between the gear and the shaft expensive, but it may also cause problems in service. Wear, fretting, or both may occur on the splines and centering surfaces, and the nuts may lose tension or even break completely. This, in turn, may lead to damage or breakage of the gear teeth.

Shaft areas damaged by fretting, as well as stress concentration areas near the splines and keyways and in the interference fits, provide the best conditions for origination of fatigue cracks that may lead to shaft failure. This caution doesn't mean that connections of all kinds should be completely avoided. But, when making a decision about gear-and-shaft design, reliability should carry weight along with other factors.

One might say that modern methods of calculation enable the design of completely reliable connections, so that a factor of reliability (or *safety factor*) need not be used. The authors are not ready to agree with this opinion. Although our knowledge in the field of machine design has increased, troubles during development of machines, as well as in service, have not declined. The point is that the intensity of machines grows relentlessly, and the demand for higher power density (reduced weight and increased capacity) grows hourly. New steels that have doubled and tripled tensile strength are extensively used, but the elasticity modulus of these steels remains the same. Therefore, the higher stresses are accompanied by greater elastic deformations, which, in turn, lead to increased micromovements in connections and hence to damage of the mating surfaces

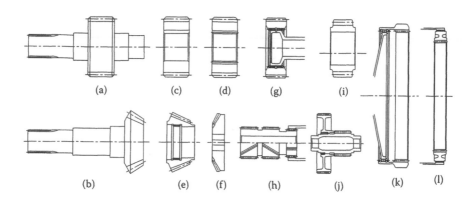

FIGURE 8.1 Common gear types.

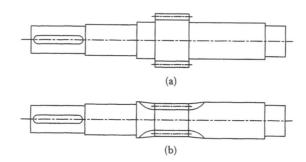

FIGURE 8.2 Design of pinion shafts.

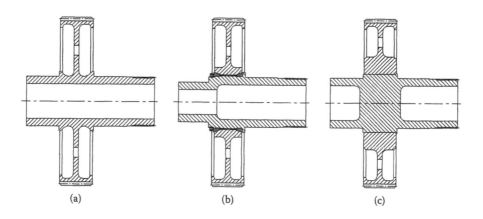

FIGURE 8.3 Connection of a gear to shaft.

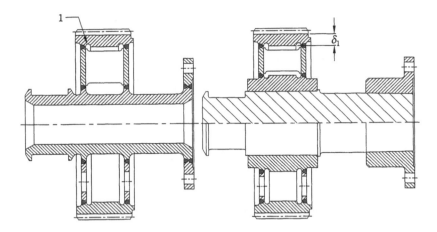

FIGURE 8.4 Gears welded to a shaft or hub.

(see Chapter 2). In addition, high-strength materials are more sensitive to stress raisers, especially to fretting. The complexity of the processes occurring in the connections doesn't enable us to predict for certain the service life of heavily loaded joints. As a result, we can't be confident in the reliability of a machine until it is duly tested.

8.2 SPUR AND HELICAL GEARS

The typical disk-type gear (Figure 8.5a) consists of toothed rim 1 and hub 2 linked to each other by disk 3. Dimensions of these elements are chosen to impart strength and stiffness to the gear. The most important in this regard are thicknesses of the rim (δ_1) and the disk (δ_2). The thinner the rim, the greater is the bending deformation of the teeth, because bending of the rim is added to bending and shear of the tooth itself (see Figure 8.6a). Because the bending compliance of the rim is not uniform across the face width (it is more rigid in the area of the disk), the tooth compliance is also not uniform. This influences the load distribution along the face width: the more rigid part of the tooth is more loaded (Figure 8.7a). The thicker the rim, the more uniform (less uneven) is the tooth rigidity, but the weight of the gear increases considerably. In usual practice, the rim thickness δ_1 is approximately equal to the height of the tooth (about $2m$).

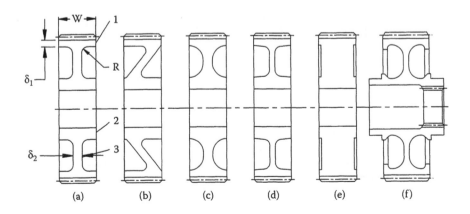

FIGURE 8.5 Shapes of gears.

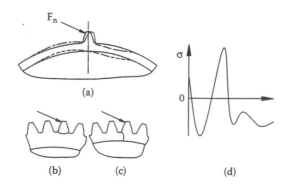

FIGURE 8.6 Bending deformation and failure of toothed rim.

Bending of the rim not only influences the teeth elastic compliance, but also may be the place of failure. If the rim is thick enough, the fatigue crack originated in the root of the tooth propagates as shown in Figure 8.6b. If the rim is thinner than $1.6m$, the crack may propagate into the rim (Figure 8.6c) and result in a very grave failure. It is more serious than just tooth breakage because it leads to immediate failure of the mechanism, and the repair involved may also be more extensive. Therefore, the rim should be made strong enough. When the weight of the gear is critical, the rim can be provided with ring ribs as shown in Figure 8.5f and Figure 8.7d.

Because of the increased rigidity of the teeth in the disk area, the maximal unit load of the tooth, as shown in Figure 8.7a, may exceed the average value by about 30% at $\delta_1 = 2m$. More uniform load distribution is achieved when the pliable parts of a tooth contact with the rigid parts of the mating tooth (Figure 8.7b and Figure 8.7c). The aforementioned ring ribs (Figure 8.7d) serve the same purpose.

Here is one example from practice. In the course of development of a propulsion gear, there was frequent tooth breakage of a gear designed as shown in Figure 8.7a. After the design was changed to that shown in Figure 8.7c, these failures ceased.

The disk thickness δ_2 for spur gears is dictated by its strength. At a static or momentary load, the weak place is the inner diameter of the disk. When the gear rotates, the stress in this place remains almost unchanged. But on the outer diameter, the disk is exposed (along with the rim) to alternating bending stress (Figure 8.6d), and a fatigue failure may completely separate the rim from the disk.

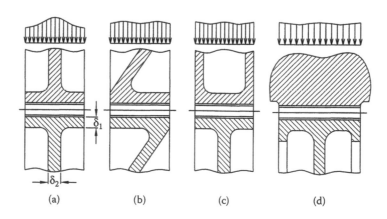

FIGURE 8.7 Load distribution across the gear face.

The thickness of the disk for spur gears usually equals 0.1–0.2 of the face width W, depending on the load level, thickness of the rim, and the fillet radius R (Figure 8.5a). Real proportions of a highly loaded spur gear are shown in Figure 8.5f. Large fillet radii and ring ribs on the rim enable a thin pliable disk, which, in turn, makes possible some self-alignment of the rim relative to the mating gear and improvement of the load distribution over the face width.

In helical gears, the thickness of the disk is limited by bending deformation under the action of axial force F_a (Figure 1.5f in Chapter 1 and Figure 8.8). This deformation results in misalignment of the teeth and uneven load distribution.

One engineer tried to increase the service life of a tractor final drive by replacing a spur gear by helical gear. The comparative strength calculation of the gears showed that the helical gear was considerably stronger. (Certainly, he didn't analyze the elastic deformations of the gears.) For the trials, he designed a set of helical gears using the same overall dimensions as the spur gears. The idea was naive: to keep all the mounting dimensions and to use the existing forged blanks made for the spur gears. It took a lot of time and money to manufacture several sets of experimental gears, but the trials were very quick and discouraging: the teeth broke within several working hours. Subsequent research disclosed that these failures were caused by bending deformation of the gear as shown in Figure 8.8. The disk was designed thin for the spur gear and had not been changed for the helical gear.

Several steel models of disk-type gear with proportions according to Figure 8.8, but different in scale, were tested for rigidity. Axial force F_a was applied to the rim, and angle φ was measured. According to this research,[1] angle φ is given by

$$\varphi = 2.09 \cdot 10^{-3} \frac{F_a}{D_0^2} \left(\frac{\delta_2}{W} \right)^{-1.5}$$

EXAMPLE 8.1

Helical gear, $m = 5$ mm, $N_1 = 20$ (pinion), $N_2 = 63$ (gear), face width $W = 60$ mm, and pitch helix angle $\psi_p = 15°10'37''$, transmits torque $T = 3.8 \cdot 10^6$ N·mm. The gear is disk-type, and the thickness of the flat disk $\delta_2 = 8$ mm. Is this gear rigid enough?

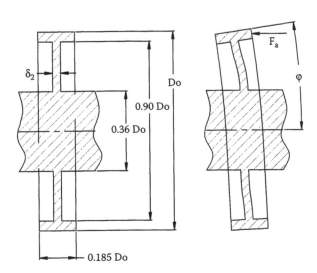

FIGURE 8.8 Deformation of disk-type gear.

Pitch diameter of the gear

$$d_P = \frac{mN}{\cos\psi} = \frac{5 \cdot 63}{\cos 15°10'37''} = 326.4 \; mm$$

Approximate outer diameter of the rim, taking into account Equation 8.3:

$$D_0 \approx d_P - 2 \cdot 1.25m + 2m(0.22 + 0.009\psi) = 326.4 - 2.5 \cdot 5 + 2 \cdot 5(0.22 + 0.009 \cdot 15.177) = 317.4 \; mm$$

The circumferential force

$$F = \frac{2T}{d_P} = \frac{2 \cdot 3.8 \cdot 10^6}{326.4} = 23284 \; N$$

The axial force

$$F_a = F \tan\psi = 23284 \tan 15°10'37'' = 6316 \; N$$

The angle of inclination of the toothed rim

$$\varphi = 2.09 \cdot 10^{-3} \frac{6316}{317.4^2} \left(\frac{8}{60}\right)^{-1.5} = 2.69 \cdot 10^{-3} \; rad$$

Now we can calculate the factor K_W (unevenness of load distribution across the face width) using Equation 7.17:

$$K_W = \sqrt{\frac{2\gamma W C}{q_a}}$$

Here, γ = angle of misalignment in the plane of action:

$$\gamma = \varphi \sin\alpha_s$$

where
α_s = pressure angle determined from the formula

$$\tan\alpha_s = \frac{\tan 20°}{\cos\psi}$$

C = rigidity of the teeth defined by a formula given in Example 7.6:

$$\frac{1}{C} = 10^{-3}\left(0.0514 + \frac{0.1425}{N_{1,e}} + \frac{0.1860}{N_{2,e}}\right)$$

where

$$N_{1,e} = \frac{N_1}{\cos^3 \psi} = \frac{20}{\cos^3 15°10'37''} = 22.25; \quad N_{2.e} = \frac{63}{\cos^3 15°10'37''} = 70.08$$

q_a = average unit load, which is

$$q_a = \frac{F}{W} = \frac{23284}{60} = 388 \; N/mm$$

From these relations, we have the following:

$$\tan \alpha_S = \frac{\tan 20°}{\cos 15°10'37''} = 0.377123; \quad \alpha_S = 20°39'45''$$

$$\gamma = 2.69 \cdot 10^{-3} \cdot \sin 20°39'45'' = 0.95 \cdot 10^{-3} \, rad$$

$$\frac{1}{C} = 10^{-3} \left(0.0514 + \frac{0.1425}{22.25} + \frac{0.1860}{70.08} \right) = 6.05 \cdot 10^{-5}; \quad C = 16530 \; N/mm^2$$

$$K_W = \sqrt{\frac{2 \cdot 0.95 \cdot 10^{-3} \cdot 60 \cdot 16530}{388}} = 2.2$$

As we can see, the thickness of the disk should be increased.

Notes: 1. If $K_W < 2$, it should be calculated using Equation 7.18.

2. This calculation is true if the helical gear proportions are as shown in Figure 8.8. In other cases, the deformation of the gear should be analyzed by numerical means.

As the disk thickness decreases, the weight of the gear also decreases, but the permissible load decreases as well. The minimal weight per unit of transferred torque may be achieved when the disk thickness is chosen from Figure 8.9a (for gray cast iron housings) or Figure 8.9b (for light housings of aluminum or magnesium alloy). These calculations were performed for some average gear proportions, and their ability to run-in was taken into account. In important cases, special

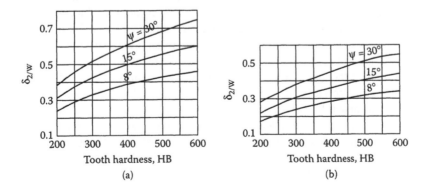

FIGURE 8.9 Recommended disk thickness of gears.

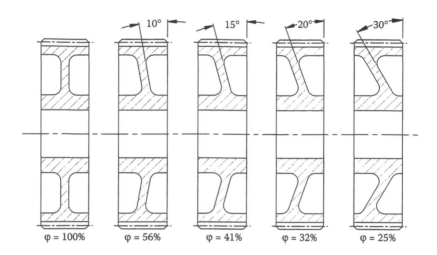

FIGURE 8.10 Rigidity of gears with conical gears.

analyses of the design should be done. Sometimes gear deformation is useful because it may compensate for deformations of the shafts and other elements (subject to direction of deformations).

Considerable increase in stiffness of the disk may be achieved by making it conical. In Figure 8.10 are given results of experiments[1] with steel models having proportions shown in Figure 8.8. (Conical disks are also very useful and widely used in wheels and pulleys in a large variety of applications.)

Holes in the disk should be made minimal in size and number, because large holes cause uneven rigidity around the circumference, as well as less rigidity overall. This unevenness also provides parametric excitation of bending and axial vibrations and may lead at high speed to resonance oscillations. In high-speed gears, the possibility of tooth frequency excitation should also be considered.

When the weight of the gear is not restricted (for instance, in stationary equipment), the design of the gears is defined by manufacturing and economic considerations. Small and middle-sized gears are forged to reduce machining (Figure 8.5c and Figure 8.5d). Gears with case-hardened teeth are usually heat-treated as massive disks (Figure 8.5e) to diminish their warpage. The steel after heat treatment is rather hard and intractable, and, if the weight is not strongly restricted, only the exact surfaces of the heat-treated gear (the joint and location surfaces and working surfaces of the teeth) are machined. If the weight must be kept minimal, the excess material is removed by machining (Figure 8.5f).

The gear shown in Figure 8.5f looks beautiful and is an excellent example of the well-known phrase, "form follows function." But it is very costly, and were it not for very practical considerations, it would not be shaped that way. After all, the gear works inside the housing, and nobody can enjoy its beautiful shape. This gear is part of a transport machine, and much effort and production costs were expended to make it both lightweight and reliable. Its shape, full of graceful curves, symmetry, and pleasing proportion, is not due to an artist's arbitrary whim, but rather is the result of careful design, stress calculation, and development.

The central part of the gear, called the hub, serves to connect it with the shaft, and its shape depends on the type of connection. A splined hub may be rather thin-walled; its outer diameter may be set equal to about 1.13–1.25 of the outer diameter of the spline. For an interference-fit connection, the hub outer diameter usually equals 1.5–1.6 of the shaft diameter (see Figure 8.3c).

At times, the hub includes bearing seats (Figure 8.5f). At other times, a hub is not used, and the disk of the gear is directly attached to the shaft by bolts (Figure 8.11). The commonly used proportions in this design are as follows:

$$d_b \approx 6 + 0.02\,D_b$$
$$h = (1.2 - 1.5)\,d_b$$

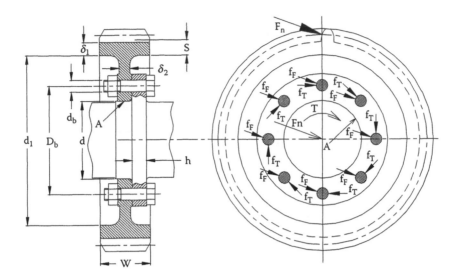

FIGURE 8.11 Flange attachment of a gear.

The load on the bolts appreciably depends on the fit of the centering surface A. The fit should be tight.

Figure 8.11 shows the load distribution between the bolts. The vector of the tooth force F_n is transferred to the center of the gear with the addition of torque T. If the centering surface is fitted loosely, the radial force is distributed between the bolts, so that each bolt carries some part, f_F of the entire force. (In this drawing the load distribution is shown as if it were equal. This assumption is not exact (see later), but for our consideration it is acceptable.) In addition, torque T loads each bolt with a force f_T. We can see that the sum of these forces is different for each bolt; vectors f_F and f_T may act in the same direction, oppositely, or at any angle within 360°. As the gear rotates, each bolt takes any place around the circle, and its load changes between the sum of vectors f_F and f_T (maximum) and the difference between them (minimum). The maximum and the minimum are directed tangentially, but the intermediate vectors are directed at different angles. If the magnitude of vector f_F is greater than that of vector f_T, then the resultant vector makes a full turn around the bolt during one full turn of the gear. But if the centering surface A is tightly fitted, it takes the radial force F_n, and the bolts transmit the torque-related forces f_T only. These forces are tangentially directed, and they reverse their direction only when the torque is reversed, meaning very seldom as compared with the number of turns. In this case, the working conditions of the bolts are much better (see, for details, Chapter 10, Subsection 10.3.2).

Because the friction in the interference fit impedes pulling the flanges together, the interference should be minimal. Grease containing molybdenum disulfide may be applied to surface A to decrease the friction force in that connection.

The load distribution between the bolts can be determined by numerical methods only. The maximal load applied to a bolt (in the area of tooth engagement)

$$F_{max} = K F_{av}$$ (8.1)

and

$$F_{av} = \frac{2T}{D_b n}$$ (8.2)

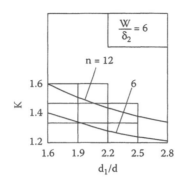
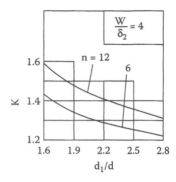

FIGURE 8.12 Factor K in Equation 8.1.

where

F_{av} = average load (N)

T = torque (N·mm)

D_b = diameter of the bolt circle (mm)

n = number of bolts

K = factor that has been calculated (see Figure 8.12)[2] for the following proportions of gear:
$D_b/d = 1.3$, $W/d = 0.6$, $S/W = 0.25$, and $\delta_2 = h$

The minimal load of a bolt

$$F_{min} \approx \frac{F_{av}}{K}$$

The thickness of the rim, S, in this case takes into consideration the influence of the teeth on the rigidity of the rim:

$$S = \delta_1 + m(0.22 + 0.009\psi) \tag{8.3}$$

where

ψ = helix angle, in degrees

m = module of the gearing

Equation 8.3 is exact enough when the teeth number 50 to 100. For more detailed recommendations, see the literature cited.[7]

8.3 BUILT-UP GEAR WHEELS

At diameters larger than 1500 mm (sometimes beginning at 800 mm), it is economically feasible to make only the rim from high-strength steel (forged as a ring) and to join it (after rough machining) to a welded or cast center made of cheaper materials. The connection between the rim and the center may be achieved by mechanical means (bolts, rivets, or interference fit) or by welding.

Examples of welded gears are shown in Figure 8.4. The thickness of the rim for such gears δ_1 is usually set equal to $(5-8)m$. Inside the rim, circular grooves are machined to form ribs 1 for welding of steel plate disks. The thickness of each disk is approximately equal to 0.1 of the gear face width. In this design, butt welds are used, because they can be checked through by x-ray photography or by ultrasonic inspection. In less important cases, in which some increased risk may

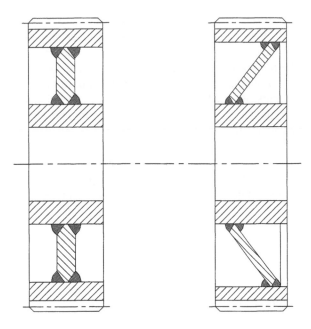

FIGURE 8.13 Welded gears.

be acceptable, cheaper design options shown in Figure 8.13 can be used. In this case, the welding can be inspected only visually. But if the welding process is well established, intermittent monitoring may guarantee the needed quality standard.

The steel grade for the rim depends on the final heat treatment: is it case hardening, nitriding, or just quenching and tempering? But in any case such steels are poorly weldable, because they are as a rule highly alloyed and often contain up to 0.4% carbon. This, in combination with the large volume of welding and necessity of subsequent cleanup, inspection, and repair of faults, makes welding very expensive. Less laborious are the mechanical connections. Figure 8.14a shows a rim of a spur gear bolted to a cast center. The centering surfaces should be tightly fitted with a minimal interference for reasons explained earlier.

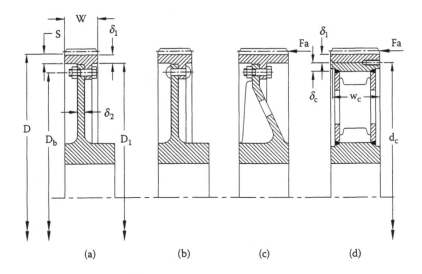

FIGURE 8.14 Connections between the rim and the center of a gear.

The rim is assembled when heated to about 100°C and bolted to the center through preliminary holes. After the parts cool, the possible gap between the flanges should be checked with a thin feeler gauge. If there is a gap, the parts should be pressed together, preferably using a manual or hydraulic press. When the gap is taken up, the holes can be finished, and the bolts fitted and torqued.

Concerning the size and number of bolts that attach the rim to the center, the designer is free to choose within certain limits. Bigger bolts need thicker and wider flanges (i.e., the weight of the flanges increases), and the tightening of such bolts requires a more powerful tool. But smaller bolts, perhaps, are not strong enough to transmit the load. Comparative calculations may help the designer to choose the right size.

The load distribution between the bolts cycles with every revolution. At each moment, the most highly loaded bolts are those that are close to the tooth engagement zone. The maximal shear load applied to a bolt (provided that the centering surfaces are tightly fitted) is approximately given by[3]

$$F_{max} = F_{av}\, n^a \tag{8.4}$$

where

F_{av} and n = symbols as in Equation 8.1 and Equation 8.2
a = exponent as in Figure 8.15

The minimal load is applied to a bolt that at this moment is diametrically opposite to the engagement zone and is approximately

$$F_{min} = \frac{F_{av}}{n^a} \tag{8.5}$$

EXAMPLE 8.2

The bolted spur gear shown in Figure 8.14a has the following dimensions: $D_1 = 794$ mm, $D_b = 740$ mm, $m = 8$ mm, $\delta_1 = 25$ mm, $\delta_2 = 22$ mm, $W = 100$ mm, and number of bolts $n = 24$.

Calculated dimensions are $S = \delta_1 + 0.22m = 25 + 0.22 \cdot 8 = 26.8$ mm (see Equation 8.3); $D = D_1 + 2S = 794 + 2 \cdot 26.8 = 847.6$ mm; $h = 0.5(D_1 - D_b) = 0.5(794 - 740) = 27$ mm; $h/S = 27/26.8 \approx 1$; $S/D = 26.8/847.6 = 0.032$; and $W/\delta_2 = 100/22 = 4.55$.

From Figure 8.15 for $S/D = 0.03$, $h/S = 1$, and $W/\delta_2 = 4.55$, we can find that $a = 0.41$. The maximal and the minimal forces equal

$$F_{max} = F_{av} \cdot 24^{0.41} = 3.68\,F_{av}; \quad F_{min} = \frac{F_{av}}{24^{0.41}} = 0.27\,F_{av}$$

Rivets may be used instead of bolts (Figure 8.14b). This connection is cheaper than the bolted joint and quite reliable when properly designed and built.

When the rivet gets cold (after hot riveting), it shortens and presses the flanges against each other with a large force. But for the same reason, the diameter of the rivet decreases and a small radial gap arises between the shank of the rivet and the hole in the flanges. Thus, the strength of the connection is provided by friction forces in the contact region of the compressed flanges.

Tightly fitted bolts also compress the flanges, and with an even larger force, because the bolt material can be much stronger than that of the rivet. In addition, the fitted bolt works as a pin, so the attainable strength of the bolt connection is much higher than a rivet connection. Besides, the quality of the bolt connection can be surely guaranteed and checked, whereas the compressive force of the rivet can't be measured. After all, the choice between riveted and bolted connections depends on the needed level of strength and reliability.

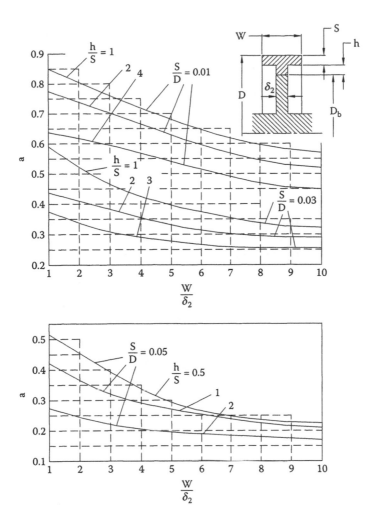

FIGURE 8.15 Exponent a in Equation 8.4.

Helical gears must have more rigid centers (Figure 8.14c) to decrease bending deformation under axial force F_a. It is preferable to place the flanges so that the axial force presses them against each other, at least under the prevailing load (for example, with the forward motion of a ship). It is also preferable to make the stiffening ribs so that they are exposed to compressive stress under the prevailing load.

Figure 8.16 explains this preference. The neutral line 0–0 is placed near the wall of the disk, and under bending the rib is much more stressed than the wall. Therefore, it is highly desirable to make the larger stress compressive.

For high-speed gears, ribs are unacceptable, because they cause unnecessary aerodynamic and fluid turbulence that result in power losses. The needed rigidity can be achieved using a thicker disk or two thin-walled disks, perhaps with ribs between them.

In cases of bolted or riveted connections, the heat treatment of the rim (and also of the center) must be performed before assembly. This is a serious drawback of these options, because the rigidity of the separate rim is insufficient. If the teeth should be case-hardened, large deformations during carbonization and quenching are very difficult to avoid, and after final grinding, the thickness of the hard layer may be very uneven and too thin here and there. Nitriding induces relatively small warpage, but in built-up designs, final grinding of the teeth may be mandatory. Because the nitrided

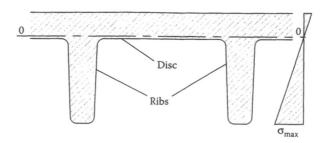

FIGURE 8.16 Stresses in a ribbed wall.

layer is relatively thin (0.4–0.5 mm), the same problem of uneven thickness may take place here as well. The most suitable type of surface hardening for the teeth of built-up gears is induction hardening: first, it can be performed after assembly, when the rim is attached to a rigid center, and second, the induction-hardened layer is thick enough, so that the possible unevenness of its depth after grinding is relatively small and doesn't impair the strength of the teeth.

Gears with a shrink-fitted rim (Figure 8.14d) have the same drawbacks. The thickness of the rim δ_1 and the thickness of the center barrel δ_c can be tentatively determined as follows:[4]

$$\delta_1 = \frac{C}{10\sqrt{(N_1+N_2)}}\left[5+\frac{W}{C}+0.001(N_1+N_2)\right] mm$$

$$\delta_c = \left(1.05+0.05\frac{W}{C}\right)\delta_1$$

(8.6)

where
   C = center distance (mm)
   W = face width (mm)
   N_1 and N_2 = number of pinion teeth and gear teeth, respectively

With this calculation, the design of the gear only begins. The point is that the friction force in the interference-fit connection (IFC) between the rim and the center may be insufficient, and the connection may become completely spinning.

The spinning torque T_S equals the torque of the friction forces:

$$T_S = 0.5\pi d_c^2 W p f_c \quad N\cdot mm$$

(8.7)

where
   d_c = diameter of the connection (mm)
   W = width of connection (mm) (In fact, it can be a bit less than the face width of the gear, but the difference usually can be neglected.)
   p = surface pressure in the connection (MPa)
   f_c = coefficient of friction in the connection

In Equation 8.7 everything is clear: the dimensions are known; surface pressure p for round parts can be calculated using the Lamé equation (see Chapter 5, Subsection 5.2.1) and, for elaborately shaped parts, using numerical methods. But the friction coefficient f in this connection is of complex nature and deserves detailed discussion. Experiments show that under static load, $f \approx 0.28$.

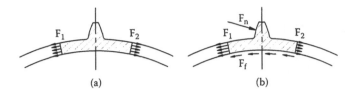

FIGURE 8.17 Local slippage of bandage.

That means, if we substitute $f = 0.28$ in Equation 8.7, we attain the breakaway torque. It follows that at some smaller load, when the entire rim is not spinning yet, there is some local micro-slip in the vicinity of the loaded tooth.

Figure 8.17 shows the very part of the shrink-fitted rim (hatched) where the local slip may occur. Before the tooth is loaded (Figure 8.17a), the tension forces F_1 and F_2 are equal, and there are no friction forces in this connection. (Please! We are aware of the pressure forces on the contacting surfaces, but we have intentionally omitted them.) When force F_n is applied to the tooth, the hatched part moves relative to the center and the neighboring parts of the rim in the direction of the force. At that F_1 becomes greater than F_2, and friction forces F_f arise at the same time.

We said "moves." This word contains the essence of the process: where there is no motion, there is no change in the stress condition. In this case, application of load to the tooth changes the stress state of the rim, and the movement of the loaded part of the rim in the direction of the load is inevitable. Another point: the movement can be elastic. Slippage between the rim and the center is not obligatory for the friction forces F_f to arise. Let's consider a simple example: A solid body is placed on a horizontal table. Until some horizontal force is applied to the body, no friction forces exist between the body and the table. To move the body, a horizontal force should be applied to it. If this force is less than the product of the weight of the body and the coefficient of friction, the body doesn't visually move. But the friction force between the body and the table exists; otherwise, why would the body not move! And if the friction force exists, there must be some kind of displacement between the body and the table. This displacement may be very small, within the elastic deformations of microasperities or even within the area of molecular forces. In this case, after the external force is removed, the body will move completely back into its initial position. Experiments show that "elastic" coefficient of friction equals about one half of the coefficient of static friction.

Let's come back to the shrink-fitted toothed rim: Until the friction force lies within the boundaries of "elastic" friction, the details stay immovable relative to each other. But if the tooth force is so large that there is some zone of microslip near the loaded area, this zone travels around the circumference of the gear once each revolution (along with the tooth force). The displacement during one revolution can be very small, but with time, it accumulates. Hence, it is clear that gradual slippage in the connection of the rim with the center may occur in a working gear at a smaller force, T_S, than is needed to break the connection away at static load. Consequently, the dynamic coefficient of friction included in Equation 8.7 should be considerably smaller.

It is also evident that the thicker the rim, the longer the local part of the rim near the loaded tooth that really participates in transmitting load to the center, and, consequently, the larger the force that can be transmitted in the elastic range. Experiments[5] have shown that at rim thicknesses of $1m$, $2m$, $4m$, and $5m$, the dynamic coefficient of friction equals 0.005, 0.017, 0.024, and 0.028, respectively. This result correlates well with the data[6] that gears with a shrink-fitted rim make no troubles in service if their maximal working force can be transmitted at $f_C = 0.016–0.033$.

Here should be underlined the peculiarity of the idea of "coefficient of friction" as applied to this case. In Equation 8.7, it is considered that the f_C value pertains to the entire surface of the IFC. But in reality, the force applied to the tooth executes the micromoving action only on

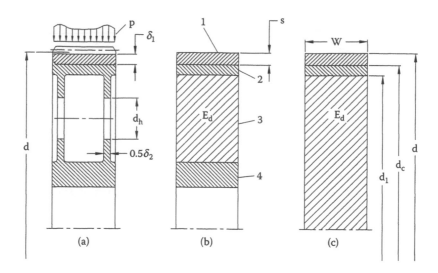

FIGURE 8.18 Surface pressure in the shrink-fit connection between the rim and the center.

a small part of the rim adjoined to the loaded tooth. More distant parts are less loaded and don't slip, although most of the connection possibly doesn't feel the load. Therefore, in the case of accumulation of microslippages (as the gear rotates), the dynamic value of the friction coefficient should be taken to calculate the strength of the rim-to-center connection in operation using Equation 8.7.

Contact pressure p is distributed nonuniformly across the width of the gear (Figure 8.18a): it is bigger where the disks prop up the barrel of the center. The real distribution of the pressure can be calculated by numerical methods, but in Equation 8.7 the average magnitude of the surface pressure should be inserted. It can be found easily by representing the design (as shown in Figure 8.18b) in the form of several cylindrical bodies: rim 1, barrel 2, disks 3, and hub 4. Because the disks are thinner than the other parts, the modulus of elasticity of this part should be decreased proportionally to the thickness of the disks:

$$E_d = E_C \frac{\delta_2}{W}$$

where
    E_C = modulus of elasticity of the center material (MPa)
    δ_2 = total thickness of the disks (mm)

Calculation of the surface pressure between parts 1 and 2 may be performed as follows:

- First, the unknown factors should be given names. For example, the surface pressure between parts 1 and 2 can be named p_{12}, between 2 and 3 — p_{23}, between 3 and 4 — p_{34}.
- Each part should then be taken separately, and the radial displacement of each interface surface to be calculated using the Lamé equation.
- Now we can write three equations of joint deformations: in connections 2–3 and 3–4, the radial displacements of both contacting surfaces are even; in connection 1–2, the difference between the displacement of the inner surface of the rim and the outer surface of the barrel equals the radial interference Δ.

As the teacher said, if we have three equations and three unknowns, we have all chances to make an error. To check your calculation, you can use the following simplified formula:

$$p = \frac{2E_1}{d_C \Omega} \Delta \ MPa \tag{8.8}$$

where
E_1 = modulus of elasticity of the rim material (MPa)
d_C = inner diameter of the rim (mm)
Δ = radial interference (difference between the inner radius of the rim and the outer radius of the barrel (mm)

$$\Omega = f_1 + \frac{E_1}{E_2} f_2$$

where
E_2 = modulus of elasticity of the center's material (MPa)
f_1 and f_2 = as given in Figure 8.19

As shown in Figure 8.18b, the thickness of the rim, S, should be taken with some addition, considering the influence of the teeth on its rigidity according to Equation 8.3.

Equation 8.8 and the charts in Figure 8.19 are calculated for a simplified model shown in Figure 8.18c. In this model, the hub is omitted, and the disks are prolonged. It doesn't usually influence considerably the calculated value of surface pressure between the rim and the disk but makes the calculation easier. When the difference between the inner and the outer radii of the disk is small, more precise calculation with the hub included shall be done.

The disks usually have round holes of moderate diameter d_h made for manufacturing reasons. These holes don't cause considerable unevenness of the pressure distribution around the circle, but they decrease the rigidity of the disks. Comparative calculations made using the finite element method (FEM) have shown that the influence of the holes on the surface pressure p can be approximately taken into account by reducing the thickness of the disks from δ_2 to δ_{2h} obtained as follows:

$$\delta_{2h} = \delta_2 \left(\frac{A_d - A_h}{A_d} \right)^2 mm$$

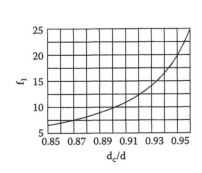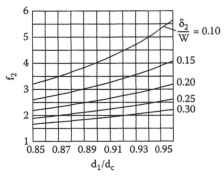

FIGURE 8.19 Functions f_1 and f_2.

where

  A_d = surface area of one disk (mm²)

  A_h = summary area of all holes in one disk (mm²)

The center of a gear made with spokes (instead of disks) has a substantial unevenness of surface pressure around the circumference. In the span between the spokes, the barrel of the center is not supported, and it bends; the connection between the rim and the center is weaker in these places. Therefore, it is preferable to design with disks.

The surface pressure q is the crucial factor in the reliability of the connection between the rim and the center. But the increase in pressure raises the tensile stress in the rim. Research[8] has revealed that at a tensile stress of 200 MPa, the bending strength of teeth without surface hardening is decreased by 17 to 23%. In operating gears, the tensile stress determined by the approximate formula

$$\sigma = \frac{p d_C}{2\delta_1} \tag{8.9}$$

usually doesn't exceed 150 MPa. If the strength of the connection requires greater surface pressure, the thickness of the rim should be increased. This results both in the increase of dynamic coefficient of friction f_C and the decrease of tensile stress.

Possible slippage is not the only problem one can face. Another problem is the separation of the rim from the center as shown in Figure 8.20. This defect is extremely harmful and may lead to quick failure of the gear. It is recommended[6] that the maximal torque that doesn't cause separation is calculated as follows:

- For the rim and center made of steel

$$T_{sep} = \frac{\pi d_p p \delta_1 W}{0.44 + 9m/\delta_1}$$

- For the steel rim and gray cast iron center

$$T_{sep} = \frac{\pi d_p p \delta_1 W}{0.2 + 6.6m/\delta_1}$$

where d_p = pitch diameter of the gear.

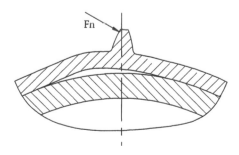

FIGURE 8.20 Local separation of the rim from the center.

EXAMPLE 8.3

Teeth numbers are $N_1 = 20$ and $N_2 = 100$; $m = 20$ mm, $\psi = 9°$, $W = 400$ mm, and output torque $T = 5 \cdot 10^8$ N $\cdot$ mm. Define the dimensions of steel rim and gray cast iron center.

Pitch diameter of the gear is

$$d_p = \frac{m N_2}{\cos \psi} = \frac{20 \cdot 100}{\cos 9°} = 2024.93 \approx 2025 \ mm$$

The center distance is

$$C = \frac{m(N_1 + N_2)}{2 \cos \psi} = \frac{20 \cdot 120}{2 \cos 9°} = 1215 \ mm$$

The root circle diameter of the teeth is

$$d_r = d_p - 2.5 m = 2025 - 2.5 \cdot 20 = 1975 \ mm$$

Preliminary estimate of the rim thickness from Equation 8.6 is

$$\delta_1 = \frac{1215}{10 \sqrt{120}} \left(5 + \frac{400}{1215} + 0.001 \cdot 120 \right) = 60.5 \ mm$$

The inner diameter of the rim (the outer diameter of the center barrel) is

$$d_C = d_r - 2 \cdot \delta_1 = 1975 - 2 \cdot 60.5 = 1854 \ mm$$

Assuming that the tensile stress in the rim should not exceed 150 MPa, the maximal permissible surface pressure in the connection can be found from Equation 8.9:

$$p_{max} = \frac{2 \delta_1 \sigma}{d_C} = \frac{2 \cdot 60.5 \cdot 150}{1854} = 9.79 \ MPa$$

This pressure may be decreased because of manufacturing tolerances. Therefore, in Equation 8.7, we will insert 80% of this value, i.e., $p = 7.8$ MPa. The coefficient of friction $f_C = 0.02$ (because $\delta_1 = 3m)^5$. Thus, the permissible torque

$$T_S = 0.5 \cdot \pi \cdot 1854^2 \cdot 400 \cdot 7.8 \cdot 0.02 = 3.37 \cdot 10^8 \ N \cdot mm$$

It is only 67% of the torque to be transmitted. To increase the strength of the connection, let's increase the rim thickness δ_1 to say, 80 mm. In that case, the inner diameter of the rim is given by

$$d_C = 1975 - 2 \cdot 80 = 1815 \ mm$$

and the maximal surface pressure

$$p_{max} = \frac{2 \cdot 80 \cdot 150}{1815} = 13.2 \; MPa$$

So, taking the same 80%, we have $p = 10.6$ MPa. Because the rim thickness now equals $4m$, the friction coefficient can be taken as $f_C = 0.024$. Thus, the permissible torque

$$T_S = 0.5 \cdot \pi \cdot 1815^2 \cdot 400 \cdot 10.6 \cdot 0.024 = 5.27 \cdot 10^8 \quad N \cdot mm$$

Now we have to check this connection for separation:

$$T_{sep} = \frac{\pi \cdot 2025 \cdot 10.6 \cdot 80 \cdot 400}{0.2 + 6.6 \cdot 20 / 80} = 11.66 \cdot 10^8 \; N \cdot mm$$

These results are satisfying, and $\delta_1 = 80$ mm is acceptable. The barrel thickness

$$\delta_C = \left(1.05 + 0.05 \frac{400}{1215}\right) 80 = 85 \; mm$$

So the inner diameter of the center barrel

$$d_1 = 1815 - 2 \cdot 85 = 1645 \; mm$$

Now we can calculate the needed interference in the connection. First, we choose the thickness of the disks: $\delta_2 = 2 \cdot 40 = 80$ mm. Each disk has 6 holes 250 mm in diameter.

$$A_d = \frac{\pi \cdot 1645^2}{4} = 2.125 \cdot 10^6 \; mm^2; \quad A_h = \frac{\pi \cdot 250^2}{4} 6 = 0.295 \cdot 10^6 \; mm^2$$

$$\delta_{2h} = 80 \left(\frac{2.125 - 0.295}{2.125}\right)^2 = 59.3 \; mm$$

$$d = d_C + 2S = 1815 + 2[80 + 20(0.22 + 0.009 \cdot 9)] = 1987 \; mm \quad \text{(see Equation 8.3)}$$

From Figure 8.19,

$$\frac{d_C}{d} = \frac{1815}{1987} = 0.913; \quad f_1 = 11.35$$

$$\frac{d_1}{d_C} = \frac{1645}{1815} = 0.906; \quad \frac{\delta_{2h}}{W} = \frac{59.3}{400} = 0.148 \approx 0.15; \quad f_2 = 3.22$$

$$E_1 = 2.06 \cdot 10^5 \; MPa; \quad E_2 = 1 \cdot 10^5 \; MPa$$

$$\Omega = 11.35 + \frac{2.06}{1} 3.22 = 18$$

Now we can obtain the maximal radial interference Δ from Equation 8.8:

$$\Delta = \frac{p_{max}d_C\,\Omega}{2\,E_1} = \frac{13.2\cdot1815\cdot18}{2\cdot2.06\cdot10^5} = 1.047\ mm$$

That means that the inner diameter of the rim should be less than the outer diameter of the barrel by about 2.1 mm maximum. If the barrel is finish-machined after the rim and according to its measured inner diameter, the tolerance may be maintained within 0.3–0.4 mm. If the minimal interference is less than the maximal by 0.4 mm (i.e., 1.7 mm instead of 2.1 mm), the minimal pressure will be, in direct proportion to the interference, 81% of the maximal.

After the design of the gear is completed (including the hub, ribs, etc.), detailed strength analysis can be performed.

The following drawbacks of gears with shrink-fitted rims should be pointed out:

- They have much greater weight as compared to welded and bolted gears.
- The rim is pretensioned, and this increases the probability of its failure.

The only benefit of such gears is their lower costs.

8.4 MANUFACTURING REQUIREMENTS AND GEAR DESIGN

Producibility considerations are an integral part of the design of anything, because it is senseless to design what can't be produced with reasonable costs. The design of gears is in many respects determined by the necessity to fasten the blank to the table of a gear-cutting machine so that it is not deformed by the attachment and weight forces, to check the datum surfaces after the blank is attached, and to provide the required clearances (runouts) for the cutting tools.

Fastening of a large double-helical gear to the table of a gear-cutting machine is shown in Figure 8.21. The gear is attached to table 1 through pedestal 2 by pressure plate 3 and stud bolts 4.

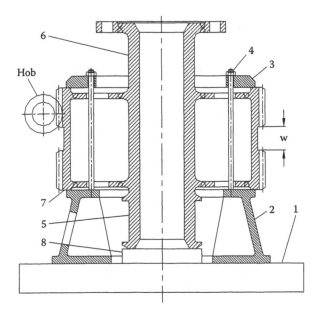

FIGURE 8.21 Double-helical gear on the table of gear hobbing machine.

Shaft journal 6 and datum face 7 are accessible to measure radial and axial runouts. Supporting device 8 is meant to take the weight of the central part of the gear.

The gear in operation is not loaded by axial forces, because they get mutually balanced in the herringbone mesh. Therefore, there are no ribs between the disks (as they are shown in Figure 8.4), and the disks are pliable in the axial direction. If the central part of the gear is not supported, its weight bends the disks elastically, and they cause the adjacent parts of the rim to deform inwardly. This results in thicker teeth in this area. When the gear is finished and installed in the horizontal position, the disks become straight, and the rim is restored to its original shape. But the gear is out of tolerances, and it can't work properly, because the load will concentrate on the thicker parts of the teeth.

In double-helical gears, the distance w between the half-gears (see Figure 8.21) needed for the hob to run out can be obtained from the equations[9]

$$w \geq \sqrt{(D_h - h)h} + 0.1km \quad (when \; \psi = 0^0)$$

$$w \geq \sqrt{(D_h - h)h} \cdot \cos\psi + km\tan\psi \quad (when \; \psi \neq 0^0)$$

where
D_h = outer diameter of the hob (mm)
h = entire height of the tooth (the standard height equals $2.25m$)
m = module of the toothing (mm)
ψ = helix angle
$k = 5$ for standard and short-addendum mesh
$k = 5 + X$ for long-addendum mesh, where X is the hob displacement from its standard position
  (see Chapter 7, Figure 7.12).

Figure 8.22a shows fastening of a smaller gear. Datum face 1 is resting against the machine table through spacing ring 2. The spacing ring provides the needed clearance space for hob 3. The gear is fastened by means of disk 4 and bolts 5 that pass through holes in the gear disk. So the holes enable this kind of attachment, which allows the alignment of datum surfaces 1 and 6 to be checked. The pliable disk of the gear has not been involved in the gear attachment.

Figure 8.22b shows an example of teeth cutting on a small two-rim gear. Mandrel 7 is attached rigidly to the table of the machine, and all the misalignments are checked on the datum surfaces of the mandrel and spacer ring 8 before the work is installed. The gear is checked after finishing the teeth.

The bigger toothing can be hobbed (hob 9) or shaped, but the smaller one can be machined only by shaping (shaper 10). Circular groove 11 is provided for the runout of the shaper cutter. The recommended width of the groove is as follows:[9]

$$w \geq 4 + 0.66m$$

The process of hobbing is more efficient, but it needs a wider groove, as shown in Figure 8.22c. If the teeth are case-hardened, they usually need grinding to improve their accuracy. The diameter of the grinding wheel 12 is greater than that of the hob, and it needs a wider clearance space (Figure 8.22d).

Figure 8.23 shows several ways of grinding teeth. Form grinders (Figure 8.23a) and conical grinders (Figure 8.23b) come out of contact with the work immediately after the center of the grinder passes the face plane of the teeth. Saucer-shaped grinders (Figure 8.23c and Figure 8.23d) stay in contact with the teeth until the grinder comes completely out of the tooth space, which requires increased width, w, of the clearance groove.

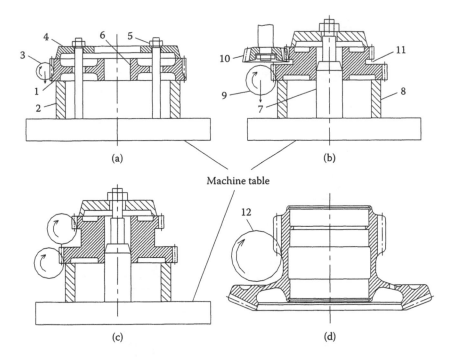

FIGURE 8.22 Clearing space for teeth machining tools.

Design innovations intended to decrease the warpage of the parts during heat treatment (to spare grinding) have brought only limited success, which is satisfying in some cases, but not acceptable in others. When high accuracy of the toothing is a must, the designer is pressed to employ any means (including separation of the gear into several parts) to ensure that grinding will be possible. Figure 8.24a shows a kind of spline connection in which the shortened teeth of the pinion serve as splines. The connection is tightly fitted, and possible axial movement is blocked by snap ring 1.

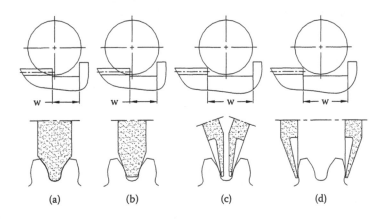

FIGURE 8.23 Types of grinding wheels.

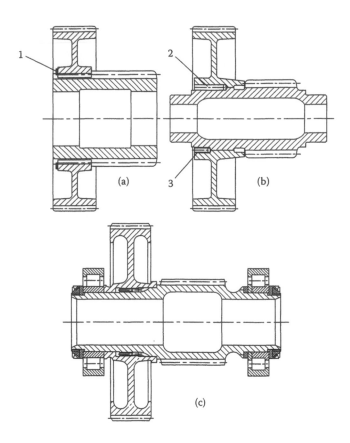

FIGURE 8.24 Assembled gears.

Figure 8.24b shows an IFC of gears with round keys (pins) 2 and screws 3 that prevent axial displacement in the connection. In both cases (Figure 8.24a and Figure 8.24b), the larger toothing is finished after assembly from the same datum surfaces that were used to align the smaller teeth. These connections are meant to be nonseparable.

Figure 8.24c shows a spline connection of gears with cylindrical centering surfaces. This connection can be dismantled, but the initial position should be marked to ensure reassembly as it was.

8.5 BEVEL GEARS

As compared with spur and helical gears, bevel gears have some peculiarities in technology and design. For example, in the middle of a shaft, a straight bevel gear can only be cut by a gear-shaping machine (Figure 8.25a). Zerol and spiral bevel gears cut by a face-mill cutter require that no part of the gear projects from the root cone (Figure 8.25b and Figure 8.25c). In the same way that helical gears have greater load capacity and work smoother than spur gears, Zerol and spiral bevel gears also have greater load capacity and work smoother than straight bevel gears. Therefore, it is worth making built-up gears (Figure 8.25d) if it is needed to use spiral bevel mesh. The separation of the gear rim also enables carrying out the heat treatment in a rigid fixture that reduces distortion.

The thickness of the rim, δ_1, usually equals the maximal tooth height. When the rim is attached to the center by bolts as shown in Figure 8.25d, the distance between the tooth root and the threaded hole, δ_2, is recommended to be not less than one third of the tooth height.

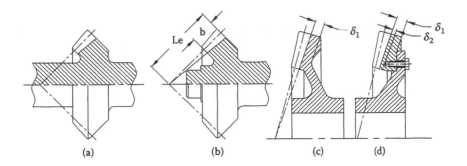

FIGURE 8.25 Bevel gears.

8.6 DESIGN OF TEETH

Part of the information related to the teeth has already been discussed in Chapter 7, Section 7.5 (for tooth root shape, see Figure 7.13) and Section 7.7 (for tooth profile modification, see Figure 7.20). There is not much to add.

The tip thickness of a tooth, S_t (Figure 8.26a) depends a little on the number of teeth (see Chapter 7, Figure 7.2), but it is very much influenced by the displacement, X, of the generating profile (Chapter 7, Figure 7.12). Therefore, when designing gears with considerable X value, close attention should be given to the thickness of the tooth tip. If the gears (without surface hardening) wear in service, it is recommended that $S_t \geq 0.5m$, because the service life of this gear may be limited by the wear until the tooth tip becomes sharp. If a gear doesn't have considerable wear, S_t of $\geq 0.25m$ is admissible. To prevent chipping (shown in Chapter 7, Figure 7.9d), case-hardened or nitrided teeth should have S_t of $\geq 2\delta_h + (0.08 - 0.1)m$, where δ_h is the case thickness after finishing. If the tooth tip is too thin, the case thickness here can be decreased by masking the top land of the teeth during the carburizing process.

The sharp edges of the tooth tip should be rounded by radius $r_1 \approx 0.1m$. That operation can be done while cutting the teeth, if the hob profile has a corresponding curvature. This rounding prevents a sharp stress concentration on the edge and may be very helpful to prevent scoring. However, finish-grinding the tooth flanks may partly or completely remove these radii. They should be restored and, when needed, polished.

The ends of a tooth should be rounded off or chamfered to make the maximal bending force more distant from its end. One of the possible options of chamfering is shown in Figure 8.26b.

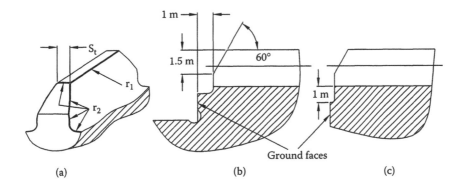

FIGURE 8.26 Tooth design.

The sharp edges of the tooth ends should be broken by chamfers or radii r_2 (Figure 8.25a). For teeth without surface hardening, it is not obligatory; deburring is usually sufficient. But for surface-hardened (case-hardened, nitrided, and such) teeth this operation must be performed before heat treatment. Usually, $r_2 \approx 0.2m$. These radii and the end surfaces of the teeth should not be machined after surface hardening, because it would diminish the bending strength of the teeth. If there is a need for an exact ground face, it is desirable (in high-strength gears) to make it at a distance of about $1m$ from the tooth in radial or axial direction, as shown in Figure 8.25b and Figure 8.25c.

The hardened case should cover surfaces at least $1m$ away from the teeth ends.

REFERENCES

1. Klebanov, B.M., *Design of Disc-Type Gears, Toothed and Worm Gears*, collection of articles, Kolchin, N.I., Ed., Mashinostroenie, Leningrad, 1974 (in Russian).
2. Kudriavtsev, V.N., Kuzmin, I.S., and Filipenkov, A.L., *Calculations and Design of Gears Handbook*, Politechnika, Sanct-Petersburg, 1993 (in Russian).
3. Barlam, D.M. and Klebanov, B.M., Calculation of bolted joints of gear rims, *Vestnik Mashinostroenija*, No. 2, 1988 (in Russian).
4. Anfimov, M.I., *Reduction Gears Design and Calculation*, Mashinostroenie, Moscow, 1972 (in Russian).
5. Arai, N., Oeda, M., and Aida, T., Study of the stress at the root fillet and safety for slipping of force-fitted helical gears, *Bulletin JSME*, Vol. 24, No. 198, 1981.
6. Safronov, Y.V., Up-to-date calculation of the rim thickness and the interference fit of large spur gears, *Transactions of the Rostov-na-Donu Agriculture University*, Rostov, 1980 (in Russian).
7. *Planetary Gears Handbook*, Kudriavtsev, V.N. and Kirdiashev, Y.N., Eds., Mashinostroenie, Leningrad, 1977 (in Russian).
8. Lechner, G., Zahnfussfestigkeit von zahnradbandagen, *Konstruktion,* 19, H.2, 1967 (in German).
9. *Manufacturing of Gears Handbook*, Taiz, B.A., Ed., Mashinostroenie, 1975 (in Russian).

9 Housings

9.1 THE FUNCTION OF HOUSINGS

Design of a housing begins when the other elements of a mechanism have been already designed in every detail: with the shafts and axles and their supports, with all dimensions of the parts and their motion paths. Then can be defined the dimensions of the bearing bosses, the possible location of partings and hatches, and other elements of the housing.

This design procedure gives the impression that the housing is a kind of envelope and can be made of any material. This impression is false; a housing is just one of the components of a mechanism, as the shell is an integral part of a tortoise's skeleton. Without this part the mechanism doesn't exist. This is the difference between a housing and a guard; the latter is just to isolate the mechanism from the surroundings.

The housing is almost always the biggest part of a mechanism. It is usually made as a closed box, so that the rest of the parts are mounted inside it. In this respect, the housing also serves as a guard.

The housing is most often (not always) the fixed part of a mechanism that serves to attach it to some foundation or substructure or to another unit.

Notice that *fixed* always means immovable relative to something. For example, let's consider the swinging mechanism of a crane shown in Figure 9.1. Platform 1 is set in motion by pinion 2, driven by electric motor 3, through worm gear 4. Pinion 2 revolves around gear 5 that is bolted to base 6, which is, in turn, bound to the ground (possibly through a movable platform, like the chassis of a vehicle). If the entire crane is considered as a mechanism, the fixed part is base 6. If the mechanism is worm gear 4, its housing 7 is the fixed part relative to the worm, the gear, and the bearings.

Though the mobility or immobility of the housing may be just a matter of definition, its movement in space should be taken into consideration. When the housing moves, all the parts of a mechanism (including the oil) are exposed to inertia forces, which can essentially change the load of the parts and the lubrication conditions.

It is appropriate to mention here that gravitation is also applied to all the parts, and it should be taken into account. For example, a purposeful increase in weight of a part may change profitably the direction and value of a load vector.

Serving as a guard, the housing isolates the mechanism from the surroundings to create conditions inside that are comfortable for the mechanism. On the other hand, the housing isolates the surroundings, including the personnel, from the mechanism, so that we are protected from hearing the noise and from being sprinkled with hot oil; besides, we can't easily put our fingers into it. This is an important function of the housing; nevertheless, it is not an envelope but one of the parts of a mechanism. Therefore, with respect to stress and strain analysis, the housing deserves attention no less than any other machine element. The problem is that housings usually have a complex spatial shape. It can be analyzed by "engineering methods" (i.e., traditional analytical methods) very approximately, under simplifying assumptions. This method, however, requires a certain level of experience and feeling of how the housing deforms under load; otherwise, the simplifications may be done incorrectly, and the results of the calculation may be far from satisfactory. More exact information of the stresses and elastic deformations of the housing can be obtained using the finite element method (FEM). But this method also requires some simplifications in most cases, for example, when defining the boundary conditions in bolted joints, when presetting the load distribution in bearings, and others. The choice of appropriate simplifications is a matter of experience with the machine elements and awareness of computer program subtleties.

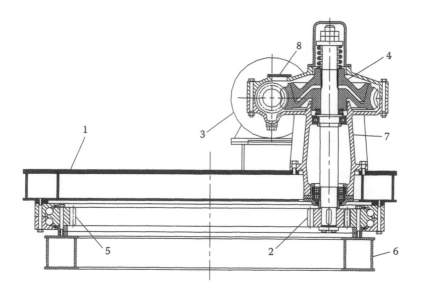

FIGURE 9.1 Swinging mechanism of a crane.

The housing is an assembly basis for the rest of the parts and units of the mechanism. Therefore, it should meet some obligatory requirements:

1. Because the mechanism is assembled inside the housing, the latter must be designed so as to enable assembly operations and the control of parameters to be checked after assembly. That means, the housing should have partings or sufficiently sized openings that enable convenient assembly and control of the mechanism. Sometimes the mechanism must be periodically inspected while in service to check the condition of wearing parts (like gear teeth or bearings). For this purpose, the housing should be provided with inspection hatches in needed places. For example, hatch 8 in Figure 9.1 enables control of the tooth contact patch during assembly, visual inspection of the teeth in service, and measurement of backlash in the worm mesh. The hatches may reduce the strength and rigidity of a housing; therefore, their dimensions should not be too large but enough to enable the actions they are made for. In relatively small mechanisms, it may be preferable to dismantle the mechanism for inspection and in that way avoid the inspection hatches.

2. As was said, the housing as an ordinary member of a mechanism should have exact dimensions where it is in contact with other parts. This accuracy should be achieved in production and then remain within admissible limits while the housing is exposed to mechanical and temperature influences. Figure 1.1 (see Chapter 1) shows frame 4, which serves as a housing (not closed in this case). Insufficient rigidity of the frame has led to increase in the distance between the lugs. Consequently, the parallel link mechanism doesn't operate properly and the weight has fallen. Figure 1.5e (see Chapter 1) shows the elastic deformations of a gear housing that lead to misalignment of the shafts.

    Dimensional stability of the housing is a mandatory requirement, as for every other member of a mechanism. Change in dimensions can relate to elastic and plastic deformations under load or to thermal expansion, or it may be caused by structural changes in the material. Metals, with their high strength, high modulus of elasticity, and stable structure (after appropriate heat treatment), are preferable to other materials. Therefore, housings are almost always made from metals.

3. In addition, some other requirements may be important, such as
   - Heat conductivity of the housing material (to enable more effective heat extraction from inside to outside)
   - Ability to damp noise and vibrations
   - low weight
   - Ability to withstand aggressive environments, such as seawater, and others, which can be achieved on any material by suitable coatings

9.2 MATERIALS FOR HOUSINGS

Because housings are intricately shaped, they are mostly cast, and usually only the surfaces that fit to other components are machined. One of the most widespread materials is gray iron. It has good castability and machinability, increased (as compared with construction steels) corrosion resistance, and good wear resistance (because it is "impregnated" with graphite). Good damping capacity enables it to reduce the noise and vibrations created in the working mechanism. But gray iron is brittle and sensitive to overload and impact load; therefore, it is not recommended in dangerous applications, for example, in devices intended for transportation of foundry ladles filled with molten metal.

The wall thickness of a gray iron housing depends on several variables, including stress, achieving adequate stiffness, method of casting, etc., and can be approximately obtained from Figure 9.2[1], curve 1. In this figure, N is the equivalent dimension of the casting which is given by

$$N = \frac{1}{3}(2L + W + H)$$

where L is the length (the largest dimension of the casting), W, the width, and H, the height of the casting, all in meters.

The recommendations given in Figure 9.2 are average. Depending on the complexity of the casting, technology used, and the temperature of the metal charged, the minimal wall thickness may be somewhat lesser or greater. For example, if a large foundry ladle feeds into several molds, the metal will cool slightly during the process of charging. So the wall thickness should be somewhat greater to be satisfactory for the last of the molds. In any case, the design of the casting should be discussed with the manufacturers and submitted for their approval.

EXAMPLE 9.1

The housing is 3000 mm long, 1500 mm wide, and 1000 mm high. $W = 1/3\ (2 \cdot 3 + 1.5 + 1) = 2.83$ m, and the practically feasible wall thickness of cast gray iron is about 20–22 mm. The strength and the rigidity of the housing should be checked afterward, and if they are insufficient, they can be increased mainly by adding ribs or, possibly, by increasing the wall thickness.

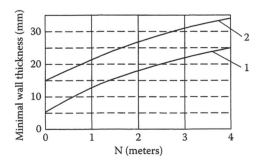

FIGURE 9.2 Practically feasible wall thickness of a casting.

The tensile strength, S_u, of gray iron is usually about 170–250 MPa, and a suitable grade of gray iron should be chosen (depending on the wall thickness) to achieve the needed strength. The Brinell hardness, HB, and the modulus of elasticity, E, can be approximately determined as follows:[1]

$$HB \approx 0.43S_u + 100; \quad E \approx \frac{9.5 \cdot 10^4 S_u}{HB} \ MPa; \tag{9.1}$$

(Here, S_u is in MPa.)

The S_u value is determined by the breaking test of cast specimens (in tension), known informally as a *tensile test* or *pull-test*. The breaking points in bending ($S_{u,b}$) and in compression ($S_{u,c}$) are appreciably greater. Approximately, they can be obtained from Figure 9.3.

When higher strength, reliability, and impact resistance are required, a steel casting is typically used. Foundry steels with a carbon content of 0.25% to 0.35% have a yield point of about 240–280 MPa and modulus of elasticity $E \approx 1.75 \cdot 10^5$ MPa. The minimal wall thickness of the steel casting can be found from Figure 9.2, using curve 2. For dimensions given in Example 9.1 the minimal wall thickness of a steel casting is about 30 mm.

Welded steel housings are lighter than cast ones, because the walls can be made thinner (by about 40%) depending on the strength and rigidity of the material. They also successfully work when subjected to impact loads. But welded housings are much more resonant, and special measures should be taken when the level of noise is restricted. Among these measures can be mentioned ribbing, and coating of the walls with vibroabsorbing materials.

Welded housings often contain forged and cast elements. Bearing seat 1 in Figure 9.4 is made of cast steel and welded to steel plates. Such a combination of intricately shaped steel castings with flat elements made of rolled plates is very common.

Welded housings have some additional advantages that shorten the production cycle:

- The dimensions of the weldment are more exact (when assembled and welded using jigs), and, consequently, machining allowances are less.
- Most of the small parts, such as pipes, flanges, and bosses, can be finish-machined, including drilling and tapping holes, on smaller machines before welding.

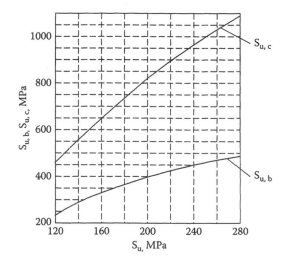

FIGURE 9.3 Dependence of breaking points of gray cast iron in bending and in compression from its tensile strength.

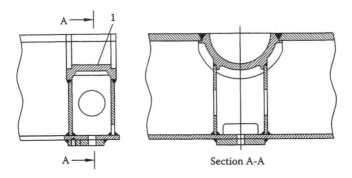

FIGURE 9.4 Fragment of a welded gear housing.

- Oil ducts can be easily welded as an integral part of the welded structure (see Chapter 6, Figure 6.39b); the incorporated pipe can be bent and bifurcated if needed, whereas in a cast housing the oil duct must be drilled after the casting is machined (Chapter 6, Figure 6.39a) and obviously must be straight.

When light weight is a mandatory requirement, aluminum and magnesium alloys are used for the housings. The specific weight of aluminum alloy is about one third, and of magnesium alloy, about one quarter, that of gray cast iron. So if the minimal wall thickness is limited by the casting technique only, the gain in weight is great. But the strength, rigidity, and temperature deformations of housings are always under consideration. The comparison characteristics of mechanical properties (yield point S_y, modulus of elasticity E) and linear expansion factors α of rolled and cast steels, gray iron, aluminum and magnesium cast alloys are given in Table 9.1. As one can see, mechanical properties of aluminum alloy can be very close to those of gray iron, but it is inferior to gray iron in wear resistance. Therefore, friction should be avoided in the design of an aluminum housing. In particular, all the seats of rolling bearings should be reinforced by steel sleeves as shown in Figure 6.19a, Figure 6.19c, and Figure 6.23 of Chapter 6. The sleeve should be tightly fitted, and the interference value should prevent joint separation at the highest service temperature. The sleeve is exposed to a turning moment, just as the bearing rings are (see Chapter 6, Subsection 6.2.5), and may rotate. This rotation may be prevented by bolts (as in Figure 6.19a) or pins (2 in Figure 6.23, Chapter 6). Otherwise, the interference should be increased. The needed value of it can be calculated in accordance with recommendations given in Chapter 6, Subsection 6.2.5.

TABLE 9.1
Mechanical Properties of Some Housing Materials

	Material	S_y (Mpa)	E (Mpa)	α 1/°C
Rolled plate	SAE 1020 annealed	250	$2.00 \cdot 10^5$	$12.2 \cdot 10^{-6}$
	SAE 4130 normalized	490	$2.00 \cdot 10^5$	$12.2 \cdot 10^{-6}$
	Hardox 400, Creusabro 4000	900	$2.06 \cdot 10^5$	$12.2 \cdot 10^{-6}$
	Cast steel	240–260	$1.70 \cdot 10^5$	$12.2 \cdot 10^{-6}$
Casting	Gray iron	$Su = 170$–250	$(0.9\text{-}1)\,10^5$ See Eq. (9.1)	$(10\text{–}12) \cdot 10^{-6}$
	High-duty cast iron	320–370	$1.70 \cdot 10^5$	$12 \cdot 10^{-6}$
	Aluminum alloys	150–180	$0.72 \cdot 10^5$	$(20\text{–}26) \cdot 10^{-6}$
	Magnesium alloys	80–120	$0.45 \cdot 10^5$	$(26\text{–}30) \cdot 10^{-6}$

The end covers of bearings that transmit axial force from the bearing to the housing (such as those shown in Figure 6.8f, Figure 6.8h, Figure 6.8i, Figure 6.11e, Figure 6.11f, and Figure 6.22 — see Chapter 6) should be made of strong, wear-resistant materials.

If the housing is split through the axes of the gears, elements of axial fixation of the bearings are sometimes installed in circular grooves (for example, ring and cover in Figure 6.8c, cup and cover in Figure 6.8f — see Chapter 6). These parts can't be installed with interference fit because it would impede the connection of two halves of the housing. So, when the outer ring of the bearing rotates (see Chapter 6, Subsection 6.2.5), it may also drag the fixing elements into rotation. (Rotation of the cover can be observed.) And if the housing or the fixing elements are made of light metal, their intensive wear is unavoidable.

The threaded connections in light-metal housings are usually made using studs with interference-fit thread of the tap end. The length of this end should be about $(1.5–2.0)d$ (Figure 9.5a) depending on the strength of the stud and of the housing materials. When needed, the diameter of the tab end can be increased as shown in Figure 9.5b. This stud is provided with hexagonal part 1 to prevent torsion of the stud shank by a heavy torque needed to assemble the tight thread.

When a standard size, reusable thread in the light-metal housing is needed, the housing wall should be reinforced by threaded inserts (2 in Figure 9.5c) made from corrosion-resistant steel and installed fast in the housing, using an insert with interference-fit outer thread, a self-tapping insert, a keyed insert, or by application of the appropriate kind of anaerobic compound such as Loctite.

All these requirements and recommendations may be surplus if the mechanism is low-loaded or designed for short-term applications.

Irrespective of the material or manufacturing method, the housing needs heat treatment to relieve internal stresses. In cast housings, these stresses are caused by uneven solidification of the casting. The more massive volumes harden later, and, as they become cooler and decrease in their dimensions, they pull their thinner neighbors that are already hard and rigid and can't deform that much. This struggle between the neighbors, each pulling in a different side, may result in a crack, or, at best, the internal stress remains in the casting.

Avoidance of massive volumes in the casting is very desirable but not always possible. Anyway, unnecessary metal should be removed from local bosses, and the transition from wall to boss should be made smooth with generously rounded corners.

Another method to prevent cracks in castings is to make pliable configurations. Figure 9.6 shows the spoked wheels, in which spokes 1 harden first, and hub 2 and rim 3, later. Depending on which of them, rim or hub, hardens later, the spokes may appear compressed or tensioned. To prevent high shrinkage stresses, the spokes can be made curved (Figure 9.6a) or slanted (Figure 9.6b).

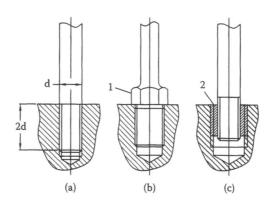

(a)   (b)   (c)

FIGURE 9.5 Reinforcement of threads in light-metal housings.

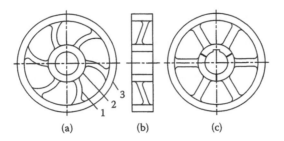

FIGURE 9.6 Prevention of cracking in the spoked wheels.

In these cases, the difference in the shrinkage of the rim and the hub leads to small bending deformations of the spokes.

Very large wheels are often made with a slit hub (Figure 9.6c). Such a wheel is slide-fitted to the shaft, and the needed press fit is achieved from massive rings mounted on the ends of the hub with a great interference. This design is archaic. Nowadays such wheels are welded of low carbon steel, and the problems of possible cracking are avoided.

The mechanism of deformation of welded housings is similar to that described previously. The welded components that are straight before welding (Figure 9.7a) are pulled by the weld while getting cold and become deformed as shown in Figure 9.7b.

The internal stresses are quite high, and, in course of time, the material little by little gets free of them owing to continuous sluggish plastic deformation in the crystals. Because the housings are relatively large and thin walled, their warpage might be unacceptably large. Therefore, to relieve these internal stresses, housings are usually heat-treated before final machining. The heat treatment consists of exposure of the casting (or the weldment) to high temperature for several hours followed by slow cooling within the oven (to prevent formation of new internal stresses because of nonuniform cooling). The temperatures depend on the kind of material and approximately range from 550 to 650°C for ferroalloys and from 170 to 250°C for aluminum and manganese alloys. At these temperatures, the material becomes pliant, so that the internal stresses deform it in a relatively short time and disappear almost completely. The same process improves the microstructure of the material, and cast materials undergo aging.

The cast and forged elements used for welded housings usually should undergo stress-relieving heat treatment before welding in order to decrease deformations during heat treatment of the entire weldment.

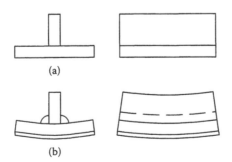

FIGURE 9.7 Deformation of welded parts.

9.3 DESIGN OF HOUSINGS

For preliminary design of the housing elements, we usually use recommendations derived from design practice. Let's illustrate housing design principles by some examples from the field of reduction gears, especially, because the gears are extremely sensitive to the rigidity and accuracy of their housings.

9.3.1 Housings Split through the Axes of Shafts

Figure 9.8 shows a housing of a three-stage gear (represented earlier in Figure 6.15 of Chapter 6 with cover off). This housing is split into two parts, base 1 and cover 2, both cast of gray iron. The wall thicknesses δ (of the base) and δ_1 (of the cover) can be chosen from Figure 9.2. The thickness of ribs δ_2 can be set equal to δ for the base and to δ_1 for the cover.

The recommended diameter d of bolts that attach the gear to a foundation or a frame depends on the center distance of the low-speed stage C_L (because the torque transmitted and the forces applied to this connection depend mainly on the dimensions of this stage):

$$d = (0.08 - 0.12)C_L$$

The mounting foot thickness h is given by

$$h = (2 - 3)d$$

Diameters of bolts connecting the two halves of the housing near the bearings (d_1) and other bolts (d_2) can be taken as follows:

$$d_1 = (0.7 - 0.8)d; \quad d_2 = (0.5 - 0.6)d$$

The shaft of the gear can be loaded with additional radial forces; for example, the power can be taken off the gear by means of a pinion mounted on the output shaft. In this case, the housing design and the bolts must accommodate the increased loads.

The height of the flange between the bearings, h_1, is dictated by the necessity to provide a place for the bolt heads and nuts between the bearing bosses. In the other areas, the flange thickness h_2 can be set equal to $1.5d_2$.

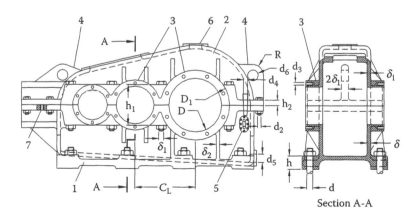

FIGURE 9.8 Housing of three-stage reduction gear with outside ribs.

The outer diameter D_1 of the bearing bosses 3 can be set equal to

$$D_1 = 1.25\,D + 10\ mm$$

where D = outer diameter of the respective bearing (mm).

The diameters of bolts d_3 that attach the bearing caps to the housing can be obtained from the equation

$$d_3 = 0.04D + 4\ mm$$

For the sake of convenience in service, the diameter of all the bolts, d_3, should be the same (but it is not a must); this may require some increase in diameter D_1 of smaller bearing bosses. These bolts are exposed to the axial force created in the gear mesh and in the bearings. Their calculation for strength should take into consideration that the axial force is distributed nonuniformly between the bolts, and the maximum-to-average ratio may be about 5 (see Chapter 6, Example 6.6).

The position of the base and the cover relative to each other is fixed by two pins; their diameter $d_4 \approx 0.8d_2$. In addition, the housing is provided with lifting eyes 4; drain hole d_5 (of about 16–42 mm in diameter, depending on the housing size), oil indicator 5, opening 6 for a breather, and jackscrew holes 7 (for disassembly).

It is desirable to place the breather in the highest point of the housing; otherwise, part of the inner volume would not be ventilated, and the steam condensate would remain in the housing. This may cause severe corrosion in nonventilated areas during warehousing.

The jackscrew hole 7 has been made in this case on the lower flange. If it is made on the upper flange, it should be closed with a greased plug to prevent filling with debris.

The relative positions of the wall and the bearing boss are important both for housing strength and bearing alignment. The minimal bending stress of the wall is achieved when it is placed symmetrically about the radial reaction of the bearing, as shown in Figure 9.9a. If, in addition, the

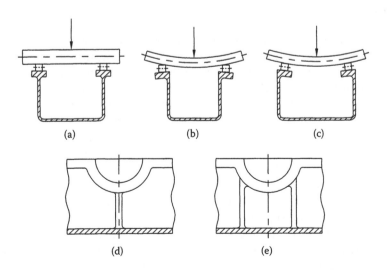

FIGURE 9.9 Deformations of the shaft and the bearing bosses.

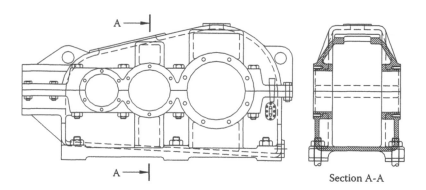

FIGURE 9.10 Housing of three-stage reduction gear with inside pockets.

bending deformations of a shaft are negligible, this arrangement is the best. If the shaft is pliable under load, the location of the wall relative to the boss shown in Figure 9.8 and Figure 9.9b is the worst (if the bearings are sensitive to misalignment). In this case, not only is the wall exposed to bending but also the deformations of the shaft and the boss are oppositely directed. The wall location shown in Figure 9.9c and Figure 9.10 is preferable when the shaft is pliable, because the deformation of the boss decreases the bearing misalignment.

Blocking the boss deformation by a central rib (Figure 9.9d) is a wrong idea because this worsens the load distribution in the bearing (see Figure 6.21c and Figure 6.21d). The ribs should be placed as shown in Figure 9.9e.

The housing design shown in Figure 9.10 offers some additional advantages, besides the favorable deformation of the bearing bosses: the walls become more rigid, and the volume of the oil reservoir increases. Above the bearing boss can be made a trap for the oil splashed by the gears so as to surely lubricate the bearing, as shown in Figure 7.26b (Chapter 7). But it also has disadvantages as against the version in Figure 9.8: the casting is more complicated, and the cleaning and painting of the inner surfaces of the casting is more difficult.

9.3.1.1 Design of Mounting Feet

The mounting feet are intended for attachment of the housing to a base or foundation. They are the most heavily stressed elements of the housing. The initial state of the mounting foot (before loading) is shown in Figure 9.11a. When there is a tensile load applied to the wall, the mounting foot is bent in two directions: crosswise (Figure 9.11b) and lengthwise (Figure 9.11c). This deformation involves the bending of the adjoined parts of the walls. If the mounting foot is thin and pliable, the load is not distributed along its length, but mostly transferred in the vicinity of the bolt. The stresses in this area may be increased considerably and result in a failure of the wall.

The thicker the mount foot, the less it bends in both directions, and the longer part of the wall is involved effectively in the load transfer between the wall and the bolt. This decreases both the bending stress and the tensile stress in the wall. Bending deformations are further reduced by adding ribs (Figure 9.11d) that also make the loading mode of the bolt closer to centric.

The maximal strength and rigidity of the mounting foot is attained in the design shown in Figure 9.11e. Here, the bolt is put into a recess, and its loading can be considered as centric. The bending of the mounting foot is relatively small. The dimensions of the recess should be as small as possible, but not less than needed to put the bolt in and then to turn it by some kind of wrench. The tolerances of the castings or weldments must take all these requirements into account.

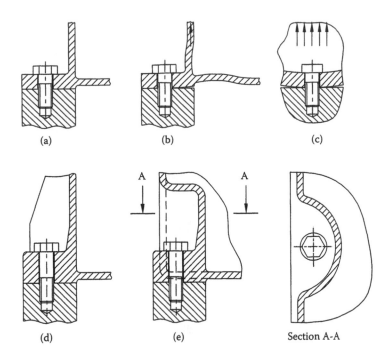

(a)   (b)   (c)

(d)   (e)   Section A-A

FIGURE 9.11 Mount feet design.

9.3.1.2 Design of Lifting Elements

Load lifting must be thoroughly planned and calculated. First, the location of the center of gravity (CG) should be defined, and the lifting elements must be so located as to prevent the load overturning. The quantity of lifting points depends on their possible location relative to the center of gravity and on the practicable load capacity of the elements used. For small- or middle-sized units, one or two lifting points may satisfy the need. In large and heavy units four or even more lifting points may be needed, and special instructions should be worked out to use them properly.

It is preferable to use elements that have standard dimensions and prescribed permissible load, such as lifting hooks (items 7 and 8 in Figure 9.12) and eyebolts (item 8 in Figure 9.14). The others, such as lifting lugs cast with the housing (item 4 in Figure 9.8), hatch covers with a lug (item 8 in Figure 9.13), and the like, must be carefully assessed for strength with safety factors generally accepted for load lifting devices.

Typical safety factors for lifting devices, such as those mentioned, may range up to 6 or even higher. These high margins are necessary to allow for manufacturing variations (undetected porosity in castings), environmental deterioration, normal wear, etc., and even possible misuse. It is very cheap insurance and allows peaceful sleep for the designer!

All the strapping elements must be controlled in production. For example, the lifting eyes and hooks attached to the wall of the housing may fail if the wall thickness in this area is thinner than indicated in the drawing (this may occur owing to displacement of the mold core).

Lifting eyes cast with the housing (item 4 in Figure 9.8) are usually twice as thick as the wall. The hole diameter $d_6 \approx 3\delta_1$, and $R = d_6$.

9.3.2 HOUSINGS SPLIT AT RIGHT ANGLE TO THE AXES OF THE SHAFTS

Examples of such design are shown in Figure 7.30, Figure 7.31 (Chapter 7), and Figure 10.26 (Chapter 10). In the first two cases, the gears are with vertical shafts. The housing of the bevel-planetary gear

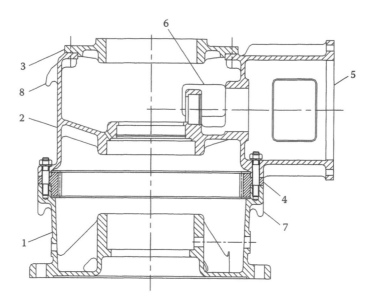

FIGURE 9.12 Housing of two-stage gear shown in Figure 7.31 (Chapter 7).

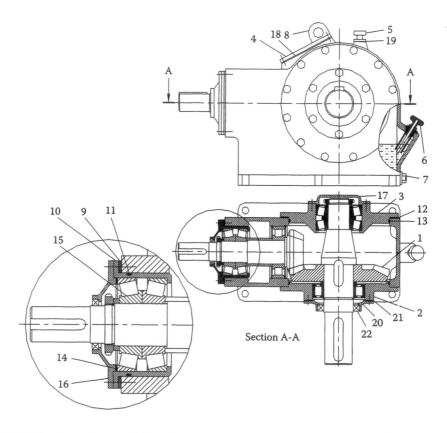

FIGURE 9.13 Bevel gear in nonsplit housing.

from Figure 7.31 is shown in Figure 9.12. It consists of base 1, upper housing 2, and cover 3. Parts 1 and 2 are centered through fixed annulus 4. Flange 5 is intended for mounting an electric motor. Inspection hatch 6 is necessary during assembly to check the contact pattern in the bevel gear. It is also useful in service to periodically inspect the bevel gear teeth.

The housings split at a right angle to the axes offer a number of advantages:

- The geometry of the bearing seats is more exact.
- The connection of two parts of the housing is not cut by the bearings, and therefore, it can be made lighter and stronger.
- The joint surface is flat and continuous, and its leakproofness can be more easily achieved.

What kind of housing design is preferable? Most general-duty gears have housings split on a horizontal or slightly slanted plane. The reason is clear: the foundation or mounting frame the gear is attached to is usually horizontal. If there is something wrong with the gear, you take off the upper part of the housing (the cover), and the lower part (the base) with the oil remains attached to the foundation. So you can inspect the whole gear mechanism, replace the damaged parts, then attach the cover and tighten the bolts. The oil filling as well as the alignment of the input and output shafts with other units remain unchanged and don't need any additional work.

9.3.3 Nonsplit Housings

Small-sized gears are often made in nonsplit housings (where possible). Figure 9.13 shows a bevel gear of such kind. Gear 1 with its shaft and bearings is installed through openings closed with covers 2 and 3. (Only one cover is enough to allow assembly of the unit. However, a second cover is provided to enable assembly with an oppositely directed output shaft without any change in details.) This housing also has inspection hatch 4, breather 5, oil gauge 6, and oil drain 7. The inspection hatch is needed during assembly to check the contact patch of the teeth and to adjust properly the axial positions of both the gear and the pinion.

A worm gear in a nonsplit housing is shown in Figure 9.14. Gear 1 with its shaft and bearings is installed through openings closed with covers 2 and 3. This housing is provided with an inspection

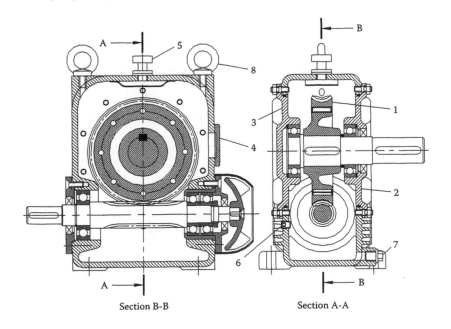

Section B-B   Section A-A

FIGURE 9.14 Worm gear in nonsplit housing.

hatch 4 that is needed while assembling to control the contact patch of the teeth and to adjust the axial position of the gear. Breather 5 is placed on the top of the housing. The oil level is controlled through plug 6. Oil drainer 7 is placed in the lowest point of the housing.

9.4 DEFORMATIONS AND STIFFNESS PROBLEMS

A housing is a relatively large and thin-walled part, and it is quite pliable. Attachment to a rigid foundation or a frame may block part of the possible deformations (not all of them). But if the abutting surfaces of the housing and foundation are not exactly flat, the very fact of tightening foundation bolts may deform the housing more than the working forces do. Moreover, the weight of the mechanism, when it is put on an uneven base, may cause noticeable deformations. Therefore, after the bolts are tightened, the tooth contact pattern should be checked again. The alignment (if needed) is achieved using jackscrews and shims.

Figure 9.15a shows the housing of a three-stage gear (see Figure 6.15 in Chapter 6 and Figure 9.8) attached to a foundation that is assumed to be infinitely rigid. Forces applied to the bearings cause deformations of bending and tension compression in the parts of the housing; in addition, some joint separation can take place.

The horizontal components of these forces are neglected here for several reasons: first, these forces are relatively small; second, their influence on tooth alignment is little; and third, the housing is very rigid in this direction, because it is loaded in tension or compression.

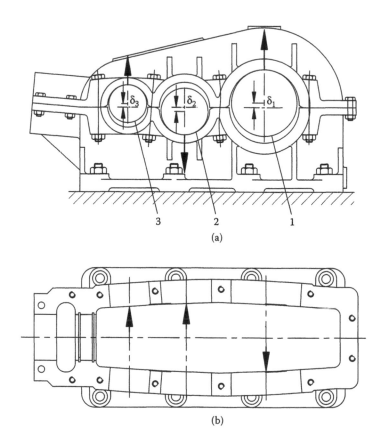

(a)

(b)

FIGURE 9.15 Gear housing deformation under load.

If shafts 1, 2, and 3 were symmetrically loaded and the bearing forces were equal, and if the housing design was also symmetrical, then displacements δ_1, δ_2, and δ_3 would be equal on both sides of the housing and would not cause any misalignment of the gear teeth. But from Figure 6.15 (Chapter 6), it is obvious that the load is distributed nonuniformly between the two bearings of each shaft, so the housing deformations (as well as the deformations of the bearings; see Chapter 6, Subsection 6.2.7) result in error in parallelism of the shafts and misalignment of the teeth.

Figure 9.15a makes clear the importance of the bolts that connect the two halves of the housing. They should be strong enough and properly tightened to prevent joint separation.

Axial forces applied to the teeth in helical gears may increase the difference in the bearing reactions, because they add bending moments to the shafts. Consequently, misalignment of the teeth may be increased. Therefore, in large gears, in which the attainment of needed teeth alignment under load presents difficulties, double-helical (i.e., herringbone) mesh is favored (for its symmetry and lack of axial forces outside the mesh).

In smaller gears, the axial forces are less harmful, and helical gears are mostly in use. Figure 9.15b shows the axial forces in a gear from Figure 6.15 (Chapter 6) and the respective deformations of the housing. These deformations may be harmful for bearings with adjustable end play, such as angular contact ball bearings (Figure 6.4b in Chapter 6) and tapered roller bearings (Figure 6.4f in Chapter 6).

Deformations of the housing are shown exaggerated in Figure 9.15, but even small deformations can't be neglected. In Chapter 7, Example 7.6, tooth misalignment of only 10 μm led to noticeably uneven load distribution across the face of the gear ($K_W = 1.14$).

Deformations caused by axial forces have been measured on two-stage helical reduction gears.[2] The maximal increase in distance between the side walls has been found as large as 0.10–0.15 mm and 0.13–0.22 mm for gears with a total center distance of 500 mm and 650 mm, respectively.

9.5   HOUSING SEALS

To prevent communication between the inside of the housing and the surroundings, all locations where this communication may take place must be sealed. The consequences of seal failure determine the requirements of the seal specification. For example, if an oil-lubricated mechanism works in a clean room, the ingress of air inside the housing is admissible. Leakage of oil from the housing is unpleasant, but if it is one drop a day, it can be bearable.

If such a mechanism works in a food production line and has contact with food, no oil leakage from the housing can be permitted, and no ingress of the product into the housing can be welcomed either. The same can be said about a submerged mechanism: oil leakage pollutes the water, and water ingress into the housing is likely to damage the mechanism.

The most stringent sealing requirements are necessary if the housing contains substances that are dangerous for the environment or personnel. In these cases, more than one stage of sealing is used as a rule, with control and alarm devices after each stage.

A principal distinction should be made between the sealing of rigid connections and that of movable connections. As the former can be usually made very reliable, the latter have more or less reliable solutions for easy applications, but still need the trial-and-error method where the working conditions are complicated. In the following subsections are discussed features of some sealing methods and devices.

9.5.1   SEALING OF RIGID CONNECTIONS (STATIC SEALS)

These connections can be divided into two groups:

> Group 1. Connections in which an exact relative position of the connected parts must be ensured: This requirement is indispensable with reference to parts that are elements of a mechanism; in other words, the mechanism can't work without any of these parts. Among

them can be mentioned connection of housing parts that contain supports of shafts, connection of bearing cap to the housing, provided that the cap adjusts the axial position of the shaft or the axial play of the bearing, and so on.

Group 2. Connection of parts that are not involved in the kinematics of the mechanism: Such parts, like covers of inspection hatches, blind flanges of manufacturing holes, pipe joints, etc., can be positioned with greater tolerances relative to the housing, provided that their displacement doesn't impede their function. That means, for example, the hatch cover must close up the opening at all tolerances, the pipe flange should be attached so that the tube is coaxial (within reasonable limits) with the duct in the housing, and the like.

Parts that belong to Group 1 must be in direct contact (metal-to-metal) with each other over their joint surfaces. These surfaces must be machined to tight tolerances (and sometimes a fine surface finish) to keep unchanged the precise geometry of the parts after assembly operations.

If the contact between two joint surfaces is imperfect, the parts become deformed while tightening the bolts, and their exactness is impaired. But this is not the only problem. After dismantling and reassembly of such parts, their deformation will be different than before, depending on the sequence of tightening the bolts, tightening forces and residual deformation of the joined parts. Therefore the geometry errors after reassembly may become increased even if the parts have been machined jointly (i.e., while assembled), and each reassembly may give different values of these errors.

It is very difficult to achieve leakproofness in plain metal-to-metal contact: the surfaces must be very exact, very smooth, and tightly pressed against each other over the entire surface. (The latter is not always easily achieved, because the bolted connection is "spotty"; see Chapter 2, Section 2.2.) Therefore, the joints are usually sealed with special compounds which are initially liquid. These compounds don't hinder the metal-to-metal contact, but they fill up all the undulations of the surfaces and later on, after the bolts are tightened, they become in small gaps semiliquid or rubberlike, vulcanized to the joint surfaces. In that way are eliminated excessive requirements to the joint surfaces and use of additional bolts where they are not needed for strength or rigidity.

To dismantle this connection, puller bolts are used. Threaded hole 7 for that bolt has been shown in Figure 9.8. (Note: if you unscrew the connecting bolts first, your attempts to separate the parts using puller bolts may be more successful.) Before reassembly the joint surfaces must be cleaned from the previous sealing material. This operation is rather time consuming, and when the mechanism is designed for repeated dismantling, the sealing effect can be achieved more efficiently using O-rings installed in milled grooves (Figure 9.16).

Special attention should be paid to the dimensions of the groove (Figure 9.16b). Its depth h must be less than the minimal diameter d of the O-ring cross-section; otherwise, it will not seal. The width of the groove b should be somewhat bigger than d to provide the needed volume for the ring's deformation. The recommendations for the groove dimensions, tolerances, and surface roughness can be found in O-ring catalogs.

O-rings, when properly installed, are very reliable and able to provide leakproof sealing for liquid and fluid medium at a pressure of hundreds and even thousands of bar.

The same can be said about bearing caps. Figure 9.13 presents several examples of this kind. Shim 9 sets up the axial position of the bevel pinion. It is made of two semirings (to install them without pulling the input shaft completely out), and it can't be sealed by applying sealing compound because there is a gap between two half-shims. Therefore, cartridge 10 is sealed with O-ring 11. For cover 3, the considerations are similar: the axial position of bevel gear 1 is adjusted by split shim 12, and cover 3 is sealed with O-ring 13. Cover 2 is not adjustable, but it is sealed in the same way as cover 3, because they are made to allow their positions to be exchanged.

Shim 14 adjusts the axial play of tapered bearings 15, and cover 16 is sealed using sealing compound. The same principle is applied to cap 17.

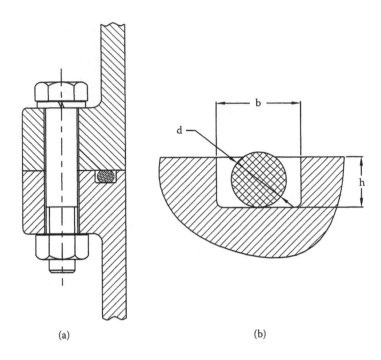

(a)   (b)

FIGURE 9.16 O-ring seal.

Parts belonging to group 2 can be connected through a deformable gasket. In Figure 9.13, to that group belong hatch cover 4 (gasket 18), breather 5 (gasket 19), and bearing cover 20 (gasket 21).

Cover 20 doesn't touch the bearing, and its axial position may change slightly (depending on the deformation of the gasket) without any influence on the moving parts. But the cover must be concentric with the bearing, because it contains shaft seal 22. For that purpose, the cover is provided with a centering spigot (also known as a pilot).

Sealing using deformable gaskets is very reliable provided that the gasket is compressed with sufficient pressure around its entire contour. In Chapter 2, it was noticed that the bolt connection is spotlike and the load distribution in the connection is uneven. Figure 9.17 shows the deformation of cover 1 pressed by bolts against housing 2 through deformable gasket 3. (Here, F is the bolt tightening force.) As is shown, the deformation of the gasket is uneven, and between the bolts there is a gap between the flange and the gasket. It is clear that to make the pressure distribution more

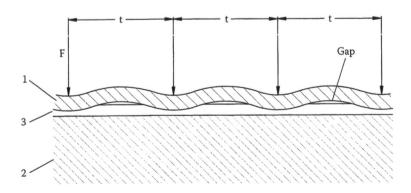

FIGURE 9.17 Deformation of a flange on a pliable gasket.

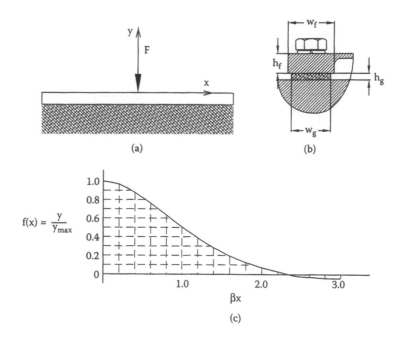

FIGURE 9.18 Deformation of a beam on elastic foundation.

even and prevent the formation of gaps, the flange should be made thicker, or the gasket more compliant, or the bolts spacing to be decreased. But how much?

Let's consider the flange on a gasket as a beam glued to the elastic foundation (Figure 9.18a).[3] Vertical movement of the beam y at a distance x from the force F is obtained from the equation:[4]

$$y = \frac{F\beta}{2k} e^{-\beta x}(\cos \beta x + \sin \beta x) \tag{9.2}$$

where

F = tightening force of the bolt (N)

k = modulus of subgrade reaction (specific stiffness of the gasket, measured in newtons per millimeter of length per millimeter of deformation = N/mm²), which is

$$k = \frac{w_g E_g}{h_g} \; N/mm^2 \tag{9.3}$$

where

w_g and h_g = width and thickness of the gasket respectively (mm)

E_g = modulus of elasticity of the gasket (N/mm²)

$$\beta = \sqrt[4]{\frac{k}{4E_f I_f}} \; mm^{-1} \tag{9.4}$$

E_f = modulus of elasticity of the flange (N/mm²)

I_f = moment of inertia of the flange cross section (mm⁴)

(See symbols in Figure 9.18b).

The maximal deformation of the gasket takes place under the bolt head (at $x = 0$). Provided that the influence of the neighboring bolts is negligible, the maximal deformation can be obtained from Equation 9.2:

$$y_{max} = \frac{F\beta}{2k} \tag{9.5}$$

In sections distant from the bolt, the gasket deformation y is less. We obtain the relative deformation y/y_{max} from Equation 9.2 and Equation 9.5:

$$\frac{y}{y_{max}} = e^{-\beta x}(\cos\beta x + \sin\beta x) = f(\beta x)$$

The graph of function $f(\beta x)$ is represented in Figure 9.18c. From the graph one can see that at $\beta x = 2.35$, deformation of the gasket equals zero, and then it becomes negative. That means, if the flange is not glued to the gasket, you should expect to have a gap between them in this area. Hence, to prevent formation of a gap, the spacing of bolts t (see Figure 9.17) should meet at least the following conditions:

$$\beta x \leq 2.35; \quad x = \frac{t}{2}$$

$$t \leq \frac{4.7}{\beta} \tag{9.6}$$

But this condition is still deficient, because zero pressure between the connected parts and the gasket is insufficient for reliable sealing effect. The minimal pressure on the gasket $p_{g,min}$ depends on the material of the gasket, the kind of sealed medium, and the pressure p_m of the medium. For example, the minimal gasket pressure for rubber gaskets that seal water can be obtained from the equation:[5]

$$p_{g,min} = (2.24 + 0.675p_m)h_g^{-0.55} \ MPa \tag{9.7}$$

where
  p_m = water pressure (MPa)
  h_g = gasket thickness (mm)

It is obvious that the bolt spacing should be less than was obtained earlier (Equation 9.6). So, we are to derive equations to calculate the gasket pressure. The gasket deformation y is given by

$$y = \frac{p_g w_g}{k}$$

From here,

$$p_g = \frac{k\, y}{w_g} \tag{9.8}$$

From Equation 9.5 and Equation 9.8 can be obtained the maximal gasket pressure $p_{g,\max}$ (at $x = 0$):

$$p_{g,\max} = \frac{k\,y_{\max}}{w_g} = \frac{F\beta}{2\,w_g}\ MPa \tag{9.9}$$

The maximal pressure is limited by the strength of the gasket. For rubber gaskets it is about 15 – 20 MPa. The minimal pressure (at $x = t/2$) is influenced by two neighboring bolts and can be obtained from Equation 9.2 and Equation 9.9:

$$p_{g,\min} = \frac{F\beta}{w_g}e^{-\beta\frac{t}{2}}\left(\cos\beta\frac{t}{2} + \sin\beta\frac{t}{2}\right) \tag{9.10}$$

EXAMPLE 9.2

The cover of an inspection hatch is attached to a relatively massive housing through a rubber gasket. Dimensions are as follows (see Figure 9.18b): $w_f = w_g = 20$ mm, $h_f = 10$ mm. Rubber gasket of Shore hardness 50, $h_g = 1$ mm. The cover is made of gray cast iron, $E_f = 0.9 \cdot 10^5$ MPa. Bolts M8-4.8, tightening force $F = 4000$ N. Sealed medium: water, $p_m = 0$.

The modulus of elasticity of rubber depends on the geometry of a specimen and degree of deformation. For deformation of 25% and less, the E_g value is[6]

$$E_g = E_{g,b}\left(1 + \frac{w_g}{2\,h_g}\right)$$

where $E_{g,b}$ = base value of the modulus of elasticity. For rubber of Shore hardness 40, 50, 60, and 70, it is recommended that $E_{g,b} = 1.5, 2.0, 3.0,$ and 4.0 MPa, respectively. In our case,

$$E_g = 2.0\left(1 + \frac{20}{2}\right) = 22\ MPa$$

It seems that now we have all we need to calculate the bolts spacing t.

From Equation 9.7, the minimal gasket pressure should equal

$$p_{g,\min} = 2.24 \cdot 1^{-0.55} = 2.24\ MPa$$

From Equation 9.3,

$$k = \frac{20 \cdot 22}{1} = 440\ MPa$$

Moment of inertia of the flange cross section

$$I_f = \frac{w_f h_f^3}{12} = \frac{20 \cdot 10^3}{12} = 1667\ mm^4$$

From Equation 9.4,

$$\beta = \sqrt[4]{\frac{440}{4 \cdot 0.9 \cdot 10^5 \cdot 1667}} = 0.0293 \ mm^{-1}$$

From Equation 9.9,

$$p_{g,max} = \frac{4000 \cdot 0.0293}{2 \cdot 20} = 2.93 \ MPa$$

From Equation 9.10,

$$p_{g,min} = \frac{4000 \cdot 0.0293}{20} f(\beta x) = 2.24 \ MPa$$

$$f(\beta x) = \frac{2.24 \cdot 20}{4000 \cdot 0.0293} = 0.382$$

From the graph of function $f(\beta x)$ (Figure 9.18c), we can find that $f(\beta x) = 0.382$ at $\beta x = 1.2$. Because for the calculated position $x = t/2$, $\beta t/2 = 1.2$, and the bolts spacing is

$$t = \frac{2 \cdot 1.2}{0.0293} = 82 \ mm$$

If the bolts spacing is not bigger than that, the hatch cover seal will be leakproof. With the increasing of the gasket thickness to 2 mm, the minimal bolt spacing grows to 118 mm.

Well, I don't like the cast cover! I want to make the cover from steel sheet of 4 mm thick with the same gasket. Let's make the calculations needed for the new case. Because the gasket is the same, the values of $p_{g,min}$ and k remain unchanged. Next, the flange, $E_f = 2 \cdot 10^5$ MPa. The moment of inertia of the flange can be approximately calculated as for a strip 20 mm wide multiplied by approximately 1.5 to consider the supporting effect of the sheet:

$$I_f = \frac{20 \cdot 4^3}{12} 1.5 = 160 \ mm^4$$

$$\beta = \sqrt[4]{\frac{440}{4 \cdot 2 \cdot 10^5 \cdot 160}} = 0.043 \ mm^{-1}$$

$$p_{g,max} = \frac{4000 \cdot 0.043}{2 \cdot 20} = 4.3 \ MPa$$

$$f(\beta x) = \frac{2.24 \cdot 20}{4000 \cdot 0.043} = 0.26$$

$$\beta \frac{t}{2} = 1.45; \quad t = \frac{1.45 \cdot 2}{0.043} = 67 \ mm$$

Increase in the gasket thickness from 1 mm to 2 mm enables the increase in bolt spacing from 67 mm to 94 mm.

If the part the hatch is bolted to is not massive, its deformation should be taken into account as follows:

$$\frac{1}{E_f I_f} = \frac{1}{E_1 I_1} + \frac{1}{E_2 I_2}$$

where symbols 1 and 2 refer to the cover and housing, respectively.

The calculations show what is obvious: the more rigid the metal parts and the more pliable the gasket, the larger can be the bolts spacing. For the greatest pliability, substitute the flat gasket with an O-ring installed in a groove.

9.5.2 SEALING MOVABLE JOINTS

What a wonderful thing it would be if we could make a mechanism with all moving parts inside the housing, as in a laundry washer! You close the hatch cover, turn it on, and it works. No moving parts can be seen outside the casing. But usually, such a pleasure is inaccessible. For example, the wheels of a car or the propeller of a ship hardly can do their work if they are hidden inside a closed housing. Even the unit-type design (for example, the motor gear) doesn't save us from the necessity to seal the moving connections between the gear housing and the input and output shafts.

But why grieve so deeply about this kind of seal? Are they problematic? Yes, they are. In comparison with the sealing of rigid connections, they are more expensive and less reliable. When the seals must be tight, they also become short-lived, and they consume quite a lot of energy to spend it for their own wear, heating, and destruction. And all we can do is decrease the deficiencies, not eliminate them.

There are many types of seals for movable connections that meet different service conditions. They can be divided into three main groups: noncontact seals, contact seals, and combined seals.

The latter may provide a satisfactory solution in some cases in which a single seal is not efficient, is too expensive, or difficult to replace when failed.

Fortunately, the seal is not a part of the mechanism, and it can be made as a separable unit, which can be easily replaced when it fails, or upgraded in order to improve its reliability, without changing anything in the mechanism itself. For this reason, the seal is one of those rare subjects in which the designer's imagination is practically unlimited; he might fight the leakage by all possible means, including such an exotic device as a saucepan under the seal. (By the way, if the saucepan is attached to the housing, it can be called "oil pan," to make it look and sound more technical.)

Here, we are going to consider only several widespread versions of seals in order to explain the main principles of their operation.

9.5.2.1 Noncontact Seals

They are usually for rotating shafts, and they combine an obligatory small gap between the stationary and rotating parts and optional elements that restrict the access of oil to the small gap. These seals are really excellent: relatively cheap, unrestrictedly long-lived, with practically no waste of energy. High speed only makes them more effective. But they are not tight, and that restricts their range of application.

Figure 9.19 represents several examples of noncontact seals. In the design shown in Figure 9.19a, the bearing is protected from inside by cover 1 and from outside by cover 2. These covers form a closed chamber filled with grease through greaser 3. Cover 1 (from the oil side) is formed so as to prevent the oil that flows down the wall to get to the shaft and enter the gap.

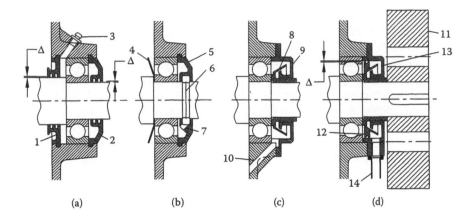

FIGURE 9.19 Noncontact seals.

The annular clearances Δ should be as small as possible, but not too small to prevent grazing of the rotating shaft against the covers.

EXAMPLE 9.3

Ball bearing 6310 has maximal internal clearance (before loading) of 18.7 μm, and its radial elastic deformation equals 60.8 μm (see Chapter 6, Example 6.7). So the possible radial displacement of the shaft in the bearing equals $60.8 + 0.5 \cdot 18.7 = 70.2$ μm. The diameter of the connection between the housing and cover equals 110 mm, the fit is *H6/h7*. The maximal diametral clearance equals 52 μm. The tolerance of concentricity of the inner and outer diameters equals, say, 10 μm, and the radial runout of the shaft in the bearing equals 10 μm as well. So the total eccentricity of the shaft relative to the bore of the cover can be as big as $70.2 + 0.5(52 + 10 + 10) = 116.2$ μm. That means, clearance Δ in this case should not be less than 0.12 mm; usually, it is greater.

In case of a sliding bearing the minimal clearance, Δ can be much bigger.

Figure 9.19b shows a seal with oil slinger 4 (placed inside) and cover 5. Sharp ridge 6 throws the oil off the shaft, cover 5 collects it in the annular groove, and then it flows through drain hole 7 into the housing. Figure 9.20a shows the oil flow in this seal. It is clear that part of the oil thrown off the ridge can come back to the shaft and leak out through clearance Δ.

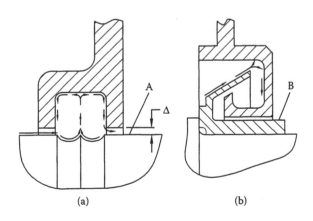

FIGURE 9.20 Sealing properties of noncontact seals.

More effective is the seal shown in Figure 9.19c. Slinger 8 throws off the oil; cover 9 is so shaped that the oil can't come back to the shaft (see also Figure 9.19b); it flows down and returns to the housing through hole 10.

The seals shown work satisfactorily on condition that they are not flooded with oil (certainly, not submerged) and the air pressure inside the housing is not greater than that outside. The last condition is usually achieved using a breather (5 in Figure 9.13 and Figure 9.14). But if there is a large rotating part, such as flywheel 11 in Figure 9.19d, it creates underpressure in the central area, which exhausts the oil mist from the housing. In such a case, the modification shown in Figure 9.19d has proved itself as an effective seal. The slinger is complemented with ring 12, which runs with a small clearance Δ against cover 13. The clearance is filled with oil and obstructs the interior of the housing. The oil that passed this clearance is thrown off the slinger, and then it drains downwards into the housing through pipe 14, which is connected to the housing under the oil level. In this way, the communication between the interior of the housing and the area of underpressure in the center of the flywheel has been hydraulically closed.

9.5.2.2 Contact Seals

They provide tightness owing to the prestressed contact between the stationary and moving elements of the connection. As in all sliding contacts, the wear and energy loss depend here on the characteristics of lubrication. As a lubricant, the products released by the seal packing (such as special grease, graphite, and polytetrafluoroethylene [PTFE]) may serve. These substances decrease the friction coefficient and the energy loss, but the friction coefficient still remains relatively high. The liquid or semiliquid friction, which is more appropriate for continuous duty and, particularly, for high-speed applications, is usually achieved by making the sealed medium serve as a lubricant. But in this case, some leakage may appear. So the requirement for a high grade of tightness results in unlubricated or boundary friction in the seal area.

Among the multitude of contact seals, we will discuss the following most widespread types: stuffing box packing, piston ring packing, O-rings, lip-type seals, and mechanical seals.

9.5.2.2.1 Stuffing Box Packing (SBP) (Figure 9.21)

This has been in use over the ages for both rotational and reciprocating services, and now it looks a bit archaic. But it has some important advantages that have promoted its use to the present day: low cost, easy adjustment (by tightening or slackening nuts), and easy replacement.

There is not only the possibility, but also the need for adjustment. This operation requires certain training and experience, or at least a torque meter (torque wrench); so it is an advantage and drawback at the same time. But when the seal starts leaking too much, the possibility to set it up (i.e., to tighten it) by just tightening a nut can be very attractive.

The SBP consists of housing 1, packing rings 2, and packing cover (gland) 3 pressed by screws. When stuffing boxes were first developed in the 18th century, the packing was usually made of felt or asbestos impregnated with oil, graphite, and other friction-reducing materials, such as animal fat. At present, diverse kinds of metallic and nonmetallic packing are produced to meet miscellaneous requirements regarding the kind of sealed medium, pressure, temperature, sliding speed, and so on. In important cases, the manufacturer should be asked about the best kind of packing for the particular application.

In effect, SBP for continuous duty is a kind of small clearance seal, as described previously, but the sealing element is made of soft, nonmetal stuff, which can be easily deformed toward the shaft, until the clearance is very small or zero. Because the packing material is compliant, the zero gap is tolerable, but at a higher speed, the heat created in the contact can burn off the contacting layer of the packing and damage the shaft. Therefore, at these conditions, the SBP should be so adjusted that there is a constant leakage so as to lubricate the contacting surfaces. The leakage can be adjusted so that both the temperature of the seal and the leakage amount are tolerable.

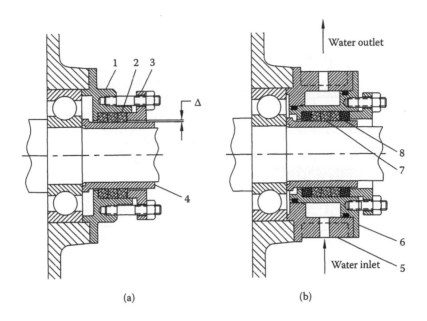

FIGURE 9.21 Stuffing box packings.

For instance, if a propeller shaft of a ship is sealed up with an SBP, there can be a leakage of several drops in a minute. It is not seen under the floor, and the bilge pump sucks the leaking water off (say, once a day) and restores it to the sea. This is one of the rare cases where a leakage is tolerable. But, usually, the leakage is unpleasant or even unacceptable, and the kind of the seal should be chosen so as to meet this requirement.

At a low speed and short-time duty, the packing can be pressed tightly to the shaft and provide dependable leakproofness at a high pressure, such as 500 bar and more.

The required hardness of the metal part (shaft, or sleeve 4 in Figure 9.21) that slides against the packing depends on the internal pressure, the sliding speed, and the needed service life. Usually, it is between HRC 40 and HRC 60. The surface roughness is critical to the service life of the packing. For heavy duty applications, it is recommended to grind and polish these surfaces to $R_a = 0.16$–0.32 μm.

The radial clearance Δ between the stuffing box and rotating part should be small, particularly at high pressure, but not smaller than needed to ensure free movement of the shaft within the box (the tolerances of size and alignment should also be taken into consideration).

The leakage through the SBP can be minimized when it is cooled from the outside, as shown in Figure 9.21b. Here, parts 5 and 6 form an annular chamber filled with running water. Packing rings 7 are supported from both sides by PTFE rings 8. They enable minimization of gap Δ and better performance of the packing rings that can't be pressed into the gap by the high pressure of the sealed medium. (At a low pressure, there is no need in rings 8.)

9.5.2.2.2 Piston Ring Packing (PRP) (Figure 9.22)

This represents another way to decrease the clearances between the stationary and rotating parts of a seal. Springy rings 1 are installed in annular grooves made in rings holder 2 with a small axial gap δ. The outer diameter d_o of rings 1 before assembly is made somewhat greater than the inner diameter d of sleeve 3 [$d_o \approx (1.02$–$1.03)d$; see Reference 7], so that the rings are tightly fitted to the female part. When the shaft rotates, the rings remain stationary or slightly sliding relative to sleeve 3. Nevertheless, to prevent local wear and formation of grooves on the inner surface of sleeve 3, its hardness should be about HRC 50–HRC 65. It is achieved by quenching, carburizing,

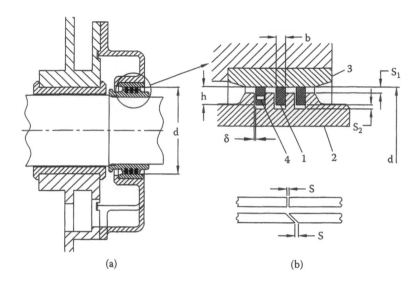

(a)   (b)

FIGURE 9.22 Piston ring packing.

or nitriding of steel. The same hardness is needed for the annular grooves on ring holder 2. In order to lower the friction between rings 1 and the steel holder and sleeve, the rings are usually made from copper alloys or gray iron of special grade.

PRP is a reliable seal; when the pressure is low or zero, it can work at very high speed (about 60–80 m/sec). It is also able to work at a rather high pressure, if the speed is not too high (say, less than 10 m/sec). The recommended dimensions are as follows[7,8] (see Figure 9.22): $\delta = 0.05$–0.1 mm, $s = 0.3$–0.5 mm (after assembly), $s_1 = 0.5$–0.7 mm, $b = (0.03$–$0.04)d$, $h = (0.05$–$0.1)d$. Gap s_2 should be sized to prevent contact between the ring holder and the inner diameter of the rings after assembly.

The radial thickness of the rings (h in Figure 9.23) depends on the strength of the ring material, because, during assembly, the ring must be elastically strained so that its inner diameter d_i becomes somewhat greater (for example, by amount $k = 0.5$ mm) than the outer diameter of the ring holder d_h.

Figure 9.23a shows the initial position, where the ring is free and its inner diameter is less than the outer diameter of the ring holder. In this position, the width of slit z should equal

$$z = \pi(d_0 - d) + s \tag{9.11}$$

That means, after the ring is installed into the female part, there must remain some small gap s.

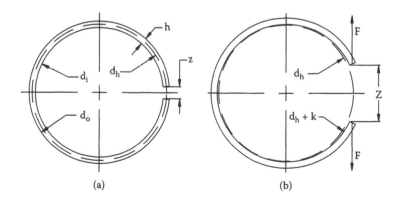

(a)   (b)

FIGURE 9.23 Ring deformation.

Figure 9.23b presents the moment of assembly, when the ring is deformed by forces F so that its inner diameter becomes large enough for assembly. The increased width of slit Z can be found from this equation:

$$Z - z = \pi(d_h + k - d_i) \qquad (9.12)$$

(Note: the last two equations result from the basic knowledge that the perimeter of a circle equals π times diameter.)

Now we can define the force F needed to deform the ring:[9]

$$F = \frac{E I \cdot \Delta z}{3 \pi r^3} \ N$$

where

E = modulus of elasticity of the ring's material (MPa)
I = moment of inertia of the ring's cross section

$$I = \frac{b h^3}{12} \ mm^4$$

$\Delta z = Z - z$ (mm)
r = mean radius of the ring (mm)

The maximal bending moment M (in the middle of the ring) equals $F \cdot 2r$ (N·mm). Consequently, the bending stress

$$\sigma = \frac{F \cdot 2r \cdot 6}{b \cdot h^2} = \frac{E I \cdot \Delta z}{3 \pi r^3} \frac{12r}{b h^2} = \frac{E h \cdot \Delta z}{3 \pi r^2} \ MPa \qquad (9.13)$$

As the height h increases, the Δz value increases as well, so the bending stress increases rapidly.

The slit of a ring is usually made as shown in Figure 9.22b: straight or at an angle of 45°. The small gap enables the media to access the groove walls and provide lubrication. If it is not enough, holes 4 (Figure 9.22) can be drilled to supply more oil to the sliding surfaces.

EXAMPLE 9.4

The inner diameter of the female part $d = 100$ mm. The outer diameter of a free ring, $d_o = 103$ mm. The cross section of the ring, $b = 4$ mm, $h = 8$ mm. The outer diameter of the spring holder, $d_h = 99$ mm. The desirable diametral clearance during assembly, $k = 0.5$ mm. The desirable slit width in assembled condition, $s = 0.5$ mm. The inner diameter of the free ring, $d_i = d_o - 2h = 103 - 2 \cdot 8 = 87$ mm. The mean radius of the ring, $r = (d_o - h)/2 = 47.5$ mm. The ring is made of beryllium copper with the following mechanical properties: tensile strength $S_u = 1130$ MPa, yield point $S_y = 960$ MPa, modulus of elasticity $E = 1.27 \cdot 10^5$ MPa.

First, we define the slit width z, using Equation 9.11:

$$z = \pi(103 - 100) + 0.5 = 10 \ mm$$

Thus, if the slit width in the free condition equals 10 mm, in the assembled condition, it will be 0.5 mm. But in operation, the temperature of the ring may be higher than that of the female part 3, and the slit may become smaller. The length L of the ring

$$L = 2\pi \cdot r - z = 2\pi \cdot 47.5 - 10 = 298.5 \ mm$$

Let's assume that the ring is warmer than sleeve 3 by 50°C. Its additional elongation ΔL caused by this difference is

$$\Delta L = L \cdot \alpha \cdot \Delta t$$

where for the beryllium copper $\alpha = 18 \cdot 10^{-6}$ 1/°C.

$$\Delta L = 298.5 \cdot 18 \cdot 10^{-6} \cdot 50 = 0.27 \ mm$$

The result shows that the ring can be warmer even by 90°C without any problem.
We obtain the Δz value from Equation 9.12:

$$\Delta z = \pi(99 + 0.5 - 87) = 39.3 \ mm$$

We obtain the maximal stress from Equation 9.13:

$$\sigma = \frac{1.27 \cdot 10^5 \cdot 8 \cdot 39.3}{3\pi \cdot 47.5^2} = 1878 \ MPa$$

Wow! It is far beyond the strength of the ring's material! Well, we must decrease the height of the cross section: say, $h = 6$ mm. In this case, $d_i = 103 - 2 \cdot 6 = 91$ mm; $r = (103 - 6)/2 = 48.5$ mm; and $\Delta z = \pi(99 + 0.5 - 91) = 26.7$ mm.
The maximal bending stress in this case is given by

$$\sigma = \frac{1.27 \cdot 10^5 \cdot 6 \cdot 26.7}{3\pi \cdot 48.5^2} = 918 \ MPa$$

This stress is less than the yield point but too close to it. Therefore, it is preferable to decrease the height of the ring, say, to $h = 5.5$ mm. At this height, the maximal stress diminishes to $\sigma = 736$ MPa.

9.5.2.2.3 O-Rings

These are relatively rigid, and the inevitable tolerances for the cross-sectional diameter of the O-ring and the depth of the groove may result in high maximal pressure in the sliding contact. This seal is not applicable in rotating connections, because no lubrication can enter the highly loaded contact zone, and the seal is doomed to overheating and quick wear. But in connections with reciprocating motion, the contact area moves, and at each moment, new surfaces covered with oil come into contact. Therefore, in such connections, O-rings are widely used.

Figure 9.24a presents the O-ring seal of a hydraulic cylinder's piston. Rubber ring 1 is installed in an annular groove 2 on the piston. The O-rings can work under high pressure, as much as hundreds of bars. This pressure may push part of the ring into the clearance, as shown in Figure 9.24b. The thin lip 3 pressed into the clearance is tenuous and can be torn off the ring by friction forces. Therefore, the clearance in the sliding joint should be as small as possible. But the minimal required

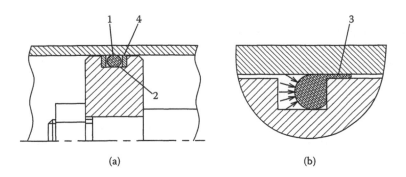

FIGURE 9.24 O-ring seal of a hydraulic piston.

clearance, together with manufacturing tolerances and radial deformation of the cylinder walls, may still result in a sizable maximal clearance, particularly at larger diameters. "Backup" rings 4 are a popular method to prevent O-rings extruding into the clearance. The width of the gap between the backup rings and the cylinder should be kept to a minimum.

9.5.2.2.4 Lip-Type Seals (LTSs)

These come in many styles, united by a common feature: their sealing element is formed as a flexible "lip," which is less sensitive to manufacturing tolerances and is pressed to the sealed surface with a relatively small and controllable force. This enables the sealed medium to enter the contact area, lubricate the surfaces, and improve the friction parameters.

Figure 9.25a shows a typical LTS for rotating shafts. Compliant lip 1 has the inner diameter less than the diameter of the shaft, so it is fitted to the shaft with a certain radial interference. In addition to that interference, the lip is pressed against the shaft by garter spring 2. Insert 3 made of brass or steel is intended to provide rigidity to the peripheral portion of the seal. Owing to that insert, the seal can be firmly pressed into cover 4, so that if the pressure difference is about zero, no other fixation is needed. When the pressure of the sealed medium is greater than zero, but less than 0.5 bar, an additional axial fixation can be provided by annular groove 5 with sharp edges. The deformed rubber layer flows slowly into the groove, and the shear resistance of the rubber augments friction to hold the seal in place.

This kind of seal can work also at a higher pressure of the medium (usually not more than 2–3 bars). Protector 6 (Figure 9.25b) backs the seal in the axial direction and prevents the lip from turning inside out. (In this case, the protector is also used to form a labyrinth seal to protect the

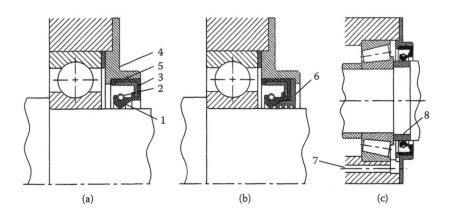

FIGURE 9.25 Lip seals for rotating shafts.

LTS against contamination from the outside.) But any increase in the pressure of the sealed medium also increases the pressing force between the lip and the shaft. This impedes oil from entering into the seal area, and increases the friction force and temperature. To achieve an acceptable service life at these pressures, the permissible sliding velocity should be decreased. Therefore, the over-pressure should be reduced as far as possible. For example, Figure 9.25c shows an LTS placed near a tapered roller bearing. Such a bearing produces overpressure on the side of bigger diameter of the cone (pumping effect), and hole 7 is made to release this pressure.

As compared with SBP, the seal area in LTS (where the lip contacts the shaft) is very small in the axial direction, and the energy loss is also much less. The easy access of the sealed medium to the small seal area enables both better lubrication of this area and also good cooling by the medium. Because temperature is the determining factor for the serviceability of rubber and polymer parts, the aforementioned combination of features in this design allows achieving both a long service life and a high grade of tightness.

Generally, the lifetime of lip seals ranges from 1000 to 5000 hours. This durability is achieved owing to the thin layer of the sealed medium that provides lubrication of the sliding contact. The friction mode (liquid, semiliquid, or boundary) depends on many factors, such as sliding velocity, pressure in the contact region, the kind and viscosity of the sealed medium, runout of the shaft, the kind of materials, and roughness of the metal surface. The latter should be made very smooth (as a rule, it is polished), but not too smooth, because the surface undulations facilitate lubricant retention. The recommended surface roughness ranges from $R_a = 0.16$ μm to 0.63 μm (the less the surface roughness, the higher is the admissible sliding velocity).

The process of polishing is important because it brings about plastic deformation of a thin surface layer in the sliding direction, blunts the asperities that are sharp after grinding, and finally makes the surface condition closer to that attained after bedding-in.

The metal surface should also be hard enough to slow down its deterioration with time. It can be made harder by appropriate heat treatment or, when needed, using carburizing, nitriding, or chrome plating. Significant improvement in lubrication conditions can be achieved by forming shallow flutes on the lip surfaces.[8,10] When the rotation is unidirectional, the flutes are directed so as to create a pumping effect and cause the media to flow inside from the sealing area, as shown in Figure 9.26a. For bi-directional operation, the flutes are shaped symmetrically (Figure 9.26b). Their pumping effect is smaller but still exists.

Better prerequisites to formation of oil film in the seal area can be also provided by making very shallow, up to 0.05 mm, flutes on the shaft (usually by knurling).[8] From the manufacturing point of view, it is more convenient to obtain all the desired features of the metal surface on a small separated part, such as sleeve 8 in Figure 9.25c, tightly fitted to the shaft. This also enables

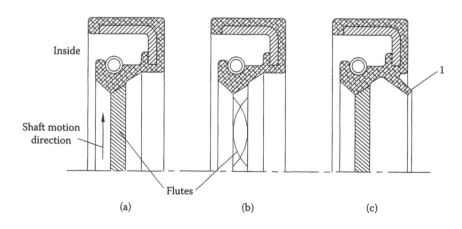

FIGURE 9.26 Improved versions of lip-type seals.

replacement of this part when it is worn out. But often the lack of space, impairment caused to the shaft strength (owing to the tight fit) and to the heat absorption from the sliding area into the shaft (despite the tight fit) make this design inappropriate.

The deterioration of the sealing surfaces can be forced by contaminations coming to the seal area from outside (dust, dirt) and from inside (wear debris, oil decomposition products). The inside-born contamination can be reduced by the following measures:

- Thoroughly clean sand and scale from cast parts, and paint them with oil-resistant, smooth, easily cleanable coatings.
- Avoid sliding where possible (for example, fixing the rings of rolling bearings against rotation in the housing).
- Increase the wear resistance of sliding surfaces by making them harder and better lubricated.
- Clean the oil by magnetic plug or by filtering.
- Replace contaminated oil in a timely manner.

Outside pollution can be prevented from entering the seal area by an additional barrier. The LTS working in polluted surroundings is usually provided with an additional lip ("wiper lip" 1, Figure 9.26c) which protects the main seal from contamination. The wiper lip runs dry, with increased friction and generation of heat. Therefore, it is subjected to increased wear, but in the long run, it works as a noncontact seal with a very small gap.

The wiper lip worsens the working conditions of the main seal because it increases the temperature in the seal area. Besides, the pumping action of the main seal creates a vacuum between the lips.[10] This vacuum presses the lips harder on to the shaft and increases the heating and wear of the lips, so that the main lip may become damaged before the wiper lip wears off enough to lose its airtightness. Therefore, seals with a wiper lip have lower permissible sliding speeds (by about 30%[8]). Sometimes the wiper lip is provided with a small groove that doesn't impair considerably its tightness but prevents creation of vacuum. LTSs with no additional lips are used, depending on the rubber type and the cooling conditions, at sliding speeds up to 10–15 m/sec, rarely up to 20 m/sec.

LTS for linear motion is shown in Figure 9.27. V-shaped deformable rubber rings are placed into annular grooves of the piston. The radial dimension W of the ring before assembly is greater than the depth of the groove, so the lips (when assembled) are pressed to the cylinder and the bottom of the groove by the elastic force of the rubber. The hydraulic pressure applies to the lips an additional force needed to make the seal perfectly leakless. These seals are available in many variants, some with energizing elements within the v-cavity, made of spring steel, polymer, or other materials (Figure 9.27b).

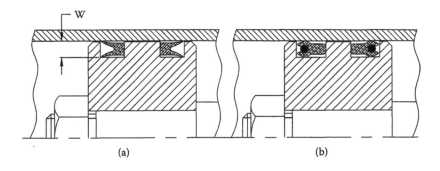

(a)   (b)

FIGURE 9.27 Lip-type seals of hydraulic pistons.

9.5.2.2.5 Mechanical Seals (MS)

These are the most reliable but also the most expensive type of seals. Their sealing element is formed by a pair of sliding rings (1 and 2 in Figure 9.28a; 13 and 14 in Figure 9.28b). One ring of each pair (1 and 14) is stationary, and the other rotates with the shaft. To decrease the friction coefficient and obtain better wear resistance, the materials of the rings are combined as in the sliding bearings: one is soft and antifriction, and another is hard. The hard ring is usually made of very hard material, like tungsten carbide or silicon carbide. The mate is usually made of artificial carbon impregnated with antifriction stuff, such as phenolic resin or babbitlike composition. The contacting surfaces of these rings are very finely finished, so that their flatness and smoothness are extremely good. This enables them to work with a very small wear and keep tightness during long-run operation.

The rings must be snug against each other independently of the tolerances and movements of the adjoined parts. This is achieved owing to the following obligatory design features:

- The rings should be so supported and attached to the adjacent parts that the forces applied to them don't cause inadmissible deformations impairing their flatness.
- One of the rings must have the possibility of self-alignment relative to the other ring.
- This ring should be pressed against its mate by a spring (or set of springs) that has a twofold purpose: it provides the pressure needed to make tight the sliding contact of the rings, and it keeps the rings aligned and compressed evenly over the circumference.

Whether the spring-loaded self-aligning ring is rotating or stationary can be important. The first variant is shown in Figure 9.28a. Rotating ring 2 is bonded to cup 3 pushed by springs 4 against stationary ring 1. Springs 4 are placed in pockets of spring holder 5. The inner surface of the spring holder is covered with a rubber layer and is attached to the shaft by the forces of interference fit with no impairment to the strength of the shaft. These friction forces between the rubber layer and the shaft are larger than that between rings 1 and 2, and they are sufficient to make the spring holder and attached parts rotate along with the shaft.

Cup 3 is fitted to spring holder 5 with a sufficient radial clearance that enables the needed self-alignment motion, whereas the tightness of this connection is provided by O-ring 6.

Stationary ring 1 is installed in cover 7, sealed with O-ring 8, and fixed against rotation with pin 9. Ring 1 is pressed against cover 7 by springs 4. So the proper position of the stationary ring

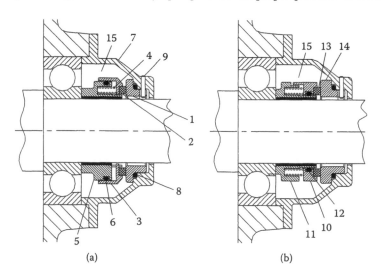

(a)   (b)

FIGURE 9.28 Mechanical seals.

depends on the perpendicularity of the supporting surface of the cover relative to the shaft axis. Thus, cover 7 should be made to close tolerances. If the angle between the face of ring 1 and the shaft axis is not exactly 90°, springs 4 afford full contact of rings over the entire seal surface, because cup 3 can move with respect to the spring holder within the clearances between them. When the shaft rotates, the motion of parts 2 and 3 becomes cyclic, wobbling with respect to the shaft. The problem is that the force needed to impart this motion to parts 2 and 3 depends on the motion amplitude and speed; if the magnitudes of these factors are too big, the springs can fail to provide the needed minimal pressure between rings 1 and 2.

The misalignment of the sealing rings can be caused not only by the manufacturing tolerances (which can be made very small), but also by elastic deformations of parts. Among them should be considered the bending deformation of the shaft, which can result in appreciable misalignment in the seal. But the problem is much easier if the spring-loaded ring is a stationary one as shown in Figure 9.29. Here, the stationary ring 1 is bonded to flanged sleeve 2, and springs 3 are mounted in cover 4. Loose fit sealed by O-ring 5 enables the self-alignment motion of parts 1 and 2. Rotating ring 6 is mounted on sleeve 7 that is sealed by O-ring 8 and fixed against the shaft by screws 9. Ring 6 is sealed in connection with the sleeve by O-ring 10 and fixed against rotation and axial movement relative to sleeve 7 by pin 11 and snap ring 12.

All these fixations can be made directly on the shaft, eliminating sleeve 7. But the designer tried to avoid stress concentrations on the shaft. Let's praise him for his effort.

In this design, as long as the sealing face of ring 6 is made perpendicular to the shaft's axle, it remains perpendicular to the local part of bent shaft independently of the load. As soon as the shaft is loaded, springs 3 align ring 1 relative to ring 6. If the load changes cyclically, the bending of the shaft changes as well, and the process of self-alignment occurs cyclically, but certainly not

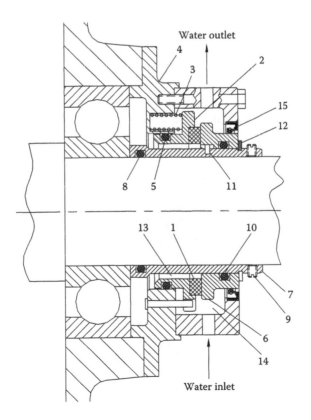

FIGURE 9.29 Mechanical seal with cooling.

in every turn of the shaft, as in the Figure 9.28 design. Therefore, the time available for accomplishing the alignment motion is much greater.

Mechanical seals are able to work under relatively high pressure of the sealed medium, even if it measures hundreds of bars. But in these cases much thought should be given to what the pressure does to the parts. For example, in the MS shown in Figure 9.28a, hydraulic pressure can overcome the spring forces, push parts 2 and 3 to the left, and disconnect rings 1 and 2. Thus, for higher pressure, this design must be changed, for example, as shown in Figure 9.28b. Here, the connection between cup 10 and spring holder 11 sealed by O-ring 12 is made on a smaller diameter, so that the medium pressure pushes cup 10 with glued ring 13 in the same direction as the springs do. The greater the medium pressure, the stronger are the sealing rings pressed together, which is exactly what is needed for better sealing effect. This additional hydraulic pressure can be made larger or smaller by changing the relevant diameters.

Stationary ring 14 is made thicker here to diminish its dishlike deformation under axial load.

Mechanical seals generate more heat than LTSs, and they need appropriate cooling, because most of their parts are sensitive to overheating. For example, the silicone carbide rings may crack; the impregnation stuff may melt and get out of the carbon ring; the O-rings can usually bear temperature up to about 200°C, but then they lose their properties. The kind of cooling depends on the pressure between the rings, but in any case, the seal should be in contact with liquid. Moreover, the liquid should have the possibility to transfer the heat from the seal area to the surroundings. In Figure 9.28, chambers 15 are filled with sealed liquid, which is turbulent because of the proximity of the rotating ball bearing. So the heat rejection from the sealing rings is quite efficient. In Figure 9.29, the rubbing rings 1 and 6 are connected with the main volume of oil through a narrow annular gap 13 that is insufficient for effective heat transfer. Therefore, this seal is cooled with running water that enters chamber 14 from below, so that chamber 14 is filled up with water. Lip seal 15 seals this chamber.

9.5.2.3 Combined Seals

When a single seal doesn't meet the requirements, a combined seal composed of two or several seals may bring the needed effect. Several options of such seals are shown in Figure 9.30. In the option as in Figure 9.30a are combined slinger 1 and labyrinth seal 2. The slinger prevents the oil leakage as described earlier (see Subsection 9.5.2.1), and the labyrinth seal (filled with grease)

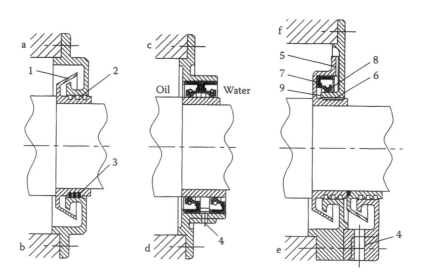

FIGURE 9.30 Combined seals.

prevents entrance of dust into the housing. Figure 9.30b presents a combination of the same slinger and piston ring packing 3. The option as in Figure 9.30c is mainly used when two media should be separated: for example, oil inside the housing and water outside it. The LTS used here, when it is single, works in one direction only; therefore, two seals should be installed as shown.

Figure 9.30d and Figure 9.30e present the combinations of two identical seals with an "informer" between them. When the sealed medium succeeds in surmounting the first seal, the leakage doesn't go out. It flows through drain holes 4 and a pipe duct to a specially appointed place, where the leakage can be detected and treated as needed before it becomes great enough to surmount the second seal. The option as in Figure 9.30f combines impeller 5 and baffle-thread 6, both of which are effective at high speed, and lip seal 7 that provides the needed tightness at a lower speed and at stop. At a higher speed, sealing element 8 of the lip seal separates from cover boss 9 under the action of centrifugal forces. The garter spring (item 2 in Figure 9.25a) can be removed here if there is a need to disconnect the lip seal at a lower speed. An additional decrease in the speed of "liftoff" can be achieved by incorporation of lead shots into the sealing lip 8.

This combination of hydrodynamic seal and a speed-controlled contact seal can provide a reasonable solution at a speed that is too high for a contact seal. If the speed-controlled seal, to be disconnected, must be shifted axially (for example, a mechanical seal), a mechanism with small rods and flyweights can be used.

REFERENCES

1. Richter, R., *Design of Manufacturable Castings*, Mashinostroenie, Moscow, 1968 (in Russian).
2. Klebanov, B.M., Problems of rigidity in design of general-purpose reduction gears, *Transactions of Leningrad Mechanical University*, No. 61, Leningrad, 1967 (in Russian).
3. Klebanov, B.M., Calculation of leak-proofness of flanged covers, *Transactions of Rostov-na Donu University of Agriculture Engineering, Mechanics of Deformable Mediums*, Rostov-na-Donu, 1983 (in Russian).
4. Timoshenko, S.P., *Strength of Materials, Part II, Advanced Theory and Problems*, 3rd ed., D. Van Nostrand Company, Princeton, NJ.
5. Burkov, V.V., Lebedev, B.I., and Mukhametshin, H.H., Investigation of leak-proofness of joints with paronite and rubber gaskets, *Chemical and Petroleum Machinery*, No. 1, 1970 (in Russian).
6. Voloshin, A.A. and Grigoryev, G.T., *Calculation and Design of Flange Connections*, Mashinostroenie, Leningrad, 1979 (in Russian).
7. Orlov, P., *Fundamentals of Machine Design*, Vol. 5, MIR Publishers, Moscow.
8. Commissar, A.G., *Seals of Rolling Bearings*, Mashinostroenie, Moscow, 1980 (in Russian).
9. Timoshenko, S., *Strength of Materials, Part I, Elementary Theory and Problems*, D. Van Nostrand Company, Princeton, NJ.
10. Johnston, D.E., Design aspects of modern rotary shaft seals, *Proceedings of the Institute of Mechanical Engineers*, Vol. 213, Part J, 1999.

10 Bolted Connections (BCs)

BCs are intended to fasten together machine components, which are manufactured singly (for technological or other reasons), but when operating they must work as a unit. The needed strength and rigidity of this connection is achieved, in the first place, by strong pressure of the connected parts against each other. The magnitude of the clamping force should be enough to prevent joint separation or any slippage in the connection.

Destruction of BCs is one of the most widespread sources of failures of machines. Despite the fact that BCs are very simple and their elements are highly standardized, there are still a lot of problems in the design, calculation, and manufacture of important and heavily loaded threaded connections. In some cases, the cost of BCs makes up a considerable part of the production cost.

EXAMPLE 10.1

Figure 10.1 shows the bolted connection of thin-walled steel tubes that are parts of the pillar of a wind-driven power plant.[1] The height of the pillar can be about 100 m, and it is built up of sections 25–30 m long. Each section consists of a tube about 3 m in diameter, with a wall 18 mm thick. The ends of the tube are provided with connection flanges that are 80 mm thick. The sections are attached to each other by 96 high-strength M36 bolts.

The first impression may be that there is some disproportion between the thin wall of the tube and the thick flanges with plenty of very strong bolts (each of them is tightened by a force of about 50 tons). The flange connection seems to be too strong in comparison to the tube. But this impression is deceptive. The pillar is bent by a wind, and at a bending moment of $3 \cdot 10^7$ N·m, the maximal bolt force is as great as 75 ton (750 kN). At that the average tension stress in the bolt shank (i.e., its body) is as great as 1000 MPa, and the stress in the thin-walled tube is only 260 MPa.

This example shows that the part becomes much weightier and more complicated, when it is split. Nevertheless, even if there is a possibility in principal to make the part as one piece, the detailed calculation of costs in production, installation, and service may reveal the profitability of a split part.

The bolt connection can be loaded in shear (when the load is to cause slippage between the joined parts in the joint plane) and in tension (when the load is to separate the joint). It is clear that as long as the shear load is fully borne by the friction between the flanges, the shear load doesn't influence the stress condition of bolts, and hence, doesn't influence their resistance to the tension load. Thus, the calculation for tension can be made irrespective of whether there is a shear load or not. It is also clear that the tension and bending loads on the whole joint assembly do change the pressing forces between the parts, and they must be taken into account when calculating the shear strength of the joint assembly.

Calculation of BC for strength is generally performed in several steps: first, the most loaded bolt should be found by analyzing the load distribution between the bolts; then, the dependence between the load and the stress in the bolt should be studied; and finally, the safety factors at stationary and cyclic load can be determined.

Finding the most loaded bolt (more exactly, *point of attachment*, but it is too long) depends on the definition of the mode of failure. For example, if the connection is loaded with a static shear force, and the load must be transmitted by friction forces between the connected parts, the initial slippage in the most loaded bolt leads immediately to redistribution of the load. In an extreme case

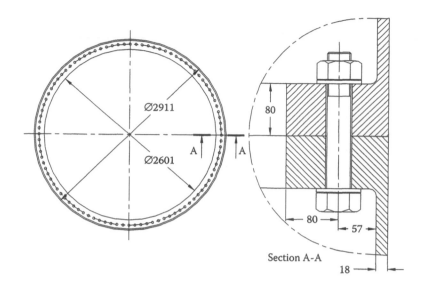

Ø2911

80

A | A

Ø2601

80

57

Section A-A

18

FIGURE 10.1 Flange connection of thin-walled tubes.

(before the connection breaks away), all the bolts become loaded with the same force that equals the product of the bolt tightening force F_t and the friction coefficient f. If the load is cyclic, any slippage in the connection is inadmissible for it may lead to fretting, and no load redistribution is expected. In this case, the most loaded bolt is the one that should be calculated for strength under its maximal load.

10.1 LOAD DISTRIBUTION BETWEEN THE BOLTS

A bolt connection usually includes more than one bolt, and the problem of determining the load distribution can be difficult and time consuming. The load can be supposed to be distributed evenly only in cases where the design of all the attachment points is invariant and the bolts are placed axisymmetrically relative to the vectors of load. This means that the load vectors must be perpendicular to the joint plane, and hence, the load can be a tension force or torque (not a bending moment because its vector is parallel to the joint plane). Figure 10.2b to Figure 10.2d, Figure 10.2g, and Figure 10.2h present the examples of such connections. The version in Figure 10.2e is wrong because bolts 1 are installed in a more rigid place (not in the corner). Bolt 2 in Figure 10.2i and bolt 3 in Figure 10.2j are the most loaded. In Figure 10.2k the most loaded bolt is 4, the less loaded is 5, and the minimal load would fall to the share of bolt 6.

The design shown in Figure 10.2g and Figure 10.2j deserves detailed consideration. In this design, rib 7 begins near the bolt. Figure 10.3a presents an element of such connection, in which rib 1 is welded to flange 2 and wall 3 to make the structure stiffer. The deformations of parts 2 and 3 loaded with force F are shown with dashed lines (exaggerated). From the deformation pattern, we can see that the rib impedes these deformations, and, hence, in areas 4 and 5 there will be high tension stresses between the rib and parts 2 and 3. These places should receive due attention of the designer. At a high load, they can be treated as shown in Figure 10.3b. Here area 4 is reinforced by boss 6 welded to rib 7 and flange 2. In area 5, the stress peak is relieved by the thin, flexible segment 8.

When the evenness of the load distribution is not obvious, or it is clear that the load distribution is uneven, it should be determined using finite element method (FEM) or simplified calculations. The deformation pattern of the joined parts is usually too complex to make confident assumptions

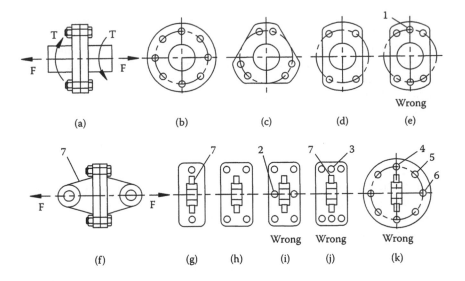

FIGURE 10.2 Some cases of even load distribution between bolts.

about it. Therefore, if we wish to calculate the bolt connection manually, we have to make greatly simplified assumptions about the deformations of the joint members. Usually, the assumptions are as follows:

- The connected parts don't deform at all; they are perfectly rigid.
- The points of connection are pliable (springy), and all of them (within one connection) are of the same rigidity.
- In connections loaded in tension–compression, the compliance of connection points is the same in tension and in compression.

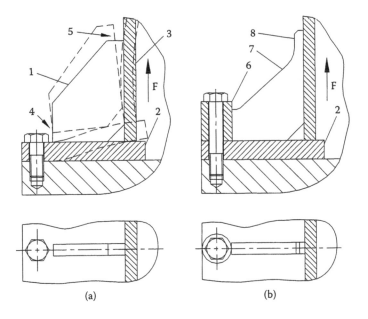

FIGURE 10.3 A bolt and a rib.

In the strictest sense, these assumptions are wide of the truth, but because they are much used, let's discuss them one by one.

When the minimum number of attachment points is used (two in a two-dimensional problem or three in a three-dimensional problem; these problems are called *statically determinate*), the reactions in the attachment points can be determined regardless of the deformations of the bodies and attachments. But when the number of attachment points is more than minimal (such problems are called *statically indeterminate*), deformations must be considered to enable calculation of reaction forces in the attachments.

For statically determinate problems, the rules of engineering mechanics are valid. That means:

- The point of application of a moment to a part is of no significance.
- A force applied at any point can be transferred to another point with the addition of an appropriate moment applied to any point.
- A force can be applied to any point coaxially with its vector with no change in the attachment reactions.

These rules are obviously invalid for statically indeterminate problems. As a case in point, in Figure 10.4 are shown two straight beams loaded with a bending moment M. One of them (Figure 10.4a) has two supports, and we can determine the reactions from the following equilibrium equations:

$$R_1 = \frac{M}{L}; \quad R_2 = -R_1$$

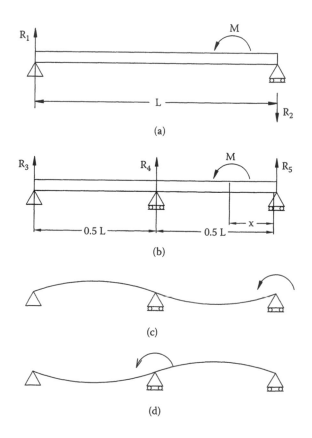

FIGURE 10.4 Examples of statically determinate and indeterminate beams.

Another beam (Figure 10.4b) is supported in three points and the number of equilibrium equations (two; both the sum of moments and the sum of forces must be zero) is less than the number of unknowns (three: R_3, R_4, and R_5). These unknowns can only be determined by applying virtual deformations. In this case, if the supports are assumed to be very rigid in comparison with the rigidity of the beam, we can use the condition that the beam sag at the supports is zero. For this purpose, we assume that the middle support doesn't exist, but instead, force R_4 is applied in this point. Then we calculate the deformations as if the beam is statically determinate:

- The sag of the beam in the middle under the action of moment M equals

$$y_M = \frac{M}{12EI}\left(3x^2 - \frac{3L^2}{4}\right)$$

- The sag of the beam in the middle under the action of force R_4 equals

$$y_R = -\frac{R_4 L^3}{48EI}$$

(Here E and I are the modulus of elasticity and moment of inertia of the beam.)

Because the beam is supported in the middle, the condition of zero sag at this point gives the following:

$$y_M + y_R = 0$$

$$R_4 = \frac{4M}{L^3}\left(3x^2 - \frac{3L^2}{4}\right)$$

Now, when force R_4 has been determined, we can easily calculate forces R_3 and R_5 as in the usual statically determinate problem. And it is obvious from these equations that the forces in the supports depend on the point of application of the moment (the x value).

This is not only a formal conclusion based on the equations. Figure 10.4c and Figure 10.4d show that there is an obvious dependence between the location of moment applied and the deformation pattern. This dependence is surely accompanied by corresponding differences in the reaction of supports represented by the preceding equations. When the bending moment is applied in the middle (as shown in Figure 10.4d), $R_4 = 0$, both obviously and mathematically.

An example of erroneous application of the rules of engineering mechanics to pliable parts is presented in Figure 10.5. No doubt, force F applied to the bearing's housing drives it into the

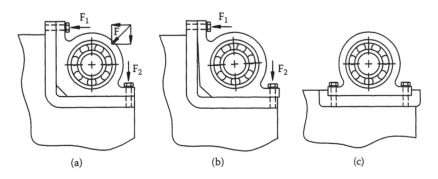

(a)   (b)   (c)

FIGURE 10.5 False use of the parallelogram of forces.

corner. But forces F_1 and F_2, which are equal to force F according to the parallelogram of forces, act differently. If force F_1 is applied first (by tightening the upper bolt), the bearing's housing takes up position shown in Figure 10.5b, and the tightening of the lower bolt (force F_2) doesn't move the housing to the left. When a horizontal force is applied to the bearing, the lower bolt becomes loaded in shear. The upper bolt is loaded in shear when a vertical force is applied to the bearing. An improved design is shown in Figure 10.5c.

The influence of deformations on the load distribution between the fasteners has been demonstrated in Chapter 8, Section 8.2 and Section 8.3, by the examples of bolted gears. If the toothed rim and the disk had been taken as perfectly rigid, the load distribution would have been even. But the FEM showed that it is far from being so; the bolts in the vicinity of the loaded tooth are stressed appreciably more than the average.

The joined parts are quite pliable, so it is not strictly correct to assume the parts are perfectly rigid. But it is very handy for developing the analytical models, because it permits using the rules of engineering mechanics as if the problem is statically determinate (see previous text). Nevertheless, deformation analysis is unavoidable in this case as well, but only the deformations of the attachment points ("springs") are considered.

For connections loaded in tension, the stiffness of the springs can be simply calculated when the load is centrically applied. From Equation 2.5 we know that the bolt "feels" only part of the load force (χ), hence the compliance of the spring, λ_{sp}, equals

$$\lambda_{sp} = \lambda_b \cdot \chi \ \ mm/N$$

where λ_b = compliance of the bolt (mm/N).

Recommendations for calculation of values λ_b and χ are given in Subsection 10.2.2 (see Example 10.7) and Subsection 10.3.1.

It can be easily shown that the centrically loaded bolt attachment has the same compliance in compression as in tension, so the "springy" presentation in this case is valid. But if the bolt is loaded eccentrically, as shown in Figure 2.9 (Chapter 2), the compliance in tension increases greatly because the bolt load F_b (at the same working load F_w) is greater, and, in addition, the bending deformation of the flange adds to the tension deformation of the bolt. But in compression, the flange is not subjected to bending, nor does the bolt feel the pressure force, so the rigidity of the attachment is much greater in compression than in tension. In this case, substituting a spring for the bolt is an inferior design but is used when no better solution can be found.

For connections loaded in shear, the stiffness of the "springs" seems to be less problematic. The shear force is transmitted mainly by friction forces between the parts compressed together and partly by pins or fitted bolts (if there are any). Although the stiffness of such attachment points can hardly be calculated, there are more reasons to assume all the "springs" to be of the same compliance. In effect, if the joined bodies are absolutely rigid, the real compliance of the attachment points doesn't play a part. The load distribution between them is influenced by the ratio of their compliances only, and if they are of the same rigidity, this difficulty doesn't exist.

It is clear from the preceding discussion that the simplifying assumptions stated previously make our calculation rather uncertain. But let's try to develop the analytical models based on these assumptions and make some comparative calculations to taste how much the error is.

10.1.1 Load Distribution in Bolted Joints Loaded in Shear

Figure 10.6a shows a beam attached to a wall by bolts and loaded with force F. Because the beam is perfectly rigid, we can transfer the force to any point (here, to the center of gravity, CG) with the addition of moment $M = F \cdot L$ as shown in Figure 10.6b. Now we can derive the equilibrium equations to calculate the reactions at supports. To ease our job, let's recall the superposition rule. It states that

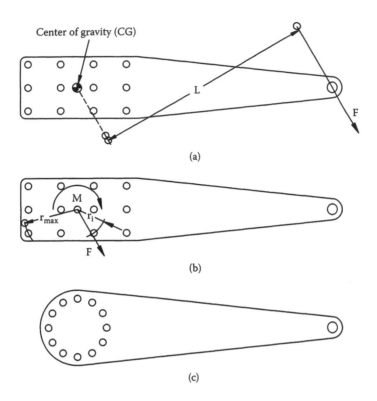

Center of gravity (CG)

(a)

(b)

(c)

FIGURE 10.6 Bolt connection loaded in shear.

the summary effect of forces on a system equals the sum of effects of each force of the system. So we can consider separately the reactions at supports caused by F and M and then sum up the results.

Force F causes linear motion of the beam relative to the wall. (As you would remember, there is a direct proportionality between the relative motion and the stress.) Because the wall and the beam are absolutely rigid, all the points of attachment become deformed by the same amount. Inasmuch as all the attachments are of the same compliance, their reactions should be equal. Hence, the load of bolt i caused by force F equals

$$R_{F,i} = \frac{F}{n} \qquad (10.1)$$

where n = the number of bolts in the connection.

Moment M turns the beam relative to the wall around the CG point, and the relative movement in each attachment is directly proportional to its distance from CG; hence, the load of bolt i caused by moment M is directly proportional to radius r_i (see Figure 10.6b). Now we can write the following equations:

$$\frac{R_{M,i}}{R_{M,max}} = \frac{r_i}{r_{max}}$$

$$R_{M,i} = R_{M,max} \frac{r_i}{r_{max}} \qquad (10.2)$$

where

$R_{M,max}$ = reaction of bolt with the maximal radius r_{max}
r_i = distance between bolt number i and the CG point
r_{max} = maximal distance between bolt and the CG point

The equilibrium equation for this case looks as follows:

$$M = \Sigma R_{M,i} \cdot r_i \tag{10.3}$$

Insertion of Equation 10.2 into Equation 10.3 gives us the maximal load caused by moment M:

$$M = \frac{R_{M,max}}{r_{max}} \Sigma r_i^2$$

$$R_{M,max} = \frac{M \cdot r_{max}}{\Sigma r_i^2} \tag{10.4}$$

As soon as we know the $R_{M,max}$ value, we can calculate the $R_{M,i}$ for each bolt using Equation 10.2, and then the summary load of each bolt can be determined.

If the bolts are equally spaced on a circle as shown in Figure 10.6c, then the former equations change as follows:

$$r_{max} = r_i = r$$

$$\Sigma r_i^2 = n \cdot r^2$$

$$R_M = \frac{M}{n \cdot r}$$

Forces $R_{F,i}$ are directed like force F, but forces $R_{M,i}$ are directed tangentially, that is, perpendicular to the line connecting the bolt i with CG. This means that each bolt has a different direction of force $R_{M,i}$, and its sum with $R_{F,i}$ should be vectorial.

You can ask, "Why on earth did you decide that the center of rotation (CR) is the CG point? In fact, we can select any other point as CR, so that the Σr_i^2 value in Equation 10.4 be greater and the force $R_{M,max}$, lesser." Unfortunately, the only way to place CR according to our wish is to make a real axle with bearings. In our case, there is no such thing, and the parts choose CR so as to minimize the resistance of the bolt joint to the rotation, i.e., to make the Σr_i^2 value minimal. There is rigorous proof that this point is placed exactly at the CG of the connection, but we want to demonstrate this by a simple example. Figure 10.7 presents a two-bolt connection subjected to torque. The CG point here is placed in the middle of the distance between them, so that $r_1 = r_2 = r$, and $\Sigma r_i^2 = 2r^2$. It is clear that if we transfer the CR horizontally (to the right or to the left), both r_1 and r_2 grow. What if we transfer the CR vertically by some amount Δ, to point O? In this case the Σr_i^2 value increases as well:

$$\Sigma r_i^2 = (r + \Delta)^2 + (r - \Delta)^2 = 2r^2 + 2\Delta^2$$

The same can be shown for any quantity of bolts. Here, we can see that any drift of CR from CG leads to increase in the Σr_i^2 value. Therefore, CR displaced from CG can't be chosen by the parts voluntarily, without some real constraint such as a spigot or an axle.

Let's make several comparative calculations.

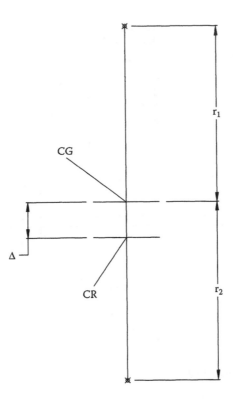

FIGURE 10.7 The center of gravity (CG) and the center of rotation (CR).

EXAMPLE 10.2 (FIGURE 10.8A)

Beam 1 is attached to wall 2 by 12 bolts and loaded with force $F = 10,000$ N. The load must be transmitted by friction forces. Dimensions are given in the sketch.

Moment of force F around the CG point: $M = 10,000 \cdot 1,000 = 10 \cdot 10^6$ N·mm. There are four bolts with $r = r_{max} = 201$ mm, two bolts with $r = 180$ mm, four bolts with $r = 108$ mm and two bolts with $r = 60$ mm. Thus,

$$\Sigma r_i^2 = 4 \cdot 201^2 + 2 \cdot 180^2 + 4 \cdot 108^2 + 2 \cdot 60^2 = 2.8 \cdot 10^5 \ mm^2$$

From Equation 10.4, the maximal bolt load from moment M equals

$$R_{M,max} = \frac{10 \cdot 10^6 \cdot 201}{2.8 \cdot 10^5} = 7180 \ N$$

The bolt load from force F equals

$$R_F = \frac{10000}{12} = 833 \ N$$

Among the 12 bolts numbers 3, 4, and 5 are obviously the most loaded for they are far from CR and both the R_M and R_F forces act in nearly the same direction. Bolts 3 and 5 are loaded equally:

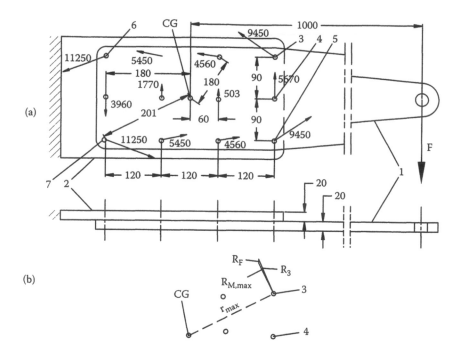

FIGURE 10.8 Sketch to Example 10.2.

vectorial sum of $R_{M,\text{max}}$ and R_F equals $R_3 = R_5 = 7930$ N (see the geometric summation in Figure 10.8b). Bolt 4 is loaded by forces $R_{M,4}$ and R_F acting in the same direction, so these two forces can be just summed. From Equation 10.2,

$$R_{M,4} = 7180 \frac{180}{201} = 6430\ N$$

Thus, the load of bolt 4 equals

$$R_4 = R_{M,4} + R_F = 6430 + 833 = 7263\ N$$

Finally, according to this calculation the most loaded bolts are 3 and 5, and $R_3 = R_5 = 7930$ N.

Calculation made using FEM gives $R_3 = R_5 = 9450$ N (the force vectors with their values are shown in Figure 10.8a), but the maximal bolt load $R_6 = R_7 = 11250$ N, that is, 1.42 times greater than was obtained under the assumption that beam 1 and wall 2 are absolutely rigid.

EXAMPLE 10.3 (FIGURE 10.9)

Part 1 loaded with force F is attached to part 2 by 12 bolts. Force F passes through the CG of the connection. So if the parts are absolutely rigid, all the bolts (see Figure 10.9a and Figure 10.9b) are loaded evenly, and the load equals

$$R = \frac{F}{12}$$

If $F = 10{,}000$ N, $R = 833$ N. But calculations using FEM show that the load distribution is uneven. The vectors of bolt forces and their magnitudes (in Newtons) are shown with arrows in Figure 10.9.

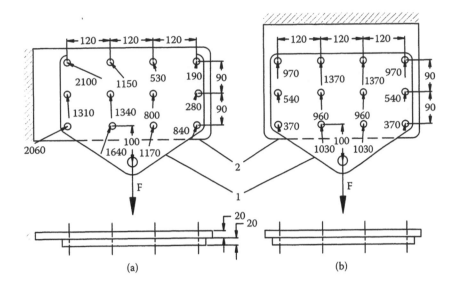

FIGURE 10.9 Sketch to Example 10.3.

We can see that in Figure 10.9a the maximal bolt load is 2,100 N, which is 2.5 times greater than the average amount of 833 N. In Figure 10.9b, the maximal bolt force (1370 N) is 1.64 times greater than the average.

How should we regard these results? The simplified calculations are very approximate, but this is better than nothing. In less important cases, the designer, in order to account for the possible difference between the calculated and real load of the bolt, just increases the safety factor and keeps in mind that after the friction forces fail to transmit the load, the shanks of bolts go into action and work as pins. In more important cases, the designer uses elements loaded in shear (such as fitted bolts, pins, or keys) in addition to the friction forces. In this case, the calculation is usually made separately for the friction forces and for the shear elements, as if each of them had to transfer the entire load with a certain factor of safety. In very important cases, special research is usually arranged, including FEM calculations and laboratory tests for strength. That is not an easy task.

10.1.2 LOAD DISTRIBUTION IN BOLTED JOINTS LOADED IN TENSION

Figure 10.10a presents a model, in which 1 and 2 are the connected parts and 3, the springs replacing bolts. In the unloaded position, all the springs are assumed to be unloaded. As mentioned, the presentation of the joined bodies as perfectly rigid means that all the forces can be transferred to the CG of the connection (with the addition of appropriate moments). It is obvious that force F applied in the center of the connection causes all the springs to be deformed (and loaded) equally, as shown in Figure 10.10b. Moment M applied to the rigid body causes the body to turn around the axis that passes through the CG of the connection and is collinear with the vector of the moment, as shown in Figure 10.10c. The deformation of each spring is directly proportional to the distance of this spring from the turning axis.

The reason for the turning axis to pass through the CG point is exactly the same as for the connections loaded in shear: this placement of the axis meets the minimal resistance of the connection to turn. Moment M is placed here in the center of connection just for the pleasure to see all the loads applied in one point. But inasmuch as the body is considered absolutely rigid, the moment can be applied at any point of the body with no influence on the spring deformation pattern.

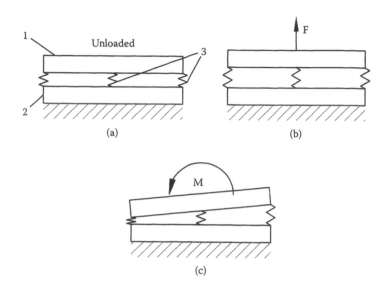

FIGURE 10.10 Elements of the analytical model for bolt joint loaded in tension and bending.

Figure 10.11 shows a scheme of a bolt connection that has two axes of symmetry. After all the forces are repositioned to the CG of the joint, we have one force F (perpendicular to the joint surface) applied in the center, and two moments M_x and M_y.

There can also be forces and moments that load the connection in shear, but they should be considered separately as shown earlier.

To derive the equilibrium equations, we consider separately the reactions at supports caused by F, M_x, and M_y and then sum up the results. We begin with F. As already mentioned, the tension (or pressure) force applied at the CG is distributed evenly between the bolts. Thus, bolt i is loaded by a force

$$R_{F,i} = \frac{F}{n} \tag{10.5}$$

where n = number of bolts.

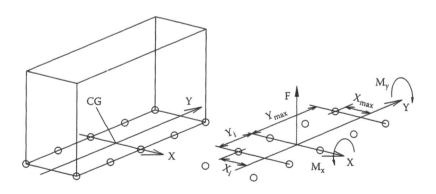

FIGURE 10.11 Schematization of bolt connection loaded in tension and bending.

The next step is to determine the reactions from the moments. Moment M_x is balanced by the total moment of the forces of bolts. Partial moment $m_{i,x}$ about axis X applied by bolt i equals

$$m_{i,x} = R_{i,x} \cdot y_i \tag{10.6}$$

where R_{ix} = reaction of bolt i caused by moment M_x.

(As said earlier, the bolt reaction is directly proportional to the distance from the axis, in this case, to the coordinate y.)

Now we can write the following:

$$\frac{R_{i,x}}{R_{\max,x}} = \frac{y_i}{y_{\max}}$$

$$R_{i,x} = R_{\max,x} \frac{y_i}{y_{\max}} \tag{10.7}$$

where $R_{max,x}$ = reaction of the bolt with the maximal distance (y_{max}) from axis X.

Insertion of Equation 10.7 into Equation 10.6 gives

$$m_{i,x} = R_{\max,x} \frac{y_i^2}{y_{\max}}; \quad M_x = \Sigma m_{i,x} = \frac{R_{\max,x}}{y_{\max}} \Sigma y_i^2$$

$$R_{\max,x} = \frac{M_x y_{\max}}{\Sigma y_i^2} \tag{10.8}$$

In the same way, we can calculate the maximal bolt reaction caused by moment M_y:

$$R_{\max,y} = \frac{M_y x_{\max}}{\Sigma x_i^2}. \tag{10.9}$$

In the case under consideration, the maximal reactions $R_{max,x}$ and $R_{max,y}$ are related to the same bolt, and the total maximal reaction equals the sum of three components presented by Equation 10.5, Equation 10.8, and Equation 10.9:

$$R_{\max} = \frac{F}{n} + \frac{M_x y_{\max}}{\Sigma y_i^2} + \frac{M_y x_{\max}}{\Sigma x_i^2} \tag{10.10}$$

The joint shown in Figure 10.11 is very handy for calculation. It is not always so. For example, in a round connection shown in Figure 10.12a, bolts 1 and 3 have the maximal reaction from moment M_x, but zero reaction from moment M_y. Conversely, bolts 2 and 4 have their maximal reaction from moment M_y and zero reaction from moment M_x. Therefore, in this case, the reaction of each bolt should be calculated using Equation 10.7 and Equation 10.8 for M_x and likewise for M_y, and then the maximal reaction can be obtained as the maximal value of their sum.

For joints that have only one symmetry axis (Figure 10.12b), the location of the neutral axis (X in this case) should be determined for the condition $\Sigma y_i = 0$.

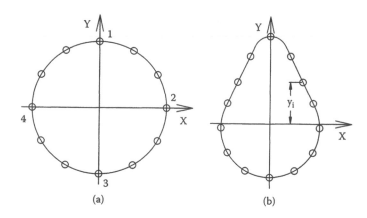

FIGURE 10.12 Bolt connection contours of greater complexity.

If the joint has no symmetry axes, the CG location can be found in the following way:

Put the X and Y axes in any position with respect to the joint; don't forget that these axes
should be perpendicular to each other.

Calculate the proper position of these axes using the condition $\Sigma x_i = 0$ and $\Sigma y_i = 0$; the
intersection of these axes is the required CG point.

Take the projections of the load vectors onto the X–Y axes (you can rotate these axes around
the CG for the sake of convenience) and calculate the reactions of each bolt using Equation
10.7 to Equation 10.10.

EXAMPLE 10.4

Figure 10.13 shows a housing attached to a rigid foundation by 10 bolts numbered from 1 to 10.
Vertical force $F = 1000$ N has been applied to the housing upward in different points numbered
from 11 to 20. In each load case, the bolt reactions have been calculated twice: using Equation 10.7
to Equation 10.10 and also by FEM.

For this connection, the following data are invariant for all ten loading options:

$$x_{max} = 150\ mm;\quad y_{max} = 450\ mm;\quad n = 10$$

$$\Sigma x_i^2 = 10 \cdot 150^2 = 2.25 \cdot 10^5\ mm^2$$

$$y_i^2 = 4 \cdot 450^2 + 4 \cdot 225^2 = 10.125 \cdot 10^5\ mm^2$$

$$\frac{x_{max}}{\Sigma x_i^2} = \frac{150}{2.25 \cdot 10^5} = 6.67 \cdot 10^{-4}\ mm^{-1};\quad \frac{y_{max}}{\Sigma y_i^2} = \frac{450}{10.125 \cdot 10^5} = 4.44 \cdot 10^{-4}\ mm^{-1}$$

To make calculations using Equation 10.10, force F is transferred from the point of
application to the CG, and the associated moments M_x and M_y are given in Table 10.1. The
maximal bolt reactions R_{max} are given in Table 10.2. In the same table are summarized the
maximal bolt reactions found using FEM. By this method, the following versions of the
connection have been calculated:

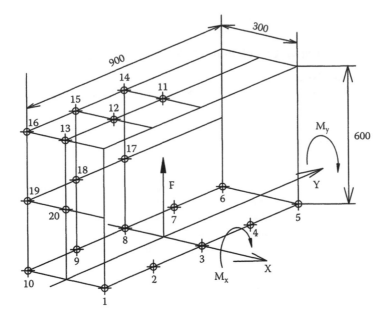

FIGURE 10.13 Sketch to Example 10.4.

- The steel housing is solid, and the compliance of the attachment point $\lambda_{sp} = 1.79 \cdot 10^{-7}$ mm/N (these reactions are labeled R_{1max}).
- The steel housing is hollow, wall thickness equals 12 mm, and $\lambda_{sp} = 1.79 \cdot 10^{-7}$ mm/N (R_{2max}).
- The housing is solid, and $\lambda_{sp} = 10.2 \cdot 10^{-7}$ mm/N (R_{3max}).
- The housing is hollow, wall thickness equals 12 mm, and $\lambda_{sp} = 10.2 \cdot 10^{-7}$ mm/N (R_{4max}).

In Table 10.2, we see that there is a certain difference in bolt reactions obtained from Equation 10.10 and FEM, but this difference is not too large and in the worst case it reaches 60%. So if we take some spare safety factor, say, 1.5 or 2, we can countervail the possible error of the simplified calculation. But if the higher safety factor is not affordable, the bolt stresses should be calculated

TABLE 10.1
Bending Moments in the Bolt Connection

Point of Load Application	M_x, kNmm	M_y, kNmm
11	0	0
12	225	0
13	450	0
14	0	150
15	225	150
16	450	150
17	0	150
18	225	150
19	450	150
20	450	0

Note: See Example 10.4.

TABLE 10.2
Comparative Calculations of Bolt Connections

Point of Load Application	R_{max}, N	R_{1max}, N	R_{2max}, N	R_{3max}, N	R_{4max}, N
11	100	123	119	120	127
12	200	197	196	177	188
13	300	274	327	266	321
14	200	263	268	234	268
15	300	328	305	292	305
16	400	409	421	357	422
17	200	270	319	237	275
18	300	334	347	295	309
19	400	410	475	357	450
20	300	277	336	268	323

Note: See Example 10.4.

more accurately and checked by direct measurements, because any calculation involves simplifications in the definition of boundary conditions.

10.2 TIGHTENING OF BOLTS

10.2.1 TIGHTENING ACCURACY

The needed pressure in a BC is achieved by tightening the bolts. The stronger the tightening, the stronger is the connection. Modern fasteners allow high tightening forces. For example, bolt M12 of property class 10.9 can be tightened by a force of 55,000 N, and bolt M24 by 230,000 N. If there are several such bolts in the connection, the connected parts may be pressed against each other by a force of hundreds of tons, provided that the bolts are tightened properly.

Until the middle of the 20th century, tightening torque was not usually specified in drawings: it was common practice to rely on a qualified fitter, except for important and highly loaded bolted connections. In the course of time, industrial production became multifarious, with machines being quickly replaced by new models that bear relentlessly increasing loads, and it has become necessary that fitters should get to know the tightening torques for each connection from the technical documentation. Therefore, the specification of this parameter is nowadays a general practice.

EXAMPLE 10.5

Once during the shop test of a marine engine, a failure occurred in the intermediate shaft that was transmitting power of 6000 HP at 250 r/min from the engine to a hydraulic brake. Shaft 1 (Figure 10.14), 250 mm in diameter, broke because of fatigue crack 2 that originated from a hole for cone bolt 3. Investigation of this case revealed that there was fretting damage in all these holes, and in all fretted areas, there were initial fatigue cracks. But what was the cause for the fretting damages and cracks?

That flange connection had been calculated for strength during design, and when the shaft failed, there was no problem ascertaining the cause of this event. In the assembly drawing there was not specified tightening torque for the bolts that are M42 in diameter. The fitters (by the way, very qualified fitters occupied with assembly of diesel engines) tightened the bolts as strongly as possible without the use of any special devices. But the friction force between the flanges of shaft 1 and resilient coupling 4 was inadequate to withstand the shear force in this connection.

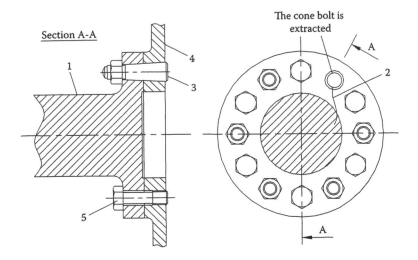

Section A-A

The cone bolt is
extracted

FIGURE 10.14 Flange connection with cone-fitted bolts.

In operation, the cone bolts transmitted most of the shear force with some micromovements caused by load fluctuations and torsional vibrations. These micromovements caused fretting damages, and the resulting origination of fatigue cracks was inevitable.

The designer rectified the error and specified the tightening torque: 6000 N·m. The failed shaft was replaced by a new one. The coupling was renewed by machining the cone holes together with the new shaft to increased diameters (to take away the fretted layer and to make sure that there were no fatigue cracks left). The cone-fitted bolts were replaced as well. The fitters built a huge lever and tightened all the bolts with the needed torque. This defect never occurred again.

How did the designer (in Example 10.5) calculate the tightening torque? First, he calculated the tightening force F_t needed to transmit the shear force (see Subsection 10.3.2). Then, he chose the quantity and diameter of bolts. And then, as he had established the diameter of thread and the tightening force, he obtained the tightening torque from this equation:[2]

$$T_t = F_t \left[\frac{d}{2} \left(\frac{p}{\pi d} + 1.15\mu_T \right) + \mu_N R_N \right] \qquad (10.11)$$

Here,

F_t = tightening force (N)
d = thread diameter (mm)
p = thread pitch (mm)
R_N = mean radius of the nut (or bolt head) bearing surface (mm)
μ_T and μ_N = coefficients of friction in the thread and on the bearing surface of the nut (or bolt head), respectively

In fact, $\mu_T > \mu_N$ because the unit pressure in the thread is higher, the oil film is thinner, and the share of metal-to-metal contact is bigger. Besides, the sliding couples may be different: the nut and the bolt are usually made of steel, and the connected parts are often made of gray cast iron, aluminum, or copper alloys that are characterized by a lower coefficient of friction. But in practice there is usually no possibility to separate these two tribological situations. Provided that $\mu_T = \mu_N$, Equation 10.11 can be converted into a simple dependence between the tightening torque T_t and the tightening force F_t. For some standard ratio of dimensions (for example, $R_N \approx 0.7d$ and $p \approx 0.125d$), it is the following:

$$T_t = k_t F_t d \tag{10.12}$$

where $k_t = 0.15$, 0.18 and, 0.20 for $\mu_T = \mu_N = \mu = 0.1$, 0.125, and 0.140, respectively.

In experiments with bolts M12 × 1.5, the following coefficients of friction were obtained:[2]

With no lubrication:
- Uncoated threads — $\mu = 0.12$–0.25 (sometimes 0.55 because of seizure)
- Cadmium-plated threads — $\mu = 0.18$–0.32
- Copper-plated threads — $\mu = 0.13$–0.32

Lubricated with machine oil:
- Uncoated threads — $\mu = 0.10$–0.17
- Cadmium-plated threads — $\mu = 0.14$–0.18
- Copper-plated threads — $\mu = 0.12$–0.19

In different experiments, the μ value may differ, but the following conclusions can be drawn:

1. Before tightening, the threads should be greased or oiled, because that prevents seizure, decreases the dispersion of the μ value, and improves the tightening accuracy.
2. Considering the dispersion of the friction coefficient, the tightening torque value can be calculated with an accuracy of ±(20–25)%.

For critically loaded BCs, such error is not acceptable, and then the tightening force should be controlled by methods that leave the friction forces out. The most exact method consists in measuring the bolt elongation. In this case, the friction forces don't influence the tightening force determination, though they do influence the bolt stress through its torsion.

For tightening large bolts, hydraulic devices are used. The hydraulic piston is connected to the thread (which is made longer to project from the nut) and applies force F_t to the bolt. Then the nut is tightened manually by a wrench, which applies to the bolt some additional force that is negligible in comparison with F_t. This method is exactly similar to the previous one, but the bolt shank is not torsioned. This means that the stress in the bolt is less by about 1.3 times, and the tightening force F_t can be increased correspondingly.

Both of these methods are laborious and used only sometimes, when the exact and reliable tightening of bolts is a must. As cases in point, the connecting rod bolts and cylinder head studs of high-power diesel engines can be mentioned.

One more method of tightening, less exact but not influenced by friction, consists in turning the nut (or the head of bolt) by a specified angle. This is aptly called the "turn-of-the-nut" method. The tightening is executed in two steps: first, the nut should be tightened by a relatively small force to take up clearances, and then it should be turned by a certain additional angle. The deformations while tightening include not only elongation of the bolts, but also contraction of the connected parts, plastic deformation and wear of the roughness (called *embedment*), and plastic deformations in the thread. Most of these factors can hardly be calculated with acceptable accuracy. Therefore, the preliminary tightening torque and the angle of additional turn should be determined experimentally on prototype models, when the real tightening force F_t can be controlled by strain gauges or by measuring bolt elongation. This method can be recommended for long bolts (more than six diameters in length). With shorter bolts, the accuracy of this method is too low because the angle of rotation of the nut is rather small.

If the bolted connection is designed with a generous safety factor, a torque wrench is used to tighten the bolts. This method is the easiest. The nominal tightening torque can be obtained from

Equation 10.11 or Equation 10.12, in which the values of F_t and the coefficients of friction should be preselected. Because the real coefficient of friction can be less than the chosen one, force F_t developed by the nominal tightening torque may be larger than planned, and the bolt may be overstressed. To prevent this possibility, the nominal torque should be decreased by about 20–25%. But the coefficient of friction may be also bigger than the preset one by 20–25%, and in this case, the real tightening force (taking into account the decreased tightening torque) may come to 65–56% of the preset value. That is the payment for the inexact method of tightening.

10.2.2 Stability of Tightening

Proper tightening of bolts is necessary but insufficient to have a reliable connection: the tightening forces may decrease in course of time, and the tightening may remain proper only on the drawing. What can be the cause of this torque retention loss? There are several such factors, as described in Subsection 10.2.2.1 and Subsection 10.2.2.2.

10.2.2.1 Self-Loosening of Bolts

All bolt threads are self-braking, or, in other words, the lead angle of the thread is less than the angle of friction. That means, if you have tightened the thread and began staring at it attentively, you are just wasting your time: it will not unscrew, even if you wait very long and watch with unrelenting interest. It will unscrew later, when the machine works. But why? Will something have changed? Yes, sure! Under the influence of vibrations and micromovements, including the variable radial deformation of the nut under variable bolt load (the nut is "breathing"), the coefficient of friction decreases. How much? It depends on the amplitude and frequency of vibrations. If, usually, $\mu = 0.10$–0.15, under vibrations it changes cyclically (in accordance with the local microslip in the contact places), and the lowest value during a cycle may be as low as 0.01 or even 0.005. During these very small periods of time, the thread ceases to be self-braking, and the fasteners begin to unscrew little by little, if they have not been locked (see Subsection 10.2.3).

10.2.2.2 Plastic Deformation of Fasteners and Connected Parts

It can be related to the following factors:

> *Creep of metal:* Creep, i.e., slow plastic deformation of a metal under load, exists always to a greater or lesser extent. But at higher temperatures, it increases by far. The temperatures when creep should be considered a possibility are 300°C for construction steels and 150°C for light alloys. Above these temperatures, high-temperature materials must be used.
>
> It should be noticed that the harmful influence of high temperature manifests itself not only in the creep of materials. In the literature, one can find information that at a temperature above 200°C the parts should not be plated with zinc, cadmium, tin, and bismuth, because these materials diffuse into the surface layer of the part (Rebinder's effect) and impair its strength. At higher temperatures copper and brass coatings that have considerably higher melting points are acceptable.
>
> *Occasional overload* that causes plastic deformations of the connection members: It is usually cheaper to replace the fasteners that have slackened because of overload than it is to examine them and incur risks using them again. The connected parts may need some additional machining, if their joint surfaces have been damaged when working with slackened fasteners.
>
> *Surface plastic deformation* of the connected parts underneath the nut or head of bolt: The bearing surface of the nut (head of bolt) nearly equals the cross section of the thread.

Both of them are exposed to the same bolt force F_b; therefore, the yield points of the materials should be comparable. Tab washers and locking plates made of mild steel are not suitable for high-strength fasteners (property class 8.8 and higher). If a high-strength bolt is tightened against a tab washer, the latter creeps out from under the nut as if it was made of plastic.

It is clear that the admissible bearing stress should not be exceeded. When the connected parts are made of softer materials (for example, aluminum or magnesium alloys), special measures should be taken to increase bearing surfaces. Figure 10.15a shows a bolt and nut made of high-strength stainless steel and designed for light metal parts. The bearing surfaces of the nut and the bolt head have been developed so as to decrease the surface pressure to a permissible value.

Figure 10.15b shows a connection with standard fasteners and thick steel washers 1 and 2 that are intended to distribute the bolt force over a bigger surface. The thickness of the washers depends on the bearing surface to be achieved on the softer material.

Figure 10.16a presents isobars in a tightened BC. It is clear that outside diameter d_2 washer 1 is separated from the flange, and increasing the washer beyond diameter d_2 is useless. Because the isobars in this area are inclined at about 45°, the effective diameter of the washer can be determined as follows (see Example 10.6):

$$d_2 = d_1 + 2h$$

where
   d_1 = diameter of the bearing surface of the nut (or head of bolt)
   h = thickness of the washer

Plastic deformation of micro-asperities: The initial plastic deformation happens during tightening. Later, the microasperities continue flattening under the influence of high pressure and micromovements between the surfaces. As a result, the fasteners lose part of their tightening force. Plastic deformation is especially significant for short bolts of low property class, because their tightening elongation is relatively small.

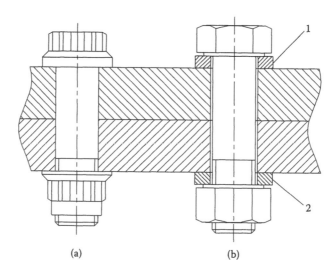

(a)   (b)

FIGURE 10.15 Bolt connection of aluminum parts.

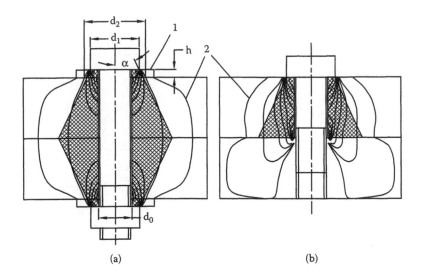

FIGURE 10.16 Stress distribution in tightened bolt connection.

EXAMPLE 10.6

Bolts M20–8.8 connect flanges made of aluminum alloy with tensile strength of 200 MPa. The permissible surface pressure for this material p = 140 MPa.[6] Tightening force for this bolt F_t = 113,500 N; hence, the bearing surface should equal

$$A = \frac{F_t}{p} = \frac{113500}{140} = 811\,mm^2$$

The diameter of the hole equals 22 mm; its surface is 380 mm², so the area inside diameter d_2 should be 811 + 380 = 1191 mm². Hence, the effective diameter of the washer

$$d_2 = \sqrt{\frac{4A}{\pi}} = \sqrt{\frac{4 \cdot 1191}{\pi}} = 39\,mm$$

The minimal diameter of the bolt head bearing surface d_1 = 28.2 mm. Hence, the needed thickness of the washer

$$h = \frac{d_2 - d_1}{2} = \frac{39 - 28.2}{2} = 5.4\,mm$$

If the bolt was of property class 10.9, the tightening force would be 159,000 N, the needed diameter of the washer d_2 = 44 mm, and its minimal thickness h = 7.9 mm.

EXAMPLE 10.7

Bolt M20–4.8 tightens two steel flanges 20 mm thick each (Figure 10.17a). Under the nut and bolt head there are installed tab washers each 1.6 mm thick. The tightening force F_t = 50,000 N. Let's first calculate the elastic deformations of the connection members using Birger's recommendations.[2]

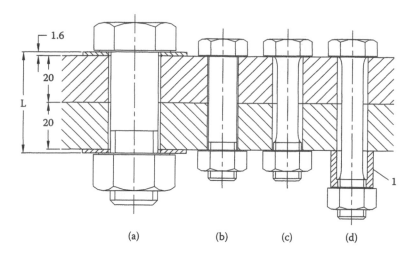

FIGURE 10.17 Comparison of bolt designs.

1. Compliance of the bolt. This includes the following:
   • Compliance of the thread:

$$\lambda_t \approx \frac{0.85}{d\,E} = \frac{0.85}{20 \cdot 2.06 \cdot 10^5} = 2.06 \cdot 10^{-7} \; mm/N$$

   where
   E = modulus of elasticity (MPa)
   d = diameter of thread (mm)

   • Compliance of the bolt head:

$$\lambda_h \approx \frac{0.15}{h\,E} = \frac{0.15}{12.5 \cdot 2.06 \cdot 10^5} = 0.58 \cdot 10^{-7} \; mm/N$$

   where h = height of the bolt head (mm).
   • Compliance of the bolt shank:

$$\lambda_{sh} = \frac{L}{E\,A_S} = \frac{43.2}{2.06 \cdot 10^5 \cdot 314} = 6.68 \cdot 10^{-7} \; mm/N$$

   Here,
   L = length of shank between the bearing surfaces of the nut and the bolt head (mm)
   A_S = cross-sectional area of the shank (mm²)

   The total compliance of the bolt equals

$$\lambda_b = \lambda_t + \lambda_h + \lambda_{sh} = (2.06 + 0.58 + 6.68)10^{-7} = 9.32 \cdot 10^{-7} mm/N$$

The elastic elongation of the bolt under tightening force equals

$$\Delta_b = F_t \lambda_b = 50000 \cdot 9.32 \cdot 10^{-7} = 0.047\,mm$$

2. The compliance of the flanges can be obtained from the following equation:[2]

$$\lambda_f = \frac{2}{E_f \pi d_0 \tan \alpha} \ln \frac{(d_1 + d_0)(d_1 + L \tan \alpha - d_0)}{(d_1 - d_0)(d_1 + L \tan \alpha + d_0)} \qquad (10.13)$$

Dimensions d_1, d_0 and angle α are illustrated in Figure 10.16. Angle α determines the so-called "cone of pressure" (hatched). There are shown also isobars, and lines 2 are the zero isobars (inside the line the stress is compressive and outside, tensile). This means that the "cone of pressure" has no physical meaning: it is just an artifice used to calculate the compliance of the connected parts. The recommended angle of the cone is $\alpha \approx 22$–$27°$. The error of this calculation usually lies within admissible limits. Because the value of λ_f directly influences the bolt force under load (see Chapter 2, Subsection 2.3.1), in important cases, it should be estimated more exactly by FEM or experimentally.

In our case, $d_1 = 28.2$ mm, $d_0 = 22$ mm, $L = 43.2$ mm, $\tan \alpha = 0.5$, and $E = 2.06 \cdot 10^5$ MPa. From Equation 10.13, $\lambda_f = 3.2 \cdot 10^{-7}$ mm/N. Now the elastic deformation of the flanges can be calculated:

$$\Delta_f = 50000 \cdot 3.2 \cdot 10^{-7} = 0.016\,mm$$

So the total elastic deformation of all connection members equals $0.047 + 0.016 = 0.063$ mm.

There are six pairs of contacting surfaces in this connection. If the mean height of microasperities equals 5 μm, and if in course of time they plastically deform by one third of their height, the elastic deformation of the connection members will decrease by 20 μm (that is 32% of the initial deformation of 63 μm), and the prestress force will decrease from 50,000 N to 34,100 N.

For the same purpose we can use bolt M12–10.9 with tightening force $F_t = 54,000$ N (Figure 10.17b). Using the same formulas and taking into account that there are no tab washers, the following values of compliances have been found:

$$\lambda_t = 3.44 \cdot 10^{-7}\,mm/N; \quad \lambda_h = 0.97 \cdot 10^{-7}\,mm/N; \quad \lambda_s = 18.56 \cdot 10^{-7}\,mm/N$$

$$\lambda_b = 22.97 \cdot 10^{-7}\,mm/N; \quad \Delta_b = 54000 \cdot 22.97 \cdot 10^{-7} = 0.124\,mm$$

$$\lambda_f = 5.73 \cdot 10^{-7}\,mm/N; \quad \Delta_f = 54000 \cdot 5.73 \cdot 10^{-7} = 0.031\,mm$$

The summary elastic deformation in this case equals $0.124 + 0.031 = 0.155$ mm. In addition, there are only four pairs of contacting surfaces, so the same plastic deformation of microasperities leads to only 13 μm of tightening loss. This comprises 8.4% of the initial tightening, and the prestress decreases from 54,000 N to 49,470 N.

If the bolt is made with a shank decreased in diameter to 90% of the outer diameter of the thread (Figure 10.17c), the loss of elastic deformation will be only 6.9%. Thus, all the above measures — using higher grade bolts, lessening the roughness of the contacting surfaces, and decreasing the quantity of connected parts — contribute to the stability of the joint.

10.2.3 LOCKING OF FASTENERS

It is not always known beforehand whether a connection needs locking or not. Therefore, the designer usually locks the fasteners using some cheap and easily available means just to feel more comfortable. The trials of the prototype may reveal insufficiency or redundancy of the locking method used.

Methods of locking fasteners are abundant, but they may be classified into two groups. In the first group, the nut is locked directly against the shank of the bolt. Among these methods, the most widespread are castle nut with a split pin (known as a cotter pin), self-locking nut and jam nut. In the second group, both the nut and the bolt are locked separately against the connected parts. That group encompasses spring lock washers, serrated lock washers, plastically deformable tab washers, nuts with a conical bearing surface, wiring, and other methods.

Each of these locking methods has its merits and weaknesses. For example, split pins (Figure 10.18a) are reliable, but they influence the tightening of the bolt: to align the holes in the bolt shank and the slots in the castle nut, the bolt will usually be a bit overtightened or undertightened. Therefore, it is unacceptable for short bolts. Besides, if the material of the pin is not soft enough, the bent area may become cracked, and in service, the bent up end may fall off, get between moving parts and damage the mechanism.

The idea of self-locking nuts is to increase the friction in the thread by a certain amount, which is independent of the tightening force and less liable to be influenced by vibrations. For this purpose, unthreaded plastic inserts have been integrated into the nut (Figure 10.18b; the bolt "threads" the plastic when screwing up). In another type, the thread of the nut is deformed in some small area, so that the bolt enters this area with interference. Both the plastic insert and the deformed threads are made in the less loaded end of the nut, which is opposite to the bearing surface. Therefore, they don't impair the load capacity of the bolt and are much less influenced by vibrations. Both of the aforementioned nut types are commonly classified as *prevailing torque*.

Self-locking nuts with a plastic insert are sensitive to increased temperature: their limit is about 80–100°C. This means that hot oil or even long exposure to the sun in a hot climate may damage the insert. Self-locking nuts with deformed thread are more reliable and can be used several times.

Using of jam nuts (Figure 10.18c) is the most questionable and old-fashioned way of locking. Nowadays it is considered unreliable. Besides, when the jam nut is tightened, it takes part of the bolt force or, possibly, the whole force. This uncertainty may lead to a failure, because the jam nut

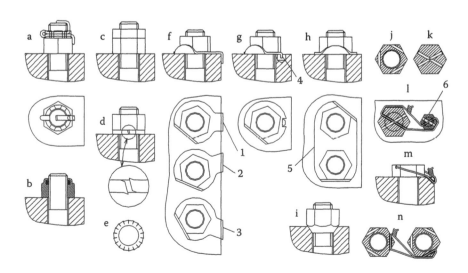

FIGURE 10.18 Locking fasteners.

is not strong enough. Also, this uncertainty may lead to loosening of the bolt, because two nuts, one of which is not loaded, are the same as a single nut.

The spring lock washer (Figure 10.18d) is, in essence, a single coil of a spring with rectangular cross section. Its slit is provided with sharp edges that project from the bearing surfaces. Under load, they indent by a small amount the bearing surfaces of both the nut and the flange. The direction of the slit is chosen so that when tightening, the sharp edges just slip, but when unscrewing, they work like chisels trying to chip the surfaces. The force needed to chip creates the locking effect. If the surfaces the spring washer contacts are too hard (more than 300 HB), the sharp edges can't do their work as needed, and the locking effect decreases. But when the material is soft (say, less than 150 HB), the slit really cuts the surfaces. Therefore, a flat washer should be placed between the spring lock washer and the soft flange (for example, one made of aluminum or magnesium alloys). In that case, the locking effect is caused only by the spring force of the washer, and it is quite low.

The projecting sharp edges of the slit cause uneven load distribution around the bearing surface of the nut (or the bolt head). This may be an important drawback for critically loaded bolts because it adds bending moment to the bolt shank. From this point of view, *serrated lock washers* (Figure 10.18e) are preferable; they are provided with many "chisels" distributed evenly around the circle. The advisable hardness of the surfaces contacting the serrated washer lies between 250 HB and 300 HB.

Tab washers (Figure 10.18f) lock the nut (or the head of bolt) in any position. The tab washer should be bent in position 1 or 2, but the nut (bolt), when tightened, often turns the washer to position 3. The fitter will then have to turn it back to position 1, but the bolt has been already tightened, and the fitter may be careless enough to leave the washer in position 3 when bending it to contact the nut. This washer doesn't lock because the nut can unscrew together with the washer. The tab washer shown in Figure 10.18g doesn't make such problems, but the friction torque developed when tightening may bend or shear off tip 4. More reliable is a locking plate 5 shown in Figure 10.18h that locks two nuts or bolts.

All the deformable locking washers and plates are made of mild steel, and they can be used only for bolts of low and middle strength (not more than grade 5 or property class 8.8).

Nuts with a conical bearing surface (Figure 10.18i) are locked by increased friction forces on the cone. This method has several disadvantages:

- Increased influence of the friction coefficient on the tightening torque.
- Additional machining of the flanges (to create the cone surfaces).
- Radial deformation inward of the most loaded part of the nut (that increases the unevenness of the load distribution between the threads and impairs the load capacity of the bolt in fatigue).
- Additional tension stresses are induced in the flange.

But despite all these disadvantages, the nuts with a conical bearing surface have found exclusive application in car wheel attachments. Here, the axial force of the bolt is nearly constant.

Wiring is one of the most reliable methods of locking fasteners. The nut (or the head of bolt) should be drilled as shown in Figure 10.18j and Figure 10.18k. When there are several fasteners, they can be wired to each other. If the fastener is single, a special (usually smaller) fastener can be used for this purpose (Figure 10.18l, bolt 6), or holes for the wire can be drilled in a rib or just in the flange (Figure 10.18m). The wire should be directed so that its tension force tightens the fastener. An example of wiring nuts is given in Figure 10.18n. The drawback of this method is that the fitter should be trained to do the wiring properly.

The strength of the bolt head can be considerably impaired by drilling holes as shown in Figure 10.18k. To diminish this influence, the holes should be as small as needed to insert the wire and placed as close as possible to the free face of the bolt head, as shown in Figure 10.18m.

Locking of studs relative to the housing is a separate problem. Studs can be locked by the following methods:

- Strong tightening against the thread runout, preferably at the bottom of the threaded opening, where the thread is less loaded; this causes a certain plastic deformation of the thread in this place, and the locking effect is similar to that of the deformed thread in the self-locking nuts.
- Using interference-fitted threads (as a rule, it is applicable for housings of nonferrous metals; a steel stud with a steel housing is prone to seizure).
- Employing adhesives (for example, Loctite); this method requires strict compliance with the adhesive manufacturer's procedures.

In practice, the designers distinguish between locking of outer BCs, which are easily accessible for periodic inspection and adjustment, and that of the inner connections, slackening of which is usually established "post mortem," when the machine is examined after failure. This difference determines the different approach to the locking of BCs of these two groups.

The outer connections are mainly locked by cheap and easy methods, such as spring lock washers and self-locking nuts. The inner connections, especially on actuated parts, should be locked surely. Here, are mainly used castle nuts with split pins, tab washers and locking plates. For stationary parts, wiring is in wide use. Under heavy vibrations, wire may break; tab washers of increased thickness are more reliable under such conditions.

10.3 CORRELATION BETWEEN WORKING LOAD AND TIGHTENING FORCE OF THE BOLT

10.3.1 Load Normal to Joint Surface

In Chapter 2, Subsection 2.2.1, forces are considered in a BC loaded coaxially. It is seen from Equation 2.5 (Chapter 2), that the working force F_W is added to the bolt force only partly, so that the bolt force under load equals

$$F_b = F_t + F_W \cdot \chi$$

where F_t = tightening force.

$$\chi = \frac{\lambda_f}{\lambda_f + \lambda_b}$$

Let's calculate the values χ for three bolt options:

1. The first option, shown in Figure 10.17a (bolt M20):

$$\chi_a = \frac{3.2 \cdot 10^{-7}}{3.2 \cdot 10^{-7} + 9.32 \cdot 10^{-7}} = 0.256$$

2. The second option, shown in Figure 10.17b (bolt M12):

$$\chi_b = \frac{5.73 \cdot 10^{-7}}{5.73 \cdot 10^{-7} + 22.97 \cdot 10^{-7}} = 0.200$$

3. The third option, shown in Figure 10.17c (Bolt M12 with thinned shank):

$$\chi_c = \frac{5.73 \cdot 10^{-7}}{5.73 \cdot 10^{-7} + 26.49 \cdot 10^{-7}} = 0.178$$

This means that in the third option, only 17.8% of the load is transmitted to the tightened bolt. In the first option, the bolt is loaded by 25.6% of the working force, i.e., 1.44 times greater. If the load changes cyclically, in the third option, the amplitude of the oscillating stress would be less by a factor of 1.44. This may be a crucial factor when choosing the parameters of a BC.

Figure 10.17d shows a BC in which the bolt length is increased by spacer 1. How is the spacer considered? Does its compliance increase the compliance of the bolt or of the flanges? The rule is easy: if the load of a connection member is increased by the applied working load, its compliance should be added to that of the bolt; otherwise it should be added to the compliance of the flanges. In this case, the load of spacer 1 rises when the working load is applied, so the compliance of the spacer increases additionally the λ_b value and reduces the sensitivity of the bolt to the working load.

In an extreme case, when the bolt with added parts is very pliable (say, the bolt is made of rubber) and the flanges with added parts are very rigid (made of metal), the bolt just doesn't feel the working load. It works at a constant load that equals the tightening force *until the joint separates*. The working load in this case is balanced by decreased contact pressure in the joint.

In the opposite extreme case, when the bolts are made of metal and the flanges of rubber, the contact pressure in the joint is constant, but the entire working load is added to the tightening force of the bolt.

As you have noticed, the decrease in the bolt load achieved by lesser χ value (good!) is accompanied by decreased contact pressure in the joint (bad!). So the initial contact pressure must be big enough to ensure the immobility in the joint at the maximal working load.

When the working load is applied eccentrically, the bolt force can be determined only using numerical methods. Figure 10.19a shows a flange connection of two vertical walls loaded by forces F_W. Bolt M20 is tightened by a force $F_t = 160$ kN. Curve 1 represents the bolt force F_b vs. load F_W. Line 2 presents the same dependence in the case of coaxial load. The calculations have been performed using FEM.[3] It is clearly seen that with eccentric loading the bolt force increases nonlinearly and may be much greater than at coaxial loading.

The F_b value depends also on the dimensions L and H (shown in Figure 10.19): the smaller these dimensions, the greater is the value of F_b. As is seen in Figure 2.9b, the bolt under eccentric load is not only tensioned but also bent. For the strength calculation of the bolt, we need not consider the bolt force F_b, but the maximal stress in the bolt. Figure 10.20[3] shows the maximal stress σ_{max} and the average stress σ_{av} in the bolt shank ($\sigma_{av} = F_b /A$, where A is the shank cross-sectional area). Curves 1, 2, and 3 correspond to dimensions $L = 20$, 30, and 40 mm, respectively. Curves 4 correspond to coaxial loading.

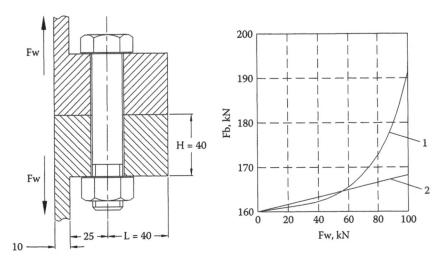

FIGURE 10.19 Bolt tension force under (a) eccentric load and (b) coaxially applied load.

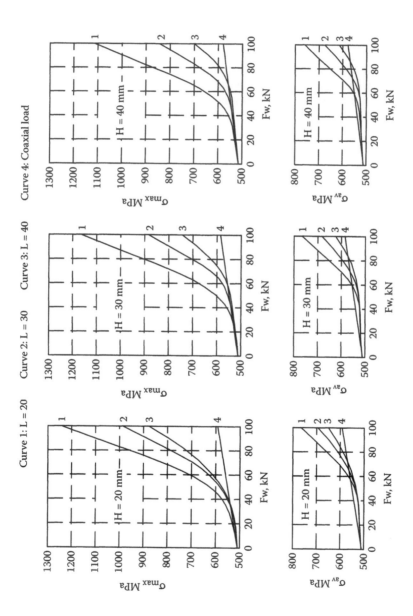

FIGURE 10.20 The maximal and the average tension stresses in the bolt shank.

These stresses are calculated in the middle of the shank, thus the stress raisers are not considered. But both the bending moment and the tension force are constant along the bolt shank, so the strength of the bolt can be calculated in all sections, considering the stress raisers where they are located.

From Figure 10.20 we find that calculation of eccentrically loaded BCs as if the load is applied coaxially may lead to a large underestimate of the bolt stresses. The error can even exceed 100%. But when the flanges are properly designed, the error may be within acceptable limits. For example, if $H = 40$ mm and $L = 40$ mm, the maximal stress obtained using FEM equals 695 MPa, whereas with coaxial application of the load, $\sigma_{max} = 576$ MPa. The difference is 20% only.

10.3.2 Shear Load

The shear load acts in the plane of the joint and tries to move the parts relative to each other. In the simplest case shown in Figure 10.21, the bolt is installed in the hole with a clearance, and the shear force Q is transmitted only by friction forces in the joint. The permissible load Q_p for this connection (of one bolt) equals

$$Q_p = \frac{F_t f}{k}$$

where
$\quad F_t =$ tightening force
$\quad f =$ friction coefficient
$\quad k =$ safety factor

Both the tightening force and the friction coefficient may have considerable spread in values; therefore, the safety factor is usually taken as $k = 3$.

A typical example of such a connection is shown in Figure 10.22a. This is a flange connection of two shafts. It transmits the torque by friction between the flanges owing to bolt tension. The

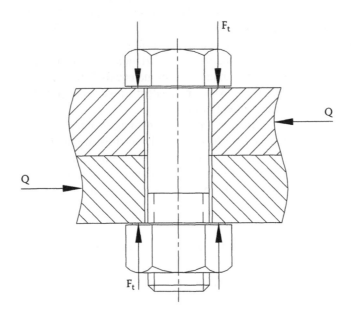

FIGURE 10.21 Bolt connection loaded with shear forces.

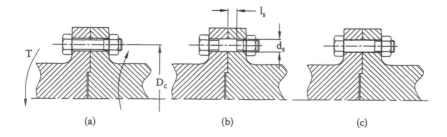

FIGURE 10.22 Types of flange connection.

shear force Q, which is transmitted to each bolt, equals

$$Q = \frac{2T}{D_C n}$$

where

T = torque transmitted by the connection
D_C = diameter of the connection (see Figure 10.22a)
n = number of bolts

According to investigations,[4,5] the friction coefficient changes with time. The initial values have been found to be $f_i = 0.12$ for turned surfaces and $f_i = 0.20$ for ground surfaces. The final values are the same for both turned and ground surfaces, $f_s = 0.20$–0.25 for connections working in oil and $f_s = 0.30$ for those working in air.

EXAMPLE 10.8

A shaft flange connection transmits power of 1600 kW at 1100 r/min. The flanges are connected by 12 bolts M20 × 1.5 of quality class 10.9. The connection diameter $D_C = 245$ mm. The bolts are greased with MoS_2 and tightened with a manual torque wrench. The flanges are ground, and the connection works in air. Does it have a satisfactory safety factor?

We start with Bossard's recommendations.[6] The friction coefficient in the thread (when MoS_2 is applied) $\mu = 0.1$, the recommended tightening torque equals 510 N·m, and the tightening force $F_t = 197,000$ N. At this strength, the maximal equivalent stress in the bolt shank (from tension and torsion) equals 90% of the material yield point.

The possible error of the specified method of tightening is about ± 20%. This means that the tightening torque should be reduced by 20% to avoid overstressing the bolt in the case of a lower coefficient of friction. But with the same probability, the coefficient of friction may be higher by 20%, and, in this case, the tightening force will be smaller by 20%. In all, the minimal tightening force may be less than tabulated by 20% because of reducing the tightening torque and by 20% because of higher friction:

$$F_{t\,min} = 197000 \cdot 0.8^2 = 126080 \; N$$

Torque that can be transmitted by friction forces when this flange connection is new equals

$$T_f = \frac{F_t f D_C}{2} n = \frac{126080 \cdot 0.20 \cdot 0.245}{2} 12 = 37068 \; N \cdot m$$

Torque to be transmitted by the connection:

$$T_w = 9555\frac{P(kW)}{n(rpm)} = 9555\frac{1600}{1100} = 13900\ N \cdot m$$

Safety factor when new:

$$k = \frac{T_f}{T_w} = \frac{37068}{13900} = 2.67$$

This is a bit less than recommended, but the holes for the bolts are made with a diameter of 20.2 mm, so, if there will be some small slippage in this connection, the bolts will work in shear (similar to fitted bolts). During prolonged service $f = 0.3$ and $k = 4$. Conclusion: this connection has a sufficient safety factor.

The increase of the coefficient of friction between the flanges in the course of working time testifies that the contacting surfaces undergo changes related to their power interaction and, undoubtedly, to micromovements in their contact spots. To make the connection more reliable and increase its load capacity, fitted bolts are often used (Figure 10.22b). The fitted bolt combines a usual bolt, which presses together the connected parts, and a pin that works in shear. The pin (the shank of the bolt) fitted with interference reduces the possibility of relative movements in the connection.

Let's consider the processes taking place in the connection with a fitted bolt. For the sake of simplicity, we assume that the interference is zero. The initial position without any load is shown in Figure 10.23a (the bolt head and the nut are not depicted, so we see only the shank of the bolt). Figure 10.23b shows this connection loaded by shear forces Q provided that the bolt is not tightened, and there is no friction force between the flanges. The shear force is taken by the pin that deforms the hole elastically as shown. After the flanges are unloaded, they return to their initial position (Figure 10.23a). Hence, during one cycle of loading and unloading, there will be a relative displacement δ of the flanges. Let's assume now that the bolt is tightened, so that the flanges are pressed against each other with force F_c (Figure 10.23c). In this case, the elastic displacement of the flanges δ_1 is less than δ, because the pin is loaded with a lesser shear force $Q_1 = Q - F_c \cdot f$. (Part of the shear force that equals $F_c \cdot f$ will be taken by the friction force between the flanges.) After the connection is unloaded, the friction force hinders the flanges from returning to their initial position. They will stop at a position in which the residual elastic deformation of the pin corresponds to the friction force $F_c \cdot f$.

It is obvious that if the friction force equals one half of the shear load ($F_c \cdot f = 0.5Q$), the flanges will be displaced relative to each other at the first load cycle by an amount corresponding to elastic deformations under load $0.5Q$, and the subsequent cycles of unload and load (in the same direction) will not cause any macromovements in the connection. That is what is needed to prevent the failure of the connection.

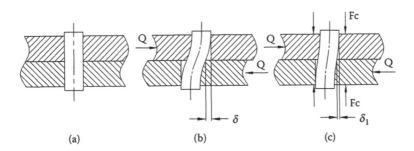

(a)   (b)   (c)

FIGURE 10.23 Connection with fitted bolts: relative movements under load.

The following conclusions can be drawn from the aforesaid:

- If there are (in the connection) fitted bolts or pins that are installed with interference and are able to take the entire shear force, the flanges still must be pressed together. This is to prevent their relative displacement when subjected to the oscillating load.
- The pressure between the flanges should be strong enough to transmit by the friction forces at least one half of the difference between the maximal and the minimal shear loads.

Translating this to the language of formulas, we can say that immobility in the bolted connection with fitted bolts, which is loaded by shear force changing from Q_{max} to Q_{min}, is achieved on the following condition:

$$F_t \cdot f \geq \frac{\bar{Q}_{max} - \bar{Q}_{min}}{2}$$

where F_t = tightening force of the bolt.

The lines over the letters Q mean that the matter concerns the vector difference. If the flange connection transmits a reversing load that is equal in both directions, the friction force must transmit the entire load, because in this case,

$$\frac{\bar{Q}_{max} - \bar{Q}_{min}}{2} = \frac{|Q|_{max} + |Q|_{min}}{2} = Q_{max}$$

The fitted bolts are ineffective in this case, and they can help only in situations of low-cycle overload.

When the bolts are tightly fitted, it should be taken into account that the pressure between the flanges (F_c) may be considerably less than the bolt tightening force F_t. This is concerned with the friction force F_S developed between the bolt shank and the connected parts. That force equals

$$F_S = \pi d_S l_S p_S f_S$$

where

d_S = diameter of the fitted shank (mm)
l_S = length of the fitted shank between the joint surface and the nut (mm) (Figure 10.22b)
p_S = surface pressure in the interference fit (MPa)
f_S = friction coefficient between the shank and the hole

Pressure p_S can be calculated by the following formula:

$$p_S = \frac{E_b \delta}{\varsigma d_S}$$

where

$$\varsigma = (1 + \mu)\frac{E_b}{E_f} + (1 - \mu)$$

E_b and E_f = modulus of elasticity of the bolt and the flange materials, respectively (MPa)
δ = diametral interference (difference in diameters of the shank and the hole in the flanges) (mm)
$\mu = 0.3$, Poisson's ratio

If the bolt and the flanges are made of materials with the same modulus of elasticity E (say, from steel), then $\zeta = 2$ and the surface pressure equals

$$p_S = \frac{E\delta}{2d_S}$$

EXAMPLE 10.9

Bolt M20 × 1.5 is fitted: $d_S = 21$ mm, $l_S = 15$ mm, and diametral interference $\delta = 0.02$ mm. The flanges are made of steel; $E = 2.06 \cdot 10^5$ MPa. Pressure in the shank fit equals

$$p_S = \frac{2.06 \cdot 10^5 \cdot 0.02}{2 \cdot 21} = 98.1 \, MPa$$

Friction force (provided that $f_S = 0.2$) equals

$$F_S = \pi \cdot 21 \cdot 15 \cdot 98.1 \cdot 0.2 = 19420 \, N$$

Taking the tightening force from Example 10.8 ($F_{t\,min} = 126{,}080$ N), we can determine the force in the contact of the flanges:

$$F_{c\,min} = F_{t\,min} - F_S = 126080 - 19420 = 106660 \, N$$

This means that the friction in the shank fit decreases the pressure between the flanges by about 15%. If the quality class (i.e., the strength and hardness) of the bolt is lower, the influence of the shank friction is greater. It is clear that the interference in the shank fit should not be large. Usually, it doesn't exceed 0.01–0.02 mm. Often, a transition fit is used for fitted bolts; for example, H6/k6. For $d_S = 21$ mm, this fit may result in a clearance of 11 μm to an interference of 15 μm. In very heavily loaded connections, all the bolts should be installed with an interference fit. This is achieved by careful machining and selection of bolts or by fitting each bolt to the measured dimensions of the holes in the flanges. When individually fitted in this way, each bolt and its hole must be marked with a unique index number to ensure they will be kept together during maintenance activities. Length l_S usually measures from $0.5d_S$ to $1d_S$.

Fitted bolts enable a great increase in the load capacity of flange connections under shear load. Such a connection is much more expensive, but when needed, the expense is justified. For instance, bolt M20 × 1.5–10.9 from Example 10.8 is tightened to $F_t = 126{,}080$ N, developing a friction force $F_f = 126{,}080 \cdot 0.3 = 37{,}820$ N. The permissible shear load for that friction force equals 37,820/3 = 12,610 N (safety factor $k = 3$ is taken here). The shank of a fitted bolt is 21 mm in diameter; its cross-section area equals 346 mm.[2] If we take the permissible shear stress to be 50 MPa (see Subsection 10.4.1), the additional shear force equals 50 · 346 = 17,300 N. So the summary shear load that is permissible for a fitted bolt equals 12,610 + 17,300 = 29,910 N (instead of 12,610 N for a nonfitted bolt, increase by 2.37 times).

If the connection with fitted bolts should be periodically dismantled, tapered bolts with taper of 1:10 to 1:20 are preferable (Figure 10.14). A problem in this case may be the attainment of needed pressure between the flanges. It is obvious that if the contact of the cones is concentrated at the bigger diameter, joint force F_c may be close to tightening force F_t. But if the contact patch is concentrated at the smaller diameter, force F_c may, in the extreme case, be even zero. Therefore, in the technical requirements for the accuracy of the cones, it is usually indicated that the contact

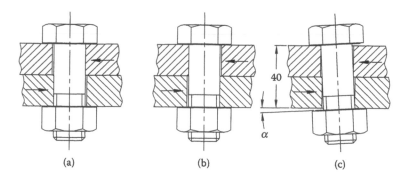

FIGURE 10.24 Slip in a flange connection.

patch should cover 70–80% of the surface, but nearer to the bigger diameter. Besides, the tapered bolts are often combined with plain bolts (see Figure 10.14) to ensure that there is pressure between the flanges.

In Figure 10.22c is shown a flange connection that is very similar to that depicted in Figure 10.22a. The only difference is that in the former the bolts are installed in smaller holes (for example, diameter 20.1 mm for bolt M20). If the load exceeds the friction forces in such a connection, the flanges slip a bit, and the bolt shanks take the overload. The nut and the head of the bolt don't move relative to their respective flanges, therefore some small bending deformation (and bending stress) is induced into the bolt. In the option as in Figure 10.22a, the slip may occur as well, but the movement is much more, and the additional bending of the bolts may be unacceptable.

EXAMPLE 10.10

Figure 10.24 shows two flanges, each 20 mm thick, connected with a bolt M20 installed in a hole of $20.1^{+0.2}$ mm in diameter. The maximal diameter of the hole can be 20.3 mm. The position of the connection members relative to each other is chosen at random, and the best case (with respect to transmission of the shear force) is shown in Figure 10.24a. No macroslip is needed here to put the shank of the bolt into action. In the worst case (Figure 10.24b), the flanges must slip by 0.6 mm before the shank becomes loaded in shear. During this motion, the nut and the head of the bolt are held in place (relative to the respective flanges) by friction forces, so that after this slippage, the bolt becomes skewed as shown in Figure 10.24c. The angle of misalignment equals

$$\alpha = \frac{0.6}{40} = 0.015 \; rad$$

This bolt is statically loaded, and experiments have shown[2] that at a static load, bolts of a middle strength (S_u <1200 MPa) are not sensitive to misalignment up to $\alpha = 8°$ (0.14 rad). Besides, this case is the worst, and in a multibolt connection, this combination of the biggest hole diameter with the worst position of the connection members will only occur at some, possibly none but almost certainly not all, of the bolt positions.

Bolts installed in smaller holes are preferred to have a full-diameter shank (not reduced in diameter) and a shorter thread that doesn't reach the joint plane. However, these bolts are not considered as fitted. When such a connection is calculated for strength, only the friction forces are taken into account. Placing the bolts in smaller holes may contribute a lot to the reliability of the connection, and will not materially affect its cost.

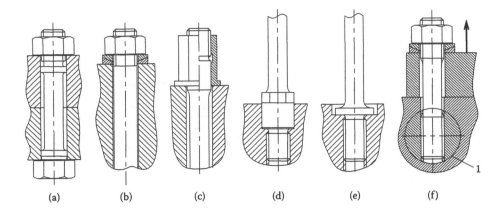

FIGURE 10.25 Preventing threads from bending.

10.3.3 BENDING LOAD

In principle, fasteners are intended for axial load only, but some bending is in many cases unavoidable. It may be related to manufacturing tolerances or deformation of parts, as shown in Figure 2.9b and Figure 2.9c (Chapter 2), or to slippage in a connection (see Example 10.10 and Figure 10.24c). Even while tightening, the bolt is bent by a force applied to the wrench. For long bolts with thinned shanks, this factor can be significant, and such bolts are often provided with centering spigots near their ends (Figure 10.25a).

Though statically loaded bolts, as pointed out earlier, are not sensitive to misalignment of their bearing surfaces (with the exception of extra-high-strength bolts), the tightening of such bolts can't be stable because of high local pressure and plastic deformation of the surfaces. Therefore, all appropriate measures should be taken to ensure that the bearing faces of the nut and the bolt head make full contact with the flanges. In the first place, these surfaces should be accurately machined. Then, the design should diminish deformations of connection members and prevent bending of the threaded parts, especially if the bolts are cyclically loaded. The illustration shows spherical washers (Figure 10.25b), a stud with a centering spigot on the nut side (Figure 10.25c) and on the housing side (Figure 10.25d), a stud with a flange (Figure 10.25e), and a stud screwed into a cylindrical insert 1 in the housing (Figure 10.25f). It should be noticed that self-alignment achieved by spherical washers and cylindrical inserts is effective during tightening. But these methods are ineffective if the bending is caused by deformations at cyclic load. High friction on the bearing surfaces prevents cyclic self-alignment. Decreasing deformations by making parts more rigid, ribbed, and so on, and making bolts longer and thinner may be helpful in decreasing bending stresses.

Besides the mentioned reasons, bending may be caused by lateral vibrations of the stud or bolt. It is relevant for relatively long fasteners.

EXAMPLE 10.11

Figure 10.26 shows the kinematic scheme of a main gear of a marine diesel engine with four crankshafts. Pinions 1, 2, 3, and 4 are connected to the crankshafts, and gear 5 is the output one.

The gear housing is welded of steel and attached to the engine case by flange 6. For assembly needs, the gear housing is split into two halves joined with flanges 7. Because the halves of the housing are rather flat and not rigid, they are additionally connected by eight studs (M36 thread) placed close to the gears. The studs are placed inside tubes 9, which take up the tightening force

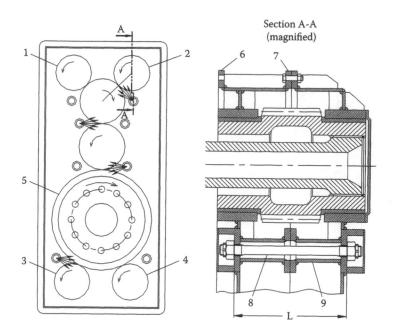

FIGURE 10.26 Vibrations of long stud bolts.

of the studs. In addition, four of the tubes are used as sprayers to lubricate the gears (they are indicated on the scheme). These tubes have small holes, and into them is supplied oil under pressure.

After the trials of the engine, it was revealed that four of the studs 8 had failed from fatigue. Because the studs are statically loaded, it was supposed that there might be lateral vibration excited by the gearing. The nominal speed of the crankshafts $n = 1000$ r/min, teeth number of the pinions $N = 32$. Consequently, the excitation frequency equals

$$f_e = \frac{nN}{60} = \frac{1000 \cdot 37}{60} = 617 \, Hz$$

The length of the stud $L = 400$ mm, diameter of the shank $d = 30$ mm. If the shank is not tightened, its first natural frequency can be obtained from Equation 4.4:

$$f_1 = 2.01 \cdot 10^6 \frac{d}{L^2} = 2.10 \cdot 10^6 \frac{30}{400^2} = 394 \, Hz$$

Because the shank is tightened, that value should be multiplied by a factor that equals[7]

$$\sqrt{1 + \frac{S L^2}{\pi^2 E I}}$$

where

$E = 2.06 \cdot 10^5$ MPa, modulus of elasticity of steel

$$I = \frac{\pi d^4}{64} = \frac{\pi \cdot 30^4}{64} = 3.98 \cdot 10^4 \, mm^4$$

$S = 4.5 \cdot 10^5$ N, tightening force of the stud.

From here the factor of the frequency increase equals

$$\sqrt{1+\frac{4.5\cdot10^5\cdot400^2}{\pi^2\cdot2.06\cdot10^5\cdot3.98\cdot10^4}}=1.375$$

Hence, the first natural frequency of the tightened stud equals $394\cdot1.375=542$ Hz. This result shows that at nearly 90% of the nominal speed there must have been a resonance oscillation.

(Note: The calculation of the stud natural frequency is not very accurate for two reasons. First, we don't know exactly the tightening force of the stud; it may differ by about ± 20%. And second, the equations we have used for the frequency estimations are correct for a rod with hinged ends, but the ends of the stud are not exactly hinged. Nevertheless, this calculation has shown clearly that the problem is very likely in the resonance oscillation. The subsequent experiments have given full confirmation of this conclusion.)

To eliminate this problem, the natural frequency of the stud should have been increased. This could have been made by increasing the diameter of the shank up to the thread diameter, i.e., from 30 mm to 36 mm. Another option was to make a centering collar in the middle of the shank slide-fitted to a machined hole in the housing (this would require some changes in the housing). In this case, the natural frequency would have been multiplied by four. But in this situation a simpler solution has been found.

It was noticed that the studs had failed only in such places where the tubes 9 were not used as sprayers, i.e., they were not filled with oil. It was supposed that the oil damps the oscillation, and this assumption was confirmed by experiments. Then all the tubes 9 were connected to the oil supply, and the stud failures were eliminated.

10.4 STRENGTH OF FASTENERS

Here we are discuss the main problems concerning the strength of fasteners. In engineering practice, the calculations for strength should be performed using appropriate reference books, standards, and methods accepted in the respective branches of industry.

The first problem, and in many cases the most difficult one, is to define the load applied to the bolt. This subject has been discussed earlier. As soon as the "personal" load of a bolt is defined, the diameter and the grade (quality class) of the bolt can be preliminarily estimated. After that, calculations should be performed for static strength while tightening and under load, and for fatigue strength if the load is periodically or cyclically changing. If the strength of the bolt is insufficient, there are several methods of getting the needed safety factors:

- Changing the tightening force (preload) of the bolt
- Using bolts of higher grade or quality class
- Reducing the load of the bolt by increasing the number of bolts and by more efficient location of the bolts relative to the working load
- Reducing the stress cycle amplitude by increasing the compliance of the bolt
- Increasing the diameter of the bolt

10.4.1 STATIC STRENGTH

As is well known, the bolt may be broken just while tightening, especially if the bolt is small, the wrench is big, and the fitter more strong than experienced. This is one of the reasons for specifying tightening torques and using torque wrenches. But a torque wrench can not always be found in service; therefore, designers of machines often avoid using bolts less than 8–10 mm in diameter. Now this tradition has become obsolete.

Bolt (or stud) failure consists mostly in shank (body) breakage. The maximal equivalent stress in the threaded part while tightening (considering both tension and torsion, but without stress raisers) equals:[2]

$$\sigma_{eq.} \approx \frac{1.3 F_t \cdot 4}{\pi d_e^2} \, MPa \tag{10.14}$$

where

F_t = tightening force (N)

d_e = equivalent diameter of the threaded part, that is, the average between the inner and the outer diameters of the thread (mm)

$$d_e = d - 0.923p$$

p = pitch of the thread (mm)

For a coarse thread $d_e = (0.85-0.90)d$, and for a fine thread it is somewhat bigger.

If the shank is thinned (as shown in Figure 10.17c and Figure 10.17d) and its diameter less than d_e, the diameter of the shank should be substituted in Equation 10.14.

The stress obtained by Equation 10.14 is valid for the static strength calculation only because it doesn't make allowance for the stress concentration in the thread.

Factor 1.3 in this equation takes into account the torsion of the shank. Sometimes, one can hear the opinion that the bolt shank is not exposed to torsion if the tightening is made by rotation of the nut while the head of bolt is held in place. This is false. The load of the shank is absolutely independent of what is rotated at tightening: the nut or the head of the bolt. Otherwise, we would have to recognize that the law of action and reaction has exceptions in the case of bolts (what a surprise for Mr. Newton!).

The torque the shank is exposed to, T_S, equals the first member of Equation 10.11:

$$T_S = F_t \frac{d}{2} \left(\frac{p}{\pi d} + 1.15\mu \right)$$

For bolt M12 of quality class 10.9 (see Example 10.7 and Figure 10.17b), the tightening force F_t = 54,000 N at μ = 0.14. The pitch of thread p = 1.75 mm. In this case,

$$T_S = 54000 \frac{12}{2} \left(\frac{1.75}{\pi \cdot 12} + 1.15 \cdot 0.14 \right) = 67200 \, N \cdot mm$$

The angle of torsion of the shank equals

$$\varphi = \frac{T_S L}{I_p G}$$

where

L = length of the shank (L = 40 mm)

$I_p = \pi d^4/32$, polar moment of inertia (mm^4)

$G = 0.8 \cdot 10^5$ MPa, modulus of elasticity in shear

It follows that

$$\varphi = \frac{67200 \cdot 40 \cdot 32}{\pi \cdot 12^4 \cdot 0.8 \cdot 10^5} = 0.0165 \; rad = 0.95°$$

If the nut is locked directly with the shank of the bolt, the torsion of the shank may disappear in a short while, because the decrease in coefficient of friction (due to vibrations) provides the nut (or the head of the bolt) the possibility to rotate relative to the housing by 1° and to release the shank from torsion. But if the nut and the bolt head are locked separately with the housing, they can't turn. In this case, when the friction coefficient diminishes, the shank may screw somewhat into the nut and decrease the residual torque, so that the tightening force must increase. In the calculated case, if the shank screws in by 0.95°, the total elastic deformation will increase by

$$1.75 \frac{0.95}{360} = 0.0046 \; mm$$

From Example 10.7, we know that the total elastic deformation of the bolt and the flanges measures 0.155 mm, so the increase in deformation (and in the tightening force) doesn't exceed 3%.

But if the bolt (or the stud) is long and thin, the torsion angle may be much larger. In this case, the shank remains under torque, and this impairs the strength of the shank. To prevent torsion in the shank, its end 1 (Figure 10.27a) that projects from the nut can be made hexagonal or square, so that while tightening, the shank can be held by another wrench. Figure 10.27b shows a design with a profiled washer 2. The washer is connected to the housing by two pins and to the shank by two protuberances that enter the slots in the thread, so that the shank can't rotate relative to the housing. This option is easier to work with, but the slots in the shank reduce its strength.

The maximal equivalent stress at tightening σ_{eq} as a rule should not exceed 90% of the yield point σ_y. Such a high preload can be used if the bolt is not exposed to additional tension at work. As an example, we refer to bolts shown in Figure 10.14. But if the working forces tension the bolt additionally, its preload should be decreased, so that at maximal load, the stress will not exceed 90% of the yield point. Exceeding the yield point may lead to plastic elongation of the bolt, loss in its tightening force, and subsequent failure of the connection.

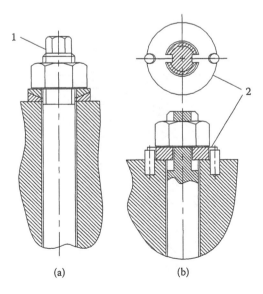

(a)   (b)

FIGURE 10.27 Preventing the stud from being twisted.

When calculating a fitted bolt, its section in the joint plane should be checked separately. It transmits the shear load, and also it is exposed to tension and torsion loads like the other parts of the shank. The shank diameter in this place (d_S) is bigger than d_e; therefore, the safety factor value here can be made not less than that in the threaded part, if the shear stress is quite moderate, say, 40–60 MPa.

The designer is not usually worried about the shear strength of the thread; the standard height of the nut that measures about $0.8d$ ensures that the strength of the thread is higher than that of the shank. This is true when the materials of the bolt and the nut are nearly of the same strength. If the nut (or the housing) is made from weaker material, the strength of the thread should be increased. In particular, bolts and studs of property class 6.8 and lower, when installed in housings of gray cast iron, bronze, or aluminum alloy, should have $(1.3–1.5)d$ of engaged thread length. For stronger bolts, to make the housing thread equal in shear strength to that of the fastener, thread engagement of $2d$ is required. For property classes 10.9 and 12.9, the thread in the housing should be reinforced by a steel insert (item 2 in Figure 9.5c), or the thread diameter should be increased (Figure 9.5b); see Chapter 9. When helical inserts are used, the static strength of the housing's thread can be increased by about 70% (Figure 10.28).[2]

Provided that the materials of the bolt and the nut are of the same strength, the shear strength of the female thread is higher by a factor of about 1.3 than that of the male thread. This is because the diameter of the nut shear surface nearly equals the outer diameter of the thread, whereas the bolt thread is sheared at its inner diameter. Therefore, the nut can be made of weaker material than

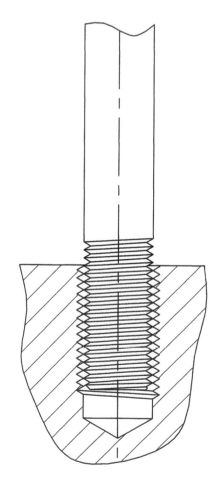

FIGURE 10.28 Reinforcement of thread using helical insert.

that of the bolt by the same factor 1.3. The weaker material of the nut enables better load distribution between the threads (because of plastic deformation of the nut material), and owing to this, it increases the fatigue strength of the bolt. The useful height of the weaker nut can be increased to $(1.4–1.6)d$.

10.4.2 FATIGUE STRENGTH

Calculation of fasteners for fatigue strength is based on the mechanical properties of the material and stress concentration factors, but the manufacturing technique also should be taken into consideration. For example, the bolt head formed by heading and the thread made by rolling are stronger than these elements formed by cutting.

The basics of calculations for fatigue strength are detailed in Chapter 12. As is shown there, the mean stress of the loading cycle affects the fatigue strength very little. As applied to fasteners, the influence of the mean stress can be neglected completely unless the maximal stress is lower than $0.9S_y$. So the only force to be taken into consideration in the calculation for endurance is the amplitude of the working (alternating or variable) load F_{aw}. Because the bolt "feels" only a portion of the working load, the amplitude of the bolt force F_{ab} can be obtained as follows:

$$F_{ab} = F_{aw} \cdot \chi \quad N$$

(about coefficient χ see Subsection 10.3.1).

The nominal stress amplitude σ_a equals

$$\sigma_a = \frac{F_{ab}}{A} \quad MPa \tag{10.15}$$

where A = area of the section under consideration (mm²).

Comparing Equation 10.15 with Equation 10.14, we notice that coefficient 1.3 has disappeared. This is not a mistake. In the endurance calculation, we consider only the variable load. But the torque induced in the bolt shank while tightening doesn't change, and it can't influence the fatigue strength of the shank. But it does influence the maximal stress, which should not exceed the limit mentioned earlier.

There are two areas in the bolt that should be checked for fatigue strength: transition from the shank to the bolt head, and also the threaded part. Usually a fatigue crack originates in the area of the most heavily loaded thread. In this place, the stress caused by tension of the shank is supplemented with stress induced by bending of the thread profiles (Figure 10.29). The latter is responsible for the low fatigue strength of the threaded shank: experiments show[2] that without bending, the fatigue limit of a tensioned threaded rod increases threefold. This factor can hardly be calculated; therefore, the admissible amplitude of the variable tensile stress is usually recommended on the basis of experiments. As a guide, for rolled threads, the admissible value of σ_a can be estimated from the following equation:[8]

$$\sigma_a = \xi S_u \frac{25+d}{25+3d}$$

where

S_u = tensile strength of the material (MPa)
d = thread diameter (mm)
ξ = 0.15 for steels with $S_u \le 1100$ MPa

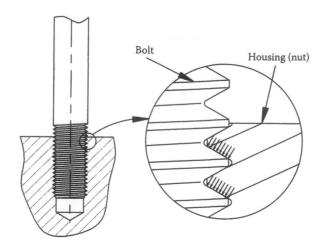

FIGURE 10.29 Bending of threads.

Example: for bolt M16–10.9 (S_u = 1040 MPa)

$$\sigma_a = 0.15 \cdot 1040 \frac{25+16}{25+3 \cdot 16} = 87.6 \, MPa$$

Another manual[9] recommends σ_a = 65–85 MPa at S_u = 600–1200 MPa, if the bolt is heat-treated before rolling the thread. With the same conditions for a cut thread, σ_a = 55–70 MPa is recommended. Because the fatigue strength of the thread displays significant dependence on manufacturing processes, in important cases, the admissible amplitude of the tensile stress should be determined experimentally. To obtain true results in the experiments, not only the bolt (or stud) but also the adjoined parts should be similar to the real ones, because they also influence the load capacity of the bolt. For instance, weaker nut materials, as well as materials with smaller modulus of elasticity, contribute to better load distribution in the thread, decrease the bending load on the most loaded threads, and increase the fatigue strength of the bolt.

A very effective idea may be to increase the radius in the root of the thread (say, to $r = 0.2p$, where p is the pitch of thread). The endurance of the threaded parts can also be improved by increasing manufacturing and assembly accuracy and decreasing clearances. Small clearances restrict the bending deformation of the threads.

Another area to be assessed for strength is the fillet that makes the transition from the shank to the head of the bolt. Increase in the fillet radius is limited by the loss in the bearing surface of the head; therefore, elliptic fillets are often used for heavily loaded bolts, as shown in Figure 10.30. Because there is only one stress raiser (fillet) with a well-known influence

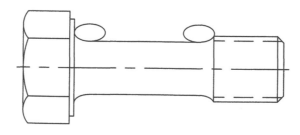

FIGURE 10.30 Bolt designed for high fatigue strength.

on the maximal stress, calculations in relation to this area are performed by conventional means (see Chapter 12).

REFERENCES

1. Seidel, M. and Schaumann, P., Measuring of Fatigue Loads of Bolts in Ring Flange Connections, EWEC 2001 Conference, Copenhagen, 2001.
2. Birger, I.A. and Iosilevich, G.B., *Threaded Joints*, Mashinostroenie, Moscow, 1973 (in Russian).
3. Gati, T., Evaluation of Stresses, Strain and Displacements in Bolted Joints, Master's thesis, Ben-Gurion University of the Negev, Beer-Sheva, 2003 (in Hebrew).
4. Gorbunov, M.N. and Saltikov, M.A., Calculation of Flange Connections of Shafts with Fitted Bolts under Varying Torsion, *Energomashinostroenie*, 1973/12 (in Russian).
5. Saltikov, M.A. and Gorbunov, M.N., *Method of Strength Evaluation of Detachable Connections*, Mashinostroenie, Moscow, 1983 (in Russian).
6. Bossard Ltd. Fasteners, Metric Fasteners for Advanced Assembly Engineering, Catalogue and technical information YB.
7. Timoshenko, S., *Vibration Problems in Engineering*, D. Van Nostrand Company, Toronto, New York, London, 1955.
8. Heywood, R.B., *Designing Against Fatigue of Metal*, Reinhold, New York, 1962.
9. Birger, I.A., Shorr, B.F., and Iosilevich, G.B., *Calculation of Machine Elements for Strength*. Manual, Mashinostroenie, Moscow, 1979 (in Russian).

11 Connection of Units

Machines are usually divided into units (subassemblies) that are manufactured separately and then linked together. This method of forming machines has many points in its favor, and some of them deserve to be mentioned.

1. Manufacturing and quality inspection of each unit can be performed separately and simultaneously, reducing the time taken to make the machine.
2. Partitioning of machines into units enables standardization and use of the standard units in different machines. As an illustration, we can refer to electric motors, gearboxes, couplings, high-pressure fuel pumps and injectors for diesel engines, and infinitely many other units. It is impossible to imagine modern technical civilization without this comprehensive unification and standardization.
3. When a machine fails, it is the work stoppage, not the repair, that usually is most expensive. The failed machine can disable a production conveyer or the entire plant. Replacing a failed unit by a new one warehoused as a spare can reduce the repair time to a minimum.
4. Use of units facilitates the task of machine designers who can compound their machines from stock-produced "cubes," linking them in the needed way.

If a machine has a part that, because of shorter service life or for some other reasons, will need to be periodically replaced, it should be segregated into an easily detachable unit to shorten the service interruption.

Sometimes the interruption time can be critical when trying to avoid very serious and even catastrophic consequences. In such cases, the equipment should be built so that when the primary machine fails, the reserve machine, which is permanently connected and kept in operating condition, starts automatically and immediately.

The stock-produced units have elements for connection to other units, like flanges, spigots, mounting feet, threads, and others. Their dimensions are given in the manufacturer's catalogs along with other technical data. In many cases, there are given several options of attachment, and sometimes a special design of the attachment elements is available, when it is economically feasible.

Connection of two units consists of housing connections and, separately, connection of parts transmitting motion, which are mostly rotating shafts.

11.1 HOUSING CONNECTIONS

Housings can be connected directly or through an intermediate part, like a frame or foundation. The direct connection is cheaper. Figure 7.30 shows a direct connection of electric motor 7 through adapter 8 to the housing of the gear. Another example of direct connection is shown in Figure 7.31. Oil pump 1 is connected to the upper bearing cup, and electric motor 8 is attached to extension 9 of the upper part of the gear housing. (Openings 10 are made to ventilate the coupling chamber and to inspect the condition of the rubber elements of the coupling.)

If the attached unit is relatively large and weighty, it can induce remarkable deformations in the housing and bring about misalignment of the teeth and bearings of the gear. The possible deformations can be calculated using FEM. The stock-produced units don't always have attachment elements that enable their direct connection. For example, pump 1 shown in Figure 11.1a is provided with mounting feet, so that electric motor 2 can be attached to it only by means of an intermediate part (frame 3 in this case). The torque transmitted from the motor shaft to the pump shaft (through coupling 4) is balanced by a torque of the same magnitude that is transmitted from the housing of the pump to the housing of the motor through the frame. If the frame was attached to a rigid foundation, its deformation would be blocked. But as long as the frame is mounted on rubber vibration dampers 5, its torsional deformation should be calculated because it leads to a certain misalignment of the shafts.

Possibly, it is not obvious that frame 3 is twisted. Let's use Newton's laws to prove this assertion. Figure 11.2 shows the same unit, but frame 3 is cut in the middle, and the halves are separated by some distance, so that the shafts of pump 1 and motor 2 are withdrawn from coupling 4. All these parts are either stationary or rotating uniformly. According to Newton's first law, each of the separated parts must be exposed to a balanced assemblage of forces. That means that if the motor applies torque $+T$ to the coupling, the other end of the coupling is loaded with torque $-T$. Now, according to Newton's law of action and reaction, the motor must be loaded with torque $-T$, and the pump with torque $+T$. Then the motor-bound half of the frame must be loaded with torque $+T$ and the pump-bound half with torque $-T$. So we have proved that the frame section between the motor and the pump is twisted with the same torque that is transmitted by the motor to the pump.

This can be confusing, as it was previously mentioned that the motor must be loaded with torque $-T$ and the pump with torque $+T$. It should be noted that these torques are applied not to the motor or to the pump, but to the shaft of the motor and to the shaft of the pump, and the shafts rotate "freely" within their housings. That is, the rotor of pump 1 is connected with its housing by forces of hydrostatic pressure that create the same torque on the rotor and on the housing. The same is true of the electromagnetic field in the motor. We can separate the rotors from the stators and apply to each of them the additional torques $+T$ or $-T$. But there's no need for such detail. From Newton's law of action and reaction, it follows that, for calculation of *external* forces, any unit (gear, pump, motor, etc.), with its shafts or other moving parts can be considered as made of one piece of metal. Frankly, Newton's laws are not always easy to understand deeply and to apply properly in practice. But they are of fundamental importance, therefore the authors took the liberty of making these (possibly annoying) explanations.

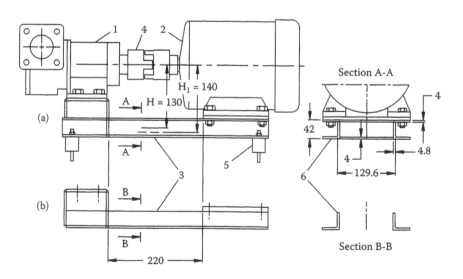

FIGURE 11.1 Design, deformations, and strength of a frame.

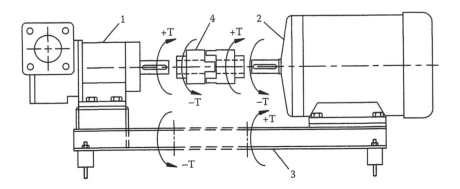

FIGURE 11.2 Twisting of the frame.

EXAMPLE 11.1

Let's calculate the deformations of the frame shown in Figure 11.1. Pump 1 is driven by electric motor 2. Their housings are connected through frame 3 and the shafts through coupling 4. The frame is installed on four dampers 5. The motor power $P = 3$ kW at $n = 1425$ r/min, so the torque T is given by:

$$T = 9555 \frac{P}{n} = 9555 \frac{3}{1425} = 20.1 \, N \cdot m$$

The frame is mainly made of two aluminum equilateral angle bars 6 ($38 \times 38 \times 4.8$ mm). Two versions of the frame are represented in Figure 11.1:

1. The angle bars are covered with plates of 4 mm thickness from above and below, along their whole length (section A-A).
2. The middle part of the frame is left without covering, and only the angles transmit the torque between the housings of the motor and the pump (section B-B).

To calculate approximately the frame deformation, we can consider the plated ends (under the motor and the pump) as completely rigid, and take into account only the middle part of the frame that is 220 mm long. The dimensions of the frame are given in Figure 11.1.

Figure 11.3a shows the simplified cross section of version a. For our calculations, we take it as a hollow rectangle, because we have formulas for this shape. The angle of torsion φ of this section is:

$$\varphi = \frac{T \cdot L}{J \cdot G}$$

where
T = torque (N·mm)
L = length of the twisted part (L = 220 mm)
G = shear modulus (G = $0.28 \cdot 10^5$ MPa for aluminum alloy)
J = polar moment of inertia (mm⁴) that equals for this section[1]

$$J = \frac{2 t_1 t_2 a^2 b^2}{a t_1 + b t_2} = \frac{2 \cdot 4.8 \cdot 4 \cdot 124.8^2 \cdot 38^2}{124.8 \cdot 4.8 + 38 \cdot 4} = 1.15 \cdot 10^6 \, mm^4$$

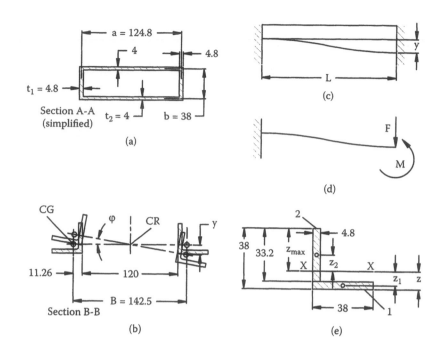

FIGURE 11.3 Calculation of deformations and stresses.

Thus, the twisting angle

$$\varphi = \frac{20.1 \cdot 10^3 \cdot 220}{1.15 \cdot 10^6 \cdot 0.28 \cdot 10^5} = 1.37 \cdot 10^{-4} \; rad$$

The misalignment of shafts Δ caused by this deformation is

$$\Delta = \varphi \cdot H$$

where H = distance between the shaft axis and the center of gravity or CG of the twisted section (H = 130 mm, see Figure 11.1a). Consequently, $\Delta = 1.37 \cdot 10^{-4} \cdot 130 = 0.018$ mm.

The maximal shear stress equals

$$\tau_{max} = \frac{T}{2abt_2} = \frac{20.1 \cdot 10^3}{2 \cdot 124.8 \cdot 38 \cdot 4} = 0.53 \; MPa$$

Now we see that the strength and rigidity of version a are satisfactory.

Let's check version b. The angle bars can be considered as fixed on both ends and turned by some angle φ around point CR (see Figure 11.3b). Each angle bar becomes both twisted by angle φ and bent (as shown in Figure 11.3c) by an amount $y = \varphi \cdot B/2$, where B is the distance between the CG points (Figure 11.3b). Our goal is to calculate the dependence between the torque T applied to the bars and the twisting angle φ.

For one equilateral angle bar, the twisting angle is given by[2]

$$\varphi = \frac{3 T_t L}{t^3 (2a - t) \cdot G}$$

where

T_t = twisting torque of one bar (N·mm)

L = length of the bar (L = 220 mm)

t = thickness of the angle's webs (i.e., its legs) (t = 4.8 mm)

a = lateral length (a = 38 mm)

G = shear modulus (G = 0.28·10^5 MPa for aluminum alloy).

From here,

$$T_t = \varphi \frac{t^3(2a-t) \cdot G}{3L} \tag{11.1}$$

The resistance forces created by bending can be calculated using a scheme shown in Figure 11.3d, where, on one side, the fixing is substituted by overhang force F and bending moment M. These forces can be defined as follows (assuming that the deflection y is known):

The bending angles on the end of a beam caused by force F (θ_F) and by moment M (θ_M) are as follows:

$$\Theta_F = -\frac{FL^2}{2EI}; \quad \Theta_M = \frac{ML}{EI}$$

where

E = modulus of elasticity (E = 0.72·10^5 MPa for aluminum alloy)

I = moment of inertia of the cross section (mm^4)

As the resultant angle at this point is zero,

$$\Theta_F + \Theta_M = -\frac{FL^2}{2EI} + \frac{ML}{EI} = 0;$$

$$M = \frac{FL}{2} \tag{11.2}$$

The deflections of the beam end caused by force F (y_F) and by moment M (y_M) are

$$y_F = \frac{FL^3}{3EI}; \quad y_M = -\frac{Ml^2}{2EI}$$

As the resultant deflection at this point equals y,

$$y_F + y_M = \frac{FL^3}{3EI} - \frac{ML^2}{2EI} = y = \varphi\frac{B}{2} \tag{11.3}$$

After insertion of Equation 11.2 into Equation 11.3, we obtain the following:

$$\varphi = \frac{FL^3}{6EIB} \tag{11.4}$$

The resistance torque (T_b) caused by the bending deformation of the angle bars equals

$$T_b = F \cdot B$$

Inserting $F = T_b/B$ into Equation 1.4, we obtain the following:

$$T_b = \varphi \, \frac{6 \, B^2 E \, I}{L^3} \tag{11.5}$$

The summary resistance torque, which equals the applied torque of the motor T, is obtained from Equation 11.1 and Equation 11.5:

$$T = 2T_t + T_b = \varphi \left(\frac{2t^3(2a - t)G}{3L} + \frac{6 \, B^2 E \, I}{L^3} \right) = \varphi \cdot \lambda \tag{11.6}$$

Here, λ is equal to the expression in parentheses.

All the terms of Equation 11.6 are already known from the preceding text except the moment of inertia I. It is well known how to determine the I value; it is given in the catalog. But the authors are glad to demonstrate this process for the case of nonstandard profile. The steps are as follows:

1. Inasmuch as the beam is bent in the vertical plane, we are to define the value of I about a horizontal axis x–x passing through the CG of the cross section (to obtain the minimal value of I).
2. The cross section of the bar is shown in Figure 11.3e. To ease our work, we divide the cross section into two rectangles 1 and 2. Each of them has its own CG obviously placed in its geometric center (marked by a small circle). The area of rectangle 1, $A_1 = 38 \cdot 4.8 = 182.4$ mm^2 and that of rectangle 2, $A_2 = (38 - 4.8)4.8 = 159.36$ mm^2.
3. To determine the position of the x–x line, let's put it at a coordinate z. Then the distances of the CGs of rectangles 1 and 2 from the x–x line are equal respectively to $z_1 = (z - 2.4)$ mm and $z_2 = 38 - z - 16.6 = (21.4 - z)$ mm.
4. The equilibrium equation is

$$A_1 z_1 = A_2 z_2;$$

$$182.4 \, (z - 2.4) = 159.36 \, (21.4 - z)$$

$$\text{where } z = 11.26 \; mm \quad z_1 = 8.86 \; mm \quad z_2 = 10.14 \; mm$$

5. Now we can calculate the I value from the following formula:

$$I = \Sigma \left(I_i + A_i z_i^2 \right)$$

$$I_1 = \frac{38 \cdot 4.8^3}{12} = 350.2 \; mm^4; \quad I_2 = \frac{4.8 \cdot 33.2^3}{12} = 14638 \; mm^4$$

$$I = 350.2 + 14638 + 182.4 \cdot 8.86^2 + 159.36 \cdot 10.14^2 = 45692 \; mm^4$$

Coming back to Equation 11.6, we obtain the λ value:

$$\lambda = \frac{2t^3(2a-t)G}{3L} + \frac{6B^2EI}{L^3} = \frac{2 \cdot 4.8^3(2 \cdot 38 - 4.8) \cdot 0.28 \cdot 10^5}{3 \cdot 220} + \frac{6 \cdot 142.5^2 \cdot 0.72 \cdot 10^5 \cdot \text{?}}{220^3}$$

$$= 6.68 \cdot 10^5 + 376.4 \cdot 10^5 = 3.83 \cdot 10^7 \; N \cdot mm;$$

$$\varphi = \frac{T}{\lambda} = \frac{20100}{3.83 \cdot 10^7} = 5.25 \cdot 10^{-4} \; rad$$

(Note that in the equation for λ, the first term is negligible as compared with the second one, which means that in this case, when the angle bars are distanced from each other, their twisting deformation is relatively small and can be neglected.)

The distance between the motor axis and the CR of the angle bars $H_1 = 140$ mm (see Figure 11.1a). Consequently, the shaft misalignments caused by the frame deformation is

$$\Delta = \varphi \cdot H_1 = 5.25 \cdot 10^{-4} \cdot 140 = 0.074 \; mm$$

The stresses in the angle bars can be calculated separately from torsion and from bending and then combined. The torque can be calculated using Equation 11.1:

$$T_t = \varphi \frac{t^3(2a-t) \cdot G}{3L} = 5.25 \cdot 10^{-4} \frac{4.8^3(2 \cdot 38 - 4.8) \cdot 0.28 \cdot 10^5}{3 \cdot 220} = 175 \; N \cdot mm$$

The maximal shear stress

$$\tau_{max} = \frac{3T_t}{t^2(2a-t)} = \frac{3 \cdot 175}{4.8^2(2 \cdot 38 - 4.8)} = 0.32 \; MPa$$

The maximal bending moment can be obtained from Equation 11.2 and Equation 11.4:

$$M = \varphi \frac{3EIB}{L^2} = 5.25 \cdot 10^{-4} \frac{3 \cdot 0.72 \cdot 10^5 \cdot 45692 \cdot 142.5}{220^2} = 15255 \; N \cdot mm$$

The maximal bending stress

$$\sigma_{max} = \frac{M}{I} z_{max} = \frac{15255}{45692} 26.74 = 8.93 \; MPa$$

(Regarding z_{max}, see Figure 11.2e.)

The maximal equivalent stress

$$\sigma_e = \sqrt{\sigma_{max}^2 + 4\tau_{max}^2} = \sqrt{8.93^2 + 4 \cdot 0.32^2} = 8.95 \; MPa$$

It is a very small stress (see, for instance, the data in Table 9.1). But the misalignment of 0.074 mm is noticeable.

Connection of units through an intermediate frame or foundation can also be forced by a heavy weight of the equipment that must be installed on a rigid horizontal foundation, as well as by greater distance between the units that makes impracticable the direct connection of the housings. But, anyway, we have to remember that the frames and foundations are not just supports. They are parts of a mechanism that participate in load transmission, and they must be checked for strength and deformations like other parts. The only difference is that these parts are stationary and made of cheaper materials, so their safety factors can be much greater and the calculations respectively more approximate.

EXAMPLE 11.2

Figure 11.4 shows a foundation made for a proving stand for escalator drive units. Drive unit 1 is linked to block brake 2 by chain 3. The forces given in Example 4.5 and Figure 4.20 are valid for an escalator in service. When testing, the weight of the hauling chains is negligible, and only the difference between forces F_{cu} and F_{cl}, which creates the torque on the main shaft, is realized:

$$F = F_{cu} - F_{cl} = 300,000 - 100,000 = 2 \cdot 10^5 \ N$$

Force F is applied to both the drive unit 1 and the brake 2 at a height of 2.5 m from the foundation surface. As the foundation height measures 2 m, the arm of force F (its distance to the neutral line (0-0) equals 3.5 m, and the bending moment M applied to the foundation is as large as

$$M = 2 \cdot 10^5 \cdot 3500 = 7 \cdot 10^8 \ N \cdot mm$$

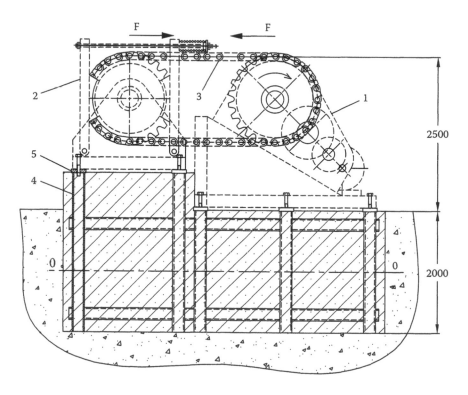

FIGURE 11.4 Foundation for proof test of escalator drive unit.

The main shaft of the escalator is provided with two sprockets (see Figure 4.20), but for the trials, only one of them (near the gear) is linked to the brake. This leads to uneven distribution of the bending moment across the width of the foundation. The factor of this unevenness can be generously taken as $k = 2$ (as only one of the two sprockets transmits the load).

Now we can calculate the stress in the foundation. Its cross section is 2 m high and 3 m wide. The bending stress

$$\sigma = M \frac{6}{b \cdot h^2} k = 7 \cdot 10^8 \frac{6}{3000 \cdot 2000^2} 2 = 0.7 \, N/mm^2$$

The ultimate bending stress of concrete (of medium quality) is about 3 MPa (on the tension side), so the safety factor S_F is given by

$$S_F = \frac{3}{0.7} = 4.3$$

As shown in Figure 11.4, the concrete foundation is reinforced by a welded structure 4 made of rectangular tubes. The vertical members of the structure are provided with thick plates 5 with threaded holes for attachment of equipment. The structure has been aligned, welded, and then installed in a pit with a form and filled with concrete.

The bending deformations of the concrete foundation, in this case, are relatively small, and they don't influence anything.

These two examples demonstrate the idea that whatever the design and the materials of the frame or the foundation, they can't be considered as completely rigid and infinitely strong. Their strength and deformations are worth checking, even if they look massive and reliable.

It is undesirable to install two connected machines on separate foundations, as in this case, the earth between the foundations is included in the chain of parts transmitting the load. The possible alteration in the relative placement of the connected units must be taken into consideration when choosing the kind of connecting couplings.

11.2 SHAFT CONNECTIONS

11.2.1 ALIGNMENT OF SHAFTS

A student of history once asked his professor: "Sir, you told us about World War II. Does it mean that there was also a World War I?" So, before discussing the alignment of shafts, we will study what misalignment is and why it may cause problems.

Misalignment of shafts is an offsetting of their axes. Figure 11.5a shows the condition called *radial misalignment*; the axes of the shafts remain parallel, but they are displaced relative to each other by an amount e in the radial direction. Shafts 1 and 2 are joined together by bolts. If the shafts and their supports were absolutely rigid, the fixed joint would not allow the shafts to rotate. But the parts are resilient, and the bearings have clearances, so the shafts can rotate by application of a certain torque. These shafts, turned by 180°, are shown in Figure 11.5b. The shafts are quite bent, and the bearings are loaded in proportion to the bending forces.

EXAMPLE 11.3

Let's calculate these forces, assuming that the shafts are identical and their dimensions are $d = 80$ mm, $C = 160$ mm, $L = 400$ mm, and $e = 0.3$ mm. To be turned by 180°, the shafts must be deformed in total by $2e = 0.6$ mm, that is, each shaft by 0.3 mm (because the shafts are identical;

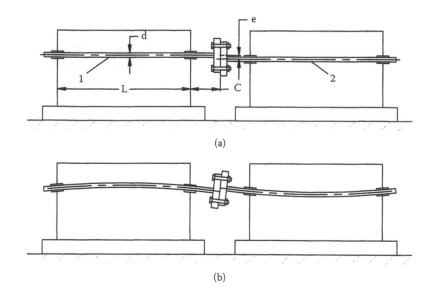

(a)

(b)

FIGURE 11.5 Effect of radial misalignment on shaft and bearing loads.

if not, the total deformation of 0.6 mm would be distributed between the shafts in proportion to their compliance). The shaft end deflection y under an overhang force F equals

$$y = \frac{F \cdot C^2}{3EI}(L+C)\ mm$$

where

    $E = 2.06 \cdot 10^5$ MPa (modulus of elasticity)
    I = moment of inertia of the shaft cross section

$$I = \frac{\pi d^4}{64} = \frac{\pi \cdot 80^4}{64} = 2.01 \cdot 10^6\ mm^4$$

From here,

$$y = F \frac{160^2}{3 \cdot 2.06 \cdot 10^5 \cdot 2.01 \cdot 10^6}(400+160) = 1.15 \cdot 10^{-5}\ F = 0.3\ mm$$

$$F = 2.6 \cdot 10^4\ N$$

The additional load F_{add} applied to the nearest bearing

$$F_{add} = F\frac{L+C}{L} = 2.6 \cdot 10^4 \frac{400+160}{400} = 3.64 \cdot 10^4\ N$$

The additional bending moment M_{add} to which the shaft is exposed due to force F is given by

$$M_{add} = F \cdot C = 3.64 \cdot 10^4 \cdot 160 = 5.82 \cdot 10^6\ N \cdot mm$$

Consequently, the additional cyclic bending stress σ_{add} is as follows:

$$\sigma_{add} = \frac{M\,d}{2\,I} = \frac{5.82 \cdot 10^6 \cdot 80}{2 \cdot 2.01 \cdot 10^6} = 116\,MPa$$

This calculation is made on the assumption that the shaft supports are gapless and absolutely rigid. If these factors were taken into account, force F would be somewhat less, but the consequences of the misalignment remain harmful. In addition to the increased bending stresses in the shaft, its elastic deformation and dislocation with respect to the housing may lead to the misalignment of the parts of the mechanism and to its premature failure.

Figure 11.6 shows *angular misalignment* characterized by angle γ. The shafts, each with two supports, are bolted together at the flange connection. As a result of these constraints, both shafts have become bent. If the two outer supports were removed and the shafts rotated by 180°, they would take up the position shown with dashed lines in Figure 11.6b.

There can also be *combined misalignment*, both radial and angular. The effect of this misalignment is similar to that described previously. These deformations and loads change cyclically, and they must be held relatively small to prevent increased noise, vibrations, and overload of the machine elements. Imparting the optimum alignment has the objective of making the shafts coaxial, thereby diminishing the additional forces and deformations that can be avoided just by fair adjustment.

In the technical literature, axial misalignment is also mentioned. Here it is not considered because the so-called *axial misalignment* is just the variation of axial distance between the connected shafts. If this distance changes within limits, admissible by the design of bearings and couplings, this has no influence on the working conditions of the mechanism, independent of the kind of coupling, whether it is rigid or resilient. The designer can increase the admissible limits of axial displacement of the shafts relative to each other, so that the procedure of installation in the axial direction doesn't cause any trouble. In some cases, the temperature expansion of the shafts should be kept in mind.

The achievable accuracy of adjustment depends not only on the instruments used, but also on the design of the shafts to be aligned and their supports. Let's consider coaxiality of two gears

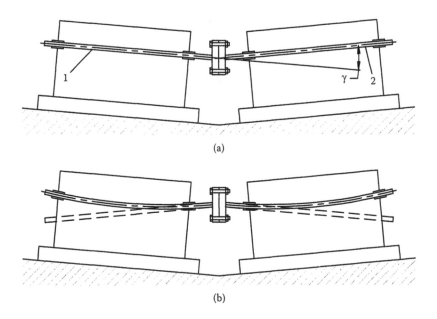

(a)

(b)

FIGURE 11.6 Effect of angular misalignment on shaft and bearing loads.

shown in Figure 5.23 or, the same, in Figure 6.43. The bearings of these gears are coaxial and the gears, when unloaded, are coaxial as well (within close manufacturing tolerances). But under a load, the gears move in the bearings within the clearances in different (possibly, in opposite) directions. If we knew exactly the position of the gears while the machine is operating, we would just displace the bearings by a calculated amount and have achieved the exact alignment. As the operating regime is varying, the position of the shafts, as well as their alignment changes in response to the variable conditions.

Another problem of alignment is the possible instability of the connection between the housings. For example, if the engine is installed on dampers (this is usual in seaborne machinery), it can move by millimeters from its nominal position depending on the torque transmitted and on the inertia forces caused by the motion of the ship in the sea. The ship's hull may also have considerable deformations, both momentary (caused by heavy seas, cargo placement, etc.) and permanent (resulting from release of the internal stresses in the welded hull). (Submarines have especially challenging alignment requirements due to hull deflections that occur due to diving and surfacing and from the effects of high-speed maneuvering.) This is the problem; alignment can only be directly performed and checked during assembly, while the shafts are coaxial under ideal running conditions. The quality of the alignment can be checked only indirectly by measuring noise and vibrations. To properly predict alignment in operation, the weight of parts, working forces, as well as the temperature and other kinds of influence on deformations must be considered when planning the alignment procedure. For example, Figure 11.7a shows a shaft of a large electric generator bent by the weight W of the rotor. In operation, this force remains and no other force is added, so the alignment of this shaft can be made just as is, i.e., the housing should be tilted as shown in Figure 11.7b.

A pinion-shaft with an overhanging flywheel is shown in Figure 11.7c unloaded (W is the weight of the flywheel and the pinion shaft together) and in Figure 11.7d under load (F_W is the working force). In this case, the angles α and β and the radial displacements in two perpendicular planes (e.g., horizontal and vertical) should be calculated. Then the alignment at rest can be optimized, taking into account the gravity load and the operating loads.

Figure 11.7e and Figure 11.7f show a less complicated case. Here, the shaft is rigid enough, but the working force F_W is directed nearly opposite to the weight. In that case, a certain calculated misalignment of the shafts must be set during the alignment procedure to take into account the vertical and lateral displacements of the shaft in the bearings during operation.

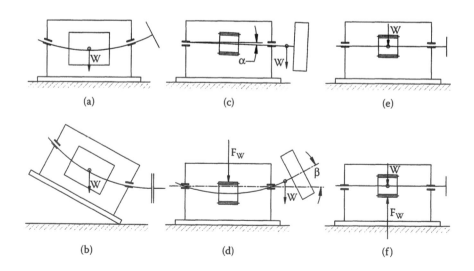

FIGURE 11.7 The influence of weight and working forces on shaft alignment.

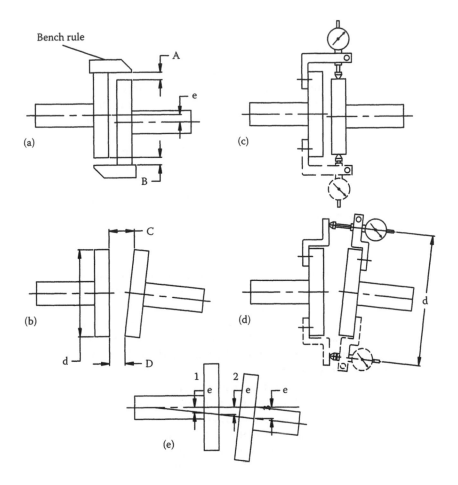

FIGURE 11.8 Alignment techniques.

In smaller machines, these difficulties don't exist as a rule, because the clearances in the rolling bearings and the elastic deformations of the parts are relatively small, and the widely used couplings have the ability to compensate for some moderate misalignment.

Figure 11.8 shows how the misalignment is usually measured. If the neighboring parts of the shafts (flanges or half-couplings) are precisely machined, the misalignment can be measured as shown in Figure 11.8a and Figure 11.8b. The radial misalignment e is given by

$$e = A = B \ mm$$

The angular misalignment γ is given by

$$\gamma = \frac{C - D}{d} \ rad$$

Dimensions A, B, C, and D can be measured by a feeler gauge or caliper (depending on the needed accuracy). Figure 11.8c and Figure 11.8d show the application of a pointer-type indicator to the alignment measurements. If the surfaces used for alignment are exact, relative to the related shafts (that means, their radial and lateral runouts are negligible as compared to the admissible misalignment), the alignment operation can be performed by the rotation of one shaft only. But if these surfaces are not exact, both shafts should be rotated together, so that the test point of the indicator touches the mating shaft nearly at the same point throughout the alignment operation.

As shown in Figure 11.8e, the angular misalignment is accompanied by radial misalignment, changing along the axis. It is often important to measure the radial misalignment in a certain place depending on the kind of coupling. For example, in the coupling shown in Figure 11.11, it is important to maintain concentricity between rubber sleeves 4 and pockets in half-coupling 2. Therefore, in this case, the radial displacement should be measured in the middle plane of the rubber sleeves. As the radial alignment changes with the angular alignment, the latter alignment should be adjusted first.

While measuring the angular misalignment, the shafts must not have any axial motion relative to each other. That should be ensured and checked.

11.2.2 RIGID COUPLINGS

The most widespread rigid coupling is just a plain flange connection. The flange can be made as an integral part of the shaft or manufactured separately and connected to the shaft by one of the numerous methods that have been partly described in Chapter 5.

Another kind of rigid coupling is a sleeve connected to both shafts by hydraulically assembled press fit (Figure 5.2a) or by key (Figure 11.9). In the last case, the sleeve should be slide fitted to the shafts (otherwise it can hardly be assembled or dismantled). Therefore, it can be applied to light-duty loads only (see Subsection 5.1.2).

The rigid couplings, in effect, convert the pair of shafts into a single four-bearing shaft, and this demands proper placement of all four bearings. In many cases, "proper placement" means that the bearings should be coaxial. But it is not always so.

One popular type of rigid connection is the male/female attachment, where one unit with a hollow shaft is mounted on the solid shaft of another unit. Figure 11.10 shows the connection of a gear motor 1 to a driven machine 2. The hollow output shaft of the gear motor is assembled with

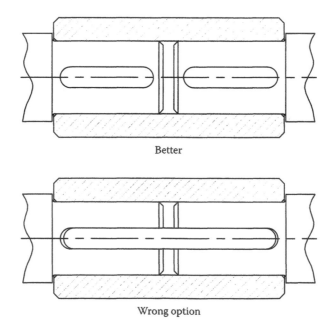

Better

Wrong option

FIGURE 11.9 Sleeve coupling.

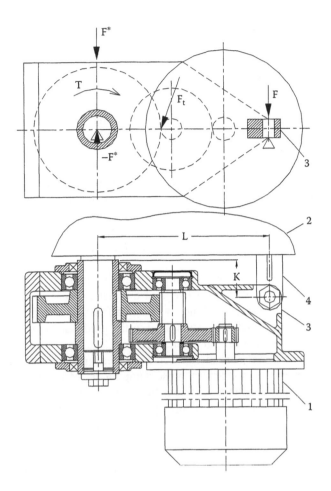

FIGURE 11.10 Shaft-mounted gear motor.

the shaft of the driven machine with a key and transition fit. The housing of the gear motor is fixed against rotation by the bolt connection of lug 3 to bracket 4. Force F applied to the lug of the gear housing equals

$$F = \frac{T}{L}$$

where T = torque on the output shaft of the gear.

Let's use our right to transfer force F from lug 3 to the point over the shaft (to position F^*) with the addition of torque T. Force F is applied to the lug that is a part of the housing. Hence, force F^* must be applied to the housing as well. The reaction of the driven machine shaft ($-F^*$) is applied to the hollow shaft and directed upwards, whereas the tooth force F_t applied to the output gear and shaft is directed nearly downwards. Thus, when determining the load on the output shaft bearings, force F^* must be subtracted from force F_t. Direction and magnitude of force F^* depends on the placement and inclination angle of lug 3, and the designer can try some options.

The bending moment M applied to the driven shaft in its most stressed section is given by

$$M = F \cdot K = \frac{T \cdot K}{L}$$

Dimension K is shown in Figure 11.10. The lesser this dimension, the smaller the bending moment.

11.2.3 RESILIENT COUPLINGS

In many cases, it is impracticable to achieve ideal alignment of the shafts because of many influences that are not exactly known, for the most part, and changing with time, like loads, temperature deformations, wear of the bearings, and others. But as is evident from the foregoing, any kind of misalignment of rigidly connected shafts leads to their cyclic deformation and possible overstress. The purpose of resilient couplings is to shift the deformation from the shafts, which are made of steel and therefore very rigid, to some intermediate resilient element that can be deformed by relatively small forces and decreases the additional load of the shafts. But this element, usually made of elastomers (rarely of steel springs), has to transmit torque from one shaft to another. Because of this, the resilient element is placed between (or attached to) two metal parts, that are in turn connected to the shafts. The unit formed by two metal parts and elastomeric intermediate parts is called a *resilient coupling*.

In addition to compensation of shaft misalignment, couplings of this kind are able to damp impact loads, vibrations, and oscillations. Change of the material and configuration of the elastomeric elements enables wide variation of the coupling stiffness; this can be useful to change the natural frequency of the entire shafting system. Due to the nonlinearity of their stiffness, resilient couplings can prevent the development of resonance.

These couplings are relatively cheap and don't need any maintenance. In some cases, their electrical insulating properties and structural noise damping are of value.

There are many resilient couplings, different in design. Figure 11.11 shows a coupling with pins and compressed rubber sleeves. Half-couplings 1 and 2 are connected to the shafts by tight fits and keys. Pins 3 with rubber sleeves 4 are attached to half-coupling 1 by their tapered ends and tightened with a nut. The rubber sleeves enter pockets in half-coupling 2, and in this way, the torque is transmitted from one shaft to another. It is clear that if the load is evenly distributed between the pins, the vector sum of the pin forces equals zero, and no shaft bending is caused by the coupling. The compliance of the pins and, mainly, of the rubber sleeves makes it possible to keep a more or less even load distribution between the pins at a moderate misalignment, and the radial forces, which bend the shaft, can be kept relatively small.

Figure 11.12 shows another design of flexible coupling. Half-couplings 1 and 2 are identical (section A-A shows them separated). The protrusions of one of them enter the recesses of the other, and rubber spider 3 is installed between them, so that the torque can be transmitted from one half-coupling to another through the spider only. Depending on the direction of torque, half of the spider's blades are loaded in compression, and the other half are idling.

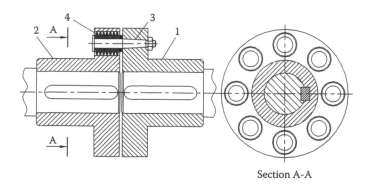

Section A-A

FIGURE 11.11 Flexible coupling with pins and rubber sleeves.

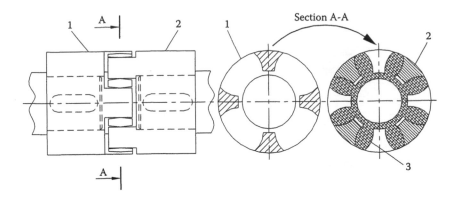

FIGURE 11.12 Flexible coupling with an elastomeric spider.

A resilient coupling with rubber pads is presented in Figure 11.13. This design is usually used for large torques, because half-couplings 1 and 2 can be made very strong and large enough in diameter, so that the number of the pads can be increased as needed. The same rubber pads can be used for different diameters due to conformable shaping of the blades. As in the previous case, only half of the pads transmit the load, whereas the other half are idling. If the torque is nonreversing, the idling pads can be made thinner, and therefore more pads can be used.

The load capacity of the couplings shown in Figure 11.11 through Figure 11.13 is limited by the permissible pressure on the elastomeric elements. Depending on the hardness of the polymer, it varies from 2–3 MPa. So the increase in quantity of the elements enables a decrease in their dimensions or a decrease in the diameter of the coupling.

The considered couplings can compensate for radial misalignment of about 0.1 mm/100 mm of their outer diameter and an angular misalignment of approximately 1°. It doesn't mean that these values of misalignment are permissible for the alignment operations. Figure 11.14 demonstrates the

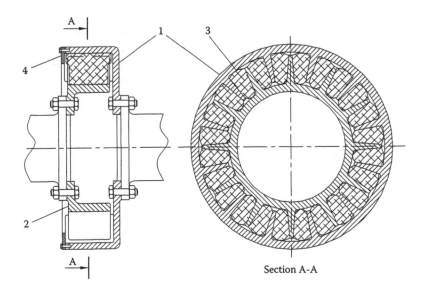

FIGURE 11.13 Flexible coupling with rubber pads.

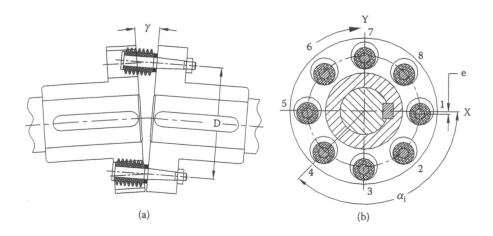

FIGURE 11.14 Effect of shaft misalignment on coupling serviceability.

effect of misalignment on coupling serviceability. Angular misalignment γ (a) leads to uneven load distribution along the elastomeric element and to its reciprocating motion relative to the mating half-coupling. For example, as the coupling rotates by 180°, element placed in position 3 moves to position 7 and shifts out of the left half-coupling. The following rotation by 180° brings this element again to position 3. If $\gamma = 1°$ and D = 200 mm, the movement from the position above the axis to the position shown below it equals 3.5 mm. This motion occurs under load with every turn of the shaft. If the shaft speed is, for example, 1500 r/min, the rubber element makes 25 end-to-end motions each second, and in the vicinity of the coupling, rubber wear debris can often be found. The increased rubbing also leads to additional heating of the elastomeric elements.

Heating of rubber elements occurs not only on account of friction energy, but also because the variable load (and variable deformation of the element) is accompanied by absorption of energy. The quantity of energy absorbed by rubber is rather large, about ten times more than that of steel. But at a temperature above 100°C, the coefficient of absorbed energy rises sharply.[3] This, in turn, increases the temperature of the part, and so on. In this way, it may lead to exponential temperature rise and quick destruction of the rubber. Therefore, the rubber elements should be well cooled (by ventilation or, when needed, by fluid cooling), and any unnecessary heating is to be avoided.

The axial motion in the region of contact of rubber elements with the pockets creates a bending moment that is applied (in opposite directions) to both half-couplings in the plane perpendicular to the plane of misalignment.

EXAMPLE 11.4

Let's calculate the bending moment M for the case shown in Figure 11.14a (angular misalignment) on the assumption that the radial displacement in the middle of the rubber elements is quite small, and the load distribution between the elements is nearly even. The data are as follows:

- Torque transmitted $T = 2100$ N·m
- Diameter over the centers of pins $D = 200$ mm
- Number of pins $N = 8$

In the position shown, pins 3 and 7 don't move in the axial direction, because they are at their "dead" locations in the vertical plane. Pins 1, 2, and 8 move to the left, and pins 4, 5, and 6 move to the right. Friction forces originated in the sliding contact create bending moment about the Y-line.

The average load of one pin equals

$$F_a = \frac{2T}{DN} = \frac{2 \cdot 2100}{0.2 \cdot 8} = 2625 \ N$$

The bending moment equals the sum of friction force moments about the Y-line:

$$M = \Sigma F_a \cdot f \ \frac{D}{2} \cos\alpha_i = 2625 \cdot 0.5 \cdot \frac{0.2}{2} (2\cos 0° + 4\cos 45° + 2\cos 90°) = 634 \ N \cdot m$$

In position turned by 22.5°, the bending moment M is a bit greater:

$$M = 2625 \cdot 0.5 \cdot \frac{0.2}{2} (4\cos 22.5° + 4\sin 67.5°) = 686 \ N \cdot m$$

In this calculation, the friction coefficient $f = 0.5$. It is worth greasing the rubber elements and pockets to decrease the f value. It is also important to decrease the angular misalignment. If it is small, the axial movement of the resilient elements can be kept within the ability of the rubber to deform elastically. The sliding will be prevented and the bending moment will be decreased.

Figure 11.14b shows the effect of radial misalignment e. If the rubber elements are driving and rotating in the arrow direction, then element 1 is the most loaded, and element 5 transmits the lowest load (if at all). This uneven load distribution between the elements brings about a lot of imperfections in the behavior of the machine, as shown in the following list:

The service life of the rubber elements decreases, because all of them (in turn) pass through the overloaded zone during one revolution of the shaft.

Because the vector sum of the forces in the coupling doesn't equal zero, the shafts become loaded with a radial force. This force results in an additional bending moment applied to both shafts and additional loads, both radial and axial, applied to their bearings. The vector of the coupling force has the same direction as the radial offset of the shafts, thus the bending moment is of constant direction, and the rotating shafts are exposed to cyclic bending.

The axial force is created when the radial misalignment is combined with angular misalignment in a perpendicular plane. This effect is similar to that discussed in Section 6.2.3 regarding the misaligned cylindrical roller bearings.

The unavoidable results of coupling misalignment are increased vibration and noise.

All the imperfections mentioned in the preceding list can occur not only due to misalignment, but also because of differences in dimensions and hardness of the elastomeric elements within one coupling. Errors in half-coupling geometry are possible as well. In these cases, the additional radial force rotates with the shaft and increases the vibrations by far. In one way or another, couplings always create some additional forces that can be larger or smaller depending on the quality and stiffness of the coupling and, mainly, on the alignment of the shafts.

EXAMPLE 11.5

Calculate the additional radial force F for the case presented in Figure 11.14b. The data are as follows:

Torque transmitted $T = 2100$ N·m
Diameter over the pin centers $D = 200$ mm

Number of pins $N = 8$
Stiffness of the rubber element $R = 2000$ N/mm
Radial misalignment $e = 0.3$ mm

The average load of one pin (assuming even load distribution between the pins) equals

$$F_a = \frac{2T}{DN} = \frac{2 \cdot 2100}{0.2 \cdot 8} = 2625\ N$$

The average deformation of a rubber element is

$$\Delta_a = \frac{F_a}{R} = \frac{2625}{2000} = 1.31\ mm$$

When there is a radial misalignment of the shafts, the deformation of any element Δ_i is given by

$$\Delta_i = \Delta_a + e \cos \alpha_i$$

where α_i = angular position of the pin relative to the most loaded one (see Figure 11.14b). From here, we can calculate the load of each element:

$F_1 = \Delta_1 R = (1.31 + 0.3 \cos 0°) 2000 = 3220\ N;$ $F_2 = (1.31 + 0.3 \cos 45°) 2000 = 3044\ N;$

$F_3 = (1.31 + 0.3 \cos 90°) 2000 = 2620\ N;$ $F_4 = (1.31 + 0.3 \cos 135°) 2000 = 2196\ N;$

$F_5 = (1.31 + 0.3 \cos 180°) 2000 = 2020\ N;$ $F_6 = (1.31 + 0.3 \cos 225°) 2000 = 2196\ N;$

$F_7 = (1.31 + 0.3 \cos 270°) 2000 = 2620\ N;$ $F_8 = (1.31 + 0.3 \cos 315°) 2000 = 3044\ N$

The vector sum of these forces F is what we want to calculate:

$$F = \Sigma F_i \cos \alpha_i = 3220 \cos 0° + 3044 \cos 45° + 2620 \cos 90° + 2196 \cos 135°$$

$$+ 2020 \cos 180° + 2196 \cos 225° + 2620 \cos 270° + 3044 \cos 315°$$

$$= 2400\ N$$

Notes:

1. When there is a radial misalignment, the deformations of the elements (and consequently, forces F_1 to F_8) are directed not exactly tangentially. But the deviation is small and can be neglected.
2. The forces are calculated for constant stiffness R. If the stiffness changes widely depending on the deformation of the element, forces F_1 to F_8 should be calculated with the changing parameter of stiffness R, and then it should be checked whether the sum of these forces multiplied by $D/2$ conforms to the given torque. If not, the deformation of all the elements must be increased or decreased by the same amount, and the forces calculated again.

As observed, force F created by this coupling is much less than what was obtained in Example 11.3 ($F = 26000$ N) for the same misalignment of shafts with a rigid connection.

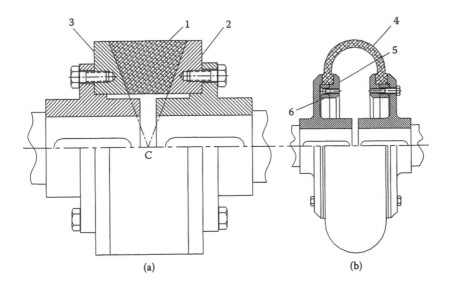

FIGURE 11.15 Flexible couplings with elastomeric elements loaded in shear.

Figure 11.15 shows couplings in which the elastomeric element is loaded in shear. In version a, rubber part 1 is glued or vulcanized to steel parts 2 and 3. This coupling is so designed that the shear stress is constant throughout the volume of the rubber part. It is achieved by placing coincident vertices of cones on the shaft axis (point C in Figure 11.15a).

If the last statement is not obvious, it can be easily proved. As is known, the dependence between shear stress τ and strain is as follows (see Figure 11.16a):

$$\tau = \gamma G = \frac{t}{l} G$$

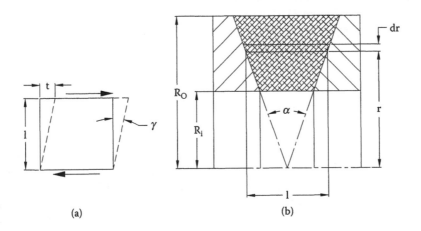

FIGURE 11.16 Shear stress distribution in a twisted rubber element.

For the rubber element, designed as shown, the following correlations are valid:

$$t = r\varphi; \quad l = 2r\tan\frac{\alpha}{2};$$

$$\tau = \frac{r\varphi}{2r\tan\dfrac{\alpha}{2}}G = \frac{\varphi}{2\tan\dfrac{\alpha}{2}}G = const$$

Here, φ is the angle of twist, which is obviously the same for all sections and layers of the rubber element. So are the angle α and the modulus of elasticity in shear G.

The infinitesimal torque created by the shear stress τ is given by

$$dT = \tau \cdot 2\pi r \cdot dr \cdot r = \tau \cdot 2\pi r^2\, dr$$

From here, the torque equals

$$T = \int_{R_i}^{R_0} dT = 2\pi\tau \int_{R_i}^{R_0} r^2 dr = \frac{2\pi}{3}\tau\left(R_0^3 - R_i^3\right)$$

From this equation, the shear stress can be obtained:

$$\tau = \frac{3T}{2\pi\left(R_0^3 - R_i^3\right)}$$

The admissible shear stress is about 0.2–0.3 MPa.[4]

Figure 11.15b shows a coupling with toroidal tirelike elastomeric element 4 connected to the steel half-couplings by means of backing rings 5. The latter rings are made of two halves each (for assembly reasons) and united through the medium of rings 6. This coupling is the most flexible among those shown here, that is, the admissible amounts of misalignment are bigger than that of any other resilient coupling. But its dimensions are relatively large. Besides, at higher speeds, this coupling creates axial forces directed inward. This behavior should be taken into consideration.

Couplings shown in Figure 11.15 have a common drawback: if the rubber (or the attachment of the rubber to half-coupling) fails, the machines suddenly disconnect. In some cases, this can be unacceptable.

11.2.4 Gear Couplings

Resilient couplings have larger dimensions because the load capacity of elastomeric elements is rather limited. In addition, they are also flexible in torsion (their torsion angles can be as large as 3° or 4°). Therefore in many cases, mechanical couplings are used, which are made mainly of surface-hardened steels, have relatively small dimensions and weight, and provide the needed compensation for shaft misalignment due to movable connections inside them. One of the most widespread couplings of this kind is the gear coupling (Figure 11.17). It consists of two toothed hubs 1 and 2 and a connecting part (sleeve) with internal toothing. The sleeve can be made of two halves (3 and 4) connected by bolts or as an integral part 5 with end covers 6. These couplings are lubricated with oil that is filled in or emptied through plug 7 or with grease. Seals 8 and O-rings in the flange connections prevent oil leakage from the coupling.

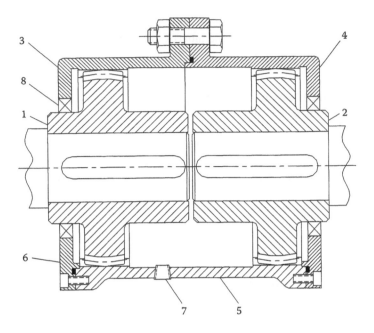

FIGURE 11.17 Gear coupling.

Power loss in gear couplings is about 0.25–0.65% at straight teeth and 0.10–0.25% at crowned teeth.[5] Therefore, at low and medium speed, the convective heat transfer suffices for cooling. But when the coupling transmits high power, oil flow through the teeth is usually needed to take up the heat. The oil is poured inside the coupling, and the parts are so designed that centrifugal forces push the oil through the mesh. To enable effective oil flow, the clearances in the mesh should be big enough.

Sleeve 1 (Figure 11.18a) is centered in the toothed connections relative to the hubs and can't move in the radial direction. Therefore, between the hub and the sleeve there can only be angular misalignment. The shafts' misalignment is compensated in the gear coupling by the pivot of the sleeve around the hub toothing, as if these connections are spherical or universal joints. But they are not, and so the ability of gear couplings to compensate for misalignment is rather limited.

Angle γ can be either the same for both connections (when the shafts are parallel, Figure 11.18a) or different (γ_1 and γ_2 in Figure 11.18b). Thus, these angles should be calculated separately for

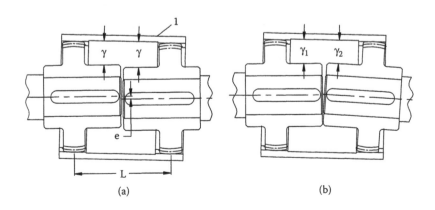

FIGURE 11.18 Gear coupling misalignment.

each of the two connections. In this case, it is preferable to measure the radial and angular misalignments of the toothed hubs and to summarize their effects.

EXAMPLE 11.6

An example of misaligned hubs is shown in Figure 11.19. The radial misalignment of the hubs measured on their outer diameters in the vertical plane is given by

$$e = A = B \ mm$$

The angular misalignment between hub 1 and sleeve 3 in the vertical plane is

$$\gamma_{1V} = \frac{e_V}{L} \ rad$$

In the same way, values e_H and γ_{1H} in the horizontal plane can be obtained, and then the total misalignment angle between hub 1 and sleeve 3 equals

$$\gamma_1 = \sqrt{\gamma_{1V}^2 + \gamma_{1H}^2}$$

Let's consider the components of misalignment between hub 2 and sleeve 3 in the vertical plane. First of all, angle γ_{1V} remains the same for that case, but there is an additional angle γ_{2V} that is given by

$$\gamma_{2V} = \frac{C - D}{M} \ rad$$

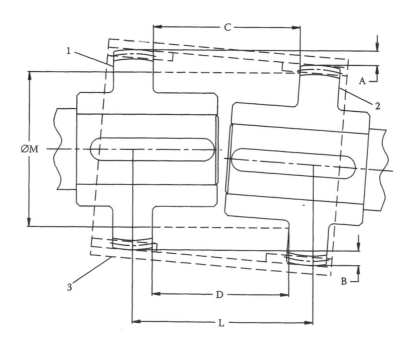

FIGURE 11.19 Misaligned hubs.

where M is the diameter at which the C and D dimensions are measured.

The combined misalignment angle in the vertical plane is given by

$$\gamma_{2VC} = \gamma_{1V} \pm \gamma_{2V}$$

In the case represented in Figure 11.19, the "minus" sign must be chosen, because both γ_{1V} and γ_{2V} are turned in the same direction. (It is obvious that if $\gamma_{1V} = \gamma_{2V}$, $\gamma_{2VC} = 0$. This is the case with hub 2.) In Figure 11.18a, where the axes of the hubs are parallel, the misalignments between the hubs and the sleeve are equal. If one of the hubs had been turned around the center of its toothing by angle γ clockwise, its misalignment relative to sleeve 1 would have been decreased to zero with no effect on the misalignment of another hub.

In the same way, the combined misalignment γ_{2HC} in the horizontal plane can be obtained, and then the total misalignment between hub 2 and the sleeve equals

$$\gamma_2 = \sqrt{\gamma_{2VC}^2 + \gamma_{2HC}^2}$$

Figure 11.20a shows the position of the hub teeth relative to the internal teeth of the sleeve when they are turned with respect to each other by angle γ. As was said, hub 1 and sleeve 2 are centered relative to each other, either by the working surfaces of the involute teeth or by the outer diameter of the toothing. The last case is typical for high-speed applications, where exact centering

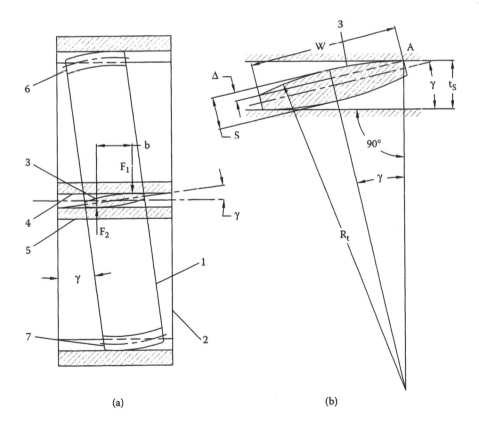

(a)   (b)

FIGURE 11.20 Angular misalignment and the shape of crowned teeth.

is needed to limit imbalance. The radial clearance in this case is very small (0.02–0.04 mm), and the top surface of the hub teeth should have the shape of a sphere to enable the unimpeded turn of the hub. But this is not enough. Let's take note of tooth 3 of the hub that lies in the space between teeth 4 and 5 of the sleeve. This is the very place where the ability of the connection members to turn relative to each other is geometrically limited. If the teeth of both the hub and the sleeve are straight and the backlash is zero, no free turning is possible in this connection. The bigger the backlash, the bigger (in direct proportion) the possible turning angle, but if straight, the teeth would contact their edges. To improve the contact of teeth, they are usually made crowned as shown in Figure 11.20. The technology of cutting such teeth is very simple in principle, but quite complicated in realization. When cutting straight teeth, the gear-cutting tool (hob) moves straight in the feed direction, (Figure 11.21a). To cut the crowned teeth, the hob moves through an arc as shown in Figure 11.21b. For that purpose, the hobbing machine must be equipped with an additional mechanism that provides the needed feed motion of the hob. Once this feature is added, this machine can be used for cutting hubs with crowned teeth with different barreling.

To calculate the needed barreling, let's return to Figure 11.20b where the section of the crowned tooth 3 is drawn in a bigger scale, so that we can make some geometric calculations. The curved profile of tooth 3 is characterized by radius R_t, and it is so built that it is tangent to the straight profile of the sleeve's tooth 4 at the point of contact A. From the geometrical construction, we can obtain the radius of the barreled tooth profile:

$$R_t = \frac{W}{2\sin\gamma} \tag{11.7}$$

If the tooth thickness in the middle equals S, on the ends, it is $S - 2\Delta$, where

$$\Delta = R_t\,(1 - \cos\gamma) \tag{11.8}$$

The nominal tooth space t_S of the sleeve needed to place tooth 3 is also obtained from the geometric calculations:

$$t_S = (S - 2\Delta)\cos\gamma + W\sin\gamma \tag{11.9}$$

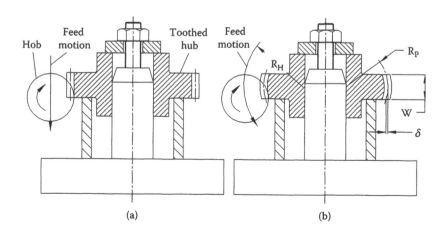

(a)   (b)

FIGURE 11.21 Hobbing of straight and crowned teeth.

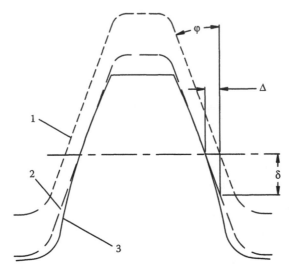

FIGURE 11.22 Relationship between hob radial motion and tooth thickness.

Now we are to define the contour of hob feed motion. Figure 11.22 shows the relation between the radial motion of the hob and the thickness of the cut tooth. When the hob is moved toward the work by δ (from position 1 to 2), the cut tooth becomes thinner by 2Δ, where

$$\Delta = \delta \cdot \tan \varphi \qquad (11.10)$$

Here, φ = pressure angle (usually $\varphi = 20°$).

From the geometry shown in Figure 11.21b, we obtain (after simple transformations) the path radius of the hob pitch line:

$$R_P = \frac{W^2}{8\delta} + \frac{\delta}{2} \approx \frac{W^2}{8\delta} \qquad (11.11)$$

To calculate radius R_H, the pitch radius of the hob should be added to R_P.

Example 11.7

Given data: gear coupling m = 5 mm, addendum height = 0.8m, pressure angle $\varphi = 20°$, teeth number N = 40, face width of the hub toothing W = 40 mm, and misalignment angle $\gamma = 3°$. The sleeve is centered by the outer diameter of the teeth.

From Equation 11.7 and Equation 11.8,

$$R_t = \frac{40}{2 \sin 3°} = 382 \; mm$$

$$\Delta = 382 \left(1 - \cos 3°\right) = 0.524 \; mm$$

This means that the tooth on the ends is thinner by 1.048 mm than in the middle. The nominal tooth thickness on the pitch circle equals (in standard case) $0.5\pi m$. Let's take the same thickness for the first approximation:

$$S = 0.5 \cdot \pi \cdot 5 = 7.854 \; mm$$

The space width of the sleeve obtained from Equation 11.9 is given by

$$t_S = (7.854 - 2 \cdot 0.524)\cos 3° + 40 \cdot \sin 3° = 8.890 \ mm$$

Now is the time to evaluate this result. The nominal width of the sleeve tooth space on the pitch circle equals $0.5\pi m = 7.854$ mm. To place the turned tooth, we have to increase the width to 8.890 mm. This means that the internal tooth of the sleeve becomes thinner by about 1 mm, that is, 6.818 mm instead of 7.854 mm. Is that OK? If we don't like this result, we can make the hub tooth thinner (by decreasing the S value) and thereby make the sleeve tooth thicker. But possibly it is OK, and the calculation of the teeth for the bending strength can only give the right answer.

To continue: The values of S and t_S have been calculated for zero backlash. But the backlash is needed to enable the oil to flow through the teeth also in position, where the teeth have the maximal inclination relative to the tooth space of the mating part. In addition, the possible temperature difference between the hub and the sleeve and the geometrical errors of the teeth must be taken into account to prevent jamming of the coupling. We will decrease the hub tooth thickness by around $0.1m = 0.5$ mm, i.e., $S = 7.35$ mm. Now all the dimensions of the hub and sleeve teeth can be calculated.

The δ value (see Equation 11.10) is

$$\delta = \frac{\Delta}{\tan \varphi} = \frac{0.524}{\tan 20°} = 1.44 \ mm$$

Then the path radius of the hob pitch line (from Equation 11.11) is given by

$$R_P = \frac{40^2}{8 \cdot 1.44} = 139 \ mm$$

In the case shown in Figure 11.20, the end point of contact is placed on the end of the tooth. To avoid that, the tooth length can be just enlarged from W to Wt (see Figure 11.23) with no change to the values calculated in the preceding text.

When the hub and sleeve are misaligned, tooth 3 (Figure 11.20) and its diametrically placed fellow sufferer are the most loaded, whereas teeth 6 and 7 are the least loaded. Figure 11.24 shows the change in clearances between the hub and the sleeve teeth over 90° (between tooth 3 and 6 in Figure 11.20). Here we can see that the teeth in the upper part of the picture are just idle.

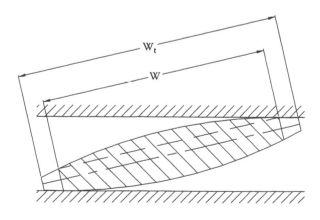

FIGURE 11.23 Enlarged tooth length.

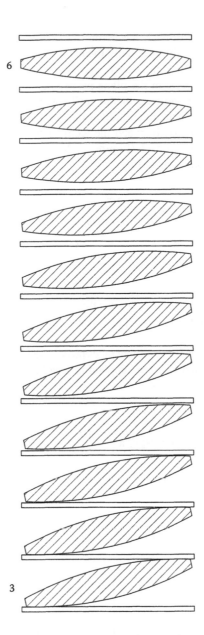

FIGURE 11.24 Change of clearance between driving and driven teeth.

This picture is made for an exaggerated misalignment angle $\gamma = 13°$. At usual misalignments of $0.5-1.5°$, the change in clearances is not so drastic, and owing to elastic deformations under load most teeth participate in the load transmission, though the load distribution is uneven. One of the authors[6] recommends the calculation of the maximal tooth force by multiplying its average value by a factor of 1.25 at $\gamma = 0.5°$, 1.5 at $\gamma = 1°$, and 1.75 at $\gamma = 1.5°$.

Actually, the load distribution depends not only on the misalignment angle, but also on the tooth shape and unit load: the greater the load, the bigger the elastic deformations and more even the load distribution. In addition, the load distribution is interesting not by itself, but as a factor that affects the load-carrying capacity and service life of the gear coupling. Quantitatively, this influence can be evaluated only experimentally and approximately. Therefore, the best way to have the needed data for calculations is to consult with the coupling's manufacturer.

As is seen from Figure 11.20, each tooth (during one revolution) moves from position 6 to position 7 and then returns to position 6. This means that each tooth performs reciprocal sliding motion relative to the mating teeth of the sleeve, and the friction forces produce a bending moment applied to both of the meshing parts in the plane perpendicular to the plane of misalignment. The mechanism of this effect is the same as in the resilient coupling with pins and rubber sleeves (discussed in the preceding text). The magnitude of the bending moment depends on the coefficient of friction f and the load distribution between the teeth. If the misalignment angle is so great that the entire load is transmitted by tooth 3 (Figure 11.20) and its diametrically opposite twin, the bending moment is maximal and is given by

$$M_{max} = \frac{2T \cdot f}{d \cdot \cos\varphi} \cdot \frac{d}{2} = \frac{T \cdot f}{\cos\varphi} \qquad (11.12)$$

where
  T = torque transmitted
  d = pitch diameter of the teeth
  φ = pressure angle on the pitch diameter (usually $\varphi = 20°$)
  f = coefficient of friction (for common applications, $f = 0.10$–0.20[5])

The minimal magnitude of bending moment takes place when the load is distributed evenly between the teeth. The infinitesimal friction force dF_f applied to small element $r \cdot d\alpha$ (see Figure 11.25) is as follows:

$$dF_f = \frac{T \cdot f}{r\cos\varphi \cdot 2\pi r} r \cdot d\alpha = \frac{T \cdot f}{2\pi r\cos\varphi} d\alpha$$

The infinitesimal bending moment

$$dM = dF_f \cdot h = \frac{T \cdot f}{2\pi r\cos\varphi} d\alpha \cdot r\sin\alpha = \frac{T \cdot f}{2\pi\cos\varphi}\sin\alpha \cdot d\alpha$$

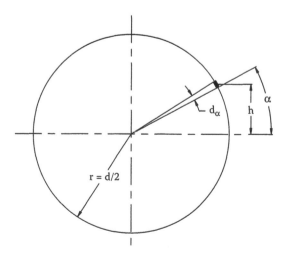

FIGURE 11.25 Computation of bending moment in a gear coupling.

From here, the minimal bending moment is given by

$$M_{min} = 2 \int_0^\pi dM = \frac{T \cdot f}{\pi \cos \varphi} \int_0^\pi \sin \alpha \cdot d\alpha = \frac{2T \cdot f}{\pi \cos \varphi}$$

Thus, the maximal and minimal bending moments differ by a factor of $\pi/2$.

There is one peculiarity in the gear coupling as compared to the coupling with pins and rubber sleeves. The sleeve of the gear coupling is loaded by two bending moments from two hubs. The directions of these moments depend on the direction of misalignment and are not known beforehand, therefore in strength calculations they should be taken as having the same direction. The doubled bending moment applied to the sleeve is balanced by radial forces applied to the hubs. In the worst case, the radial force F_r equals

$$F_r = \frac{2M_{max}}{L} \tag{11.13}$$

where L = distance between the toothed rims (see Figure 11.19).

EXAMPLE 11.8

A diesel engine is connected to the drive shaft by a gear coupling. The engine power is 2000 kW, rotational speed n = 1000 r/min, torque fluctuation evaluated by calculation of torsional vibrations equals ±15%, tooth profile angle $\varphi = 20°$, the friction coefficient $f = 0.15$, and the distance between the hubs meshing $L = 300$ mm.

The maximal torque transmitted is

$$T = 9555 \frac{2000}{1000} 1.15 = 22000 \, N \cdot m$$

The maximal bending moment applied to the shaft (from Equation 11.12) is

$$M_{max} = \frac{22000 \cdot 0.15}{\cos 20°} = 3510 \, N \cdot m$$

The possible maximal radial force (from Equation 11.13) equals

$$F_r = \frac{2 \cdot 3510}{0.3} = 23400 \, N$$

An additional bending moment may be caused by the displacement of the point of contact along the teeth. Look at tooth 3 in Figure 11.20a. We will assume that it contacts the sleeve tooth with its upper profile, and the force applied to it is F_1. Then its diametrically opposite twin, which looks exactly as tooth 3, must work with its lower profile, and the force applied to it (F_2) is of the same magnitude, but oppositely directed. The bending moment applied by these two forces to both the hub and the sleeve is

$$M_b = F_1 b$$

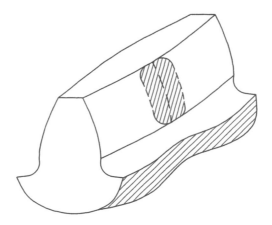

FIGURE 11.26 Contact patch on a crowned tooth.

Provided that the misalignment angle is so great that the entire torque is transmitted by two teeth under consideration, the following relations are valid:

$$F_1 = F_2 = \frac{T}{d}; \quad b = R_t \sin 2\gamma \approx 2\,\gamma\,R_t;$$

$$M_{b,max} = \frac{2T}{d}\,\gamma\,R_t$$

Here γ is the misalignment angle in rad.

This bending moment acts in the plane of misalignment and may be of remarkable magnitude at a large misalignment. When the misalignment is small, the load is distributed more evenly over the circumference, and the b value is small so that this moment can be neglected.

Another peculiarity of gear couplings with crowned teeth is that they don't like working at zero misalignment. If the misalignment is zero or very small, the contact patch (hatched, see Figure 11.26) remains immovable or moves only slightly, so that most of the contact area remains constantly compressed with no access for the lubricant. Combined with high pressure and micromovements, this creates suitable conditions for fretting (see also Chapter 2). Moderate misalignment, which brings about sliding motion in the teeth contact, enables renewal of the oil film on the contact surfaces and also facilitates creation of a hydrodynamic oil wedge. Experiments[7] have shown that the gear coupling with a certain radius of the crowned tooth profile has an optimal misalignment angle, at which the friction coefficient f is minimal. At greater misalignment, the f value rises because of growing unevenness of load distribution and sliding velocity. At smaller misalignment, the f value rises because of worsening of the lubrication conditions in the contact regions.

REFERENCES

1. Roark, R.J. and Young, W.C., *Formulas for Stress and Strain*, McGraw-Hill Kogakusha.
2. Timoshenko, S., *Strength of Materials. Part II*, D. Van Nostrand Company, Princeton, NJ.
3. Poturaev, V.N. and Dirda, V.I., *Rubber Elements of Machines*, Mashinostroenie, Moscow, 1977 (in Russian).

4. Poljakov, V.S., Barbash, I.D., and Rjakhovsky, O.A., *Couplings Handbook*, Leningrad, Mashinostroenie, 1974 (in Russian).
5. Popov, A.P., *Gear Couplings in Seaborne Machinery*, Sudostroenie, Leningrad, 1985 (in Russian).
6. Reshetov, D.N., *Machine Elements*, Mashinostroenie, Moscow, 1989 (in Russian).
7. *Handbuch der Schadenverhütung*, Herausgabe von der Allianz Versicherungs AG, Berlin, 1976 (in German).

Part III

Life Prediction of Machine Parts

Calculations for strength are incorporated into almost all phases of the development process of a mechanism. During preliminary specifications, when the designs diagram is to be chosen, its feasibility must be proven by strength calculations. Here, approximate calculations are mostly used, such as estimation of shaft diameter using Equation 4.1 and Equation 4.2 or gear dimensions from Equation 7.11. Thereupon, in the course of detailed design, the strength calculations are made in detail. When the mechanism is on trials or in service, it may fail for various reasons, and the investigation of the failures mostly involves additional calculations for strength and durability. Later on, if the customer wants some upgrade of the mechanism (for example, to a greater capacity), the strength calculations are indispensable.

Who is to make these calculations? At a design office there is usually one or a group of engineers specializing in strength calculations. To do his work, the stress engineer must receive from the designer the detailed drawings of the mechanism, with all dimensions, tolerances, materials, and so on. To prepare these drawings, the designer himself has to make all the strength calculations needed to choose the dimensions, materials, etc. Thus, the final calculation made by the stress engineers is verifying only, but if it reveals that the designer was mistaken and the project must be done anew ... oh, no!

In practice, the designer can consult with the stress engineer about the accepted method of calculation. He also can order calculation of some parts that are critically stressed or that need more sophisticated mathematical tools or computer programs in which the designer is not an expert. But the rest, which amounts to about 90% or 100% of the strength calculations, the designer has to do himself. So he is not able to do his job if he is not friendly with the strength calculations.

The methods of such calculations are usually taken from the technical literature, or they are recommended by professional societies. (As an example of the latter can be mentioned Lloyd's standards in ship building, ASME and SAE approved calculations in automobile construction, aircraft construction, and others.) Unfortunately, the correct substitution of numbers for letters in the equations recommended doesn't ensure the reliable operation of the mechanism.

Once, at a plant manufacturing marine engines, a meeting was held concerning the fracture of a crank shaft. One of the main participants in this meeting was the guru in this field, the author of the method of strength calculations for crank shafts. One young engineer asked him:

"How could the crankshaft break, if the safety factor equals 5?!"
"If the shaft broke, that means, there was no safety margin," said the professor calmly, and
everybody appreciated his answer.

This distant case obviously shows that sometimes it is very difficult to develop a strength
calculation method that takes into account quite precisely both the complicated shape of the machine
part and the intricate loading conditions, associated sometimes with unknown operation and maintenance factors. In such cases the researcher develops an approximate method and uses it to calculate
the stresses in a number of machines of the same type that are known to be safely working. The
quotient obtained when the admissible stress of material the part is made from is divided by the
stress calculated using the approximate method is defined as the safety factor. Hence, the safety
factor is essentially an index of the lack of knowledge attached to a certain method of calculation.
It is clear that in the design of a similar mechanism, such a calculation gives only the first
approximation. Later on, the reliability of the mechanism, and possibly the real stresses, have to
be checked during trials.

Nowadays the calculation methods have improved drastically, but the real reliability and service
life of the machine parts remains an estimate. Now, as well as in the distant past, the failure at a
safety factor of 5 is sometimes possible. The authors are going to discuss in detail the basics of
calculation for strength and durability. They also want to inspire in the reader's mind an understanding of some conventions and, to some extent, the uncertainty of these calculations to urge him
to be cautious about the results obtained.

"The world rests on three things," the ancient sages said. Each of them had his own version of
what these were. The authors, being in agreement with their subject's requirements, can state
unequivocally that the calculation for strength is based on three data units:

- The strength parameters of the material
- The stress magnitude and pattern of change with time
- The safety factor

Each of these data units deserves detailed discussion.

12 Strength of Metal Parts

Strength is the ability of a material to withstand loading forces without damage. In Section 3.1, mechanisms of destruction of the crystal lattice in a metal under load have been discussed. At present, the knowledge in this field is still insufficient to determine the admissible stresses for different loading conditions. This purpose is reached more safely by processing experimental data.

Engineers noticed long ago that the repeated application of a moderate load may in time bring about the failure of a part, whereas the static application of the same load doesn't cause any visible change to the material. The first systematic research of what had been called "fatigue of metal" was undertaken by August Wöhler during 1857–1870. He tested until failure railroad axles loaded with a bending moment in a rotating-beam testing machine, the plotted results yielding the first σ–N (or S–N) curves. (Here σ = bending stress, N = number of cycles to failure). Such a plot is shown in Figure 12.1a.

These plots are usually drawn in log-log coordinates, (see Figure 12.1b). In these coordinates, the σ–N curve can be plotted as a broken line. Its sloping part 1 is expressed by the equation

$$N \cdot \sigma^m = Constant \tag{12.1}$$

Here $m = 1/\tan\alpha$ = the slope angle. The greater the m is, the less the slope of the line and the stronger the dependence of the number of cycles to failure on the stress magnitude. For smooth parts, the m value may differ from 4 to 10, and for parts with stress raisers (the vast majority are such) from 4 to 6.[1] Another source[2] recommends taking the m value as approximately 13.6 for smooth shafts and 9.3 for notched shafts. As we see, the stress raisers (also known as stress *risers* or stress *concentrations*) decrease the m value. It is also known that stronger materials have greater m values. To consider these two factors, the following formula has been offered:[7]

$$m = \frac{C}{K}$$

where
  C = some constant coefficient ($C = 12$–15 for carbon steels, and $C = 20$–35 for alloy steels)
  K = factor of fatigue strength decrease that integrates the harming influences of the stress raiser, surface properties, and size factor

If, for example, the stress concentration factor $K_t = 2$, surface factor $K_S = 0.9$, and size factor $K_d = 0.9$, then $K = 2/(0.9 \cdot 0.9) = 2.47$, and for an alloy steel $m = 8$–14. We can see that the m value is very undependable.

The horizontal part 2 of the σ–N curve corresponds to the maximal stress of the repeated load cycle S_{lim} that doesn't lead to a failure at an infinite number of cycles. This value is often referred to as the "fatigue strength" or "fatigue limit."

At a relatively short life span (up to 10^4 cycles) portion 1 of the σ–N curve gives an inflated strength estimate, and so the σ–N curve in this area is represented more truly by portion 3. Here the metal undergoes significant plastic deformations, and the admissible stress amplitude must be calculated using the strain–life method (see Section 12.1.3). This case is rare in machinery.

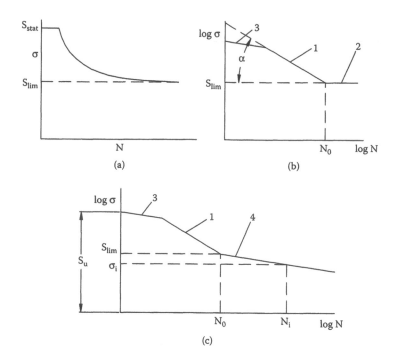

FIGURE 12.1 The σ–N curves.

The number of cycles N_0 to the breakpoint of the σ–N curve lies in the range 10^6–$3 \cdot 10^6$. Therefore endurance tests usually last $5 \cdot 10^6$–10^7 cycles: if the specimen doesn't fail within this period, it will not fail at this stress magnitude at all.

Occasionally, the failure may occur after 10^7 cycles as well, because fatigue failure is essentially probabilistic. But it doesn't influence appreciably the accepted S_{lim} magnitude because the occasional exception occurs too infrequently to be statistically significant.

The σ–N plot shown in Figure 12.1b is typical for steels of low and middle strength. High-strength alloy steels and light alloys don't have the horizontal part of the σ–N curve; that is, the admissible stress magnitude decreases continuously with the increase in number of load cycles (Figure 12.1c). Nevertheless, the breakpoint of the σ–N curve at N_0 cycles remains. Exponent m is approximately 10 times greater at portion 4 than at portion 1.[1]

Let's see the effect of the stress on the service life before and after the breakpoint. Let's assume that at portion 1, $m_1 = 10$, $N_0 = 10^6$ cycles and the stress $\sigma_1 = 1.1 \, S_{lim}$. From Equation 12.1,

$$N_0 \cdot S_{lim}^{10} = N_1 \cdot \sigma_1^{10}$$

$$N_1 = N_0 \left(\frac{S_{lim}}{\sigma_1} \right)^{10} = 10^6 \cdot \left(\frac{1}{1.1} \right)^{10} = 0.39 \cdot 10^6 \; cycles$$

Let's see now what happens at portion 4 of the σ–N curve, where $m_3 = 100$, provided that $\sigma_3 = 0.9 \, S_{lim}$:

$$N_0 \cdot S_{lim}^{100} = N_3 \cdot \sigma_3^{100}$$

$$N_3 = N_0 \left(\frac{S_{lim}}{\sigma_3} \right)^{100} = 10^6 \left(\frac{1}{0.9} \right)^{100} = 3.76 \cdot 10^{10} \; cycles$$

So we see that the increase of stress by 10% above S_{lim} has led to decrease of durability by a factor of 2.6, whereas the decrease of stress by 10% below S_{lim} has increased the durability by a factor of 3760. Thus, the breakpoint parameters (S_{lim} and N_0) are very important also in a case in which the material has no infinite life stress.

By the way, is $3.76 \cdot 10^{10}$ cycles a lot or not? For example, the crankshaft of a car engine makes in the mean about 2000 r/min. During 30 years of work 6 h daily the crankshaft makes $7.9 \cdot 10^9$ turns. Because of the four-stroke engine, the combustion stroke takes place once in two turns of the crankshaft, so the number of cycles with high bending stress equals $3.95 \cdot 10^9$ only.

Almost all machines perform cyclic work, and their parts are exposed to a load that changes in conformance with a certain program. Therefore, calculation for fatigue strength has become the dominant method. Nevertheless, against the background of repeated load cycles, an accidental overload with a very small number of cycles may occur. The admissible stress in this case is higher than at a high-cycle load, and it is most often set equal to the stress that is bearable under static load.

12.1 STRENGTH OF METALS

12.1.1 STRENGTH AT A STATIC LOAD

For static strength calculations, we use the mechanical properties of materials that are specified in standards, the manufacturers catalogs, or in reference books. In important cases, the supplier provides the results of the test for static strength along with the material.

The most used strength characteristics are the ultimate strength S_u (also called *tensile strength*), yield point S_y, and hardness. Among the required data are the parameters of plasticity (tensile strain δ and necking ψ, both taken at the fracture point of the specimen) and impact toughness (determined by the Charpy V-Notch test). These parameters are always paid attention to: The steel for engineering purposes must not be brittle. It is desirable to have $\delta \geq 6\%$, $\psi \geq 10\%$, and the impact toughness not less than 50 J/cm². At the same time, these three parameters don't appear in the strength calculation. That indicates that we lack knowledge in this field.

The mechanical properties of metals change with temperature: as the temperature increases, the strength of a metal reduces, and the plasticity increases. The strength parameters obtained at room temperature are valid up to approximately 300°C for steels and 100–150°C for light metals. For higher temperatures, the possible decrease in mechanical properties and creep resistance should be considered.

At low temperatures, the strength of a metal increases (the change is not big and usually is not taken into consideration), but the plasticity decreases considerably, and at a certain low temperature called the *transition temperature*, it falls sharply, making the metal brittle similar to glass. (The curious observer is encouraged to research maritime shipping disasters.) Normally, structural steels can be used safely up to −20°C. When the temperature lowers to −40°C, the impact toughness of the material, as well as the starting conditions of the mechanism, should be checked. For example, if a combustion engine must be warmed up before starting, the real working temperatures can be much higher than that of the environment.

At −60°C, we usually use alloy steels that retain satisfactory impact toughness at this temperature. Among the specified strength parameters for steels intended for use at low temperatures must be the impact toughness at these temperatures.

The mechanical properties of metals are obtained by tensile test of specimens that have standard dimensions to get comparable results from different tests. Typical design of a specimen is shown in Figure 12.2a ($d = 5$–10 mm). The specimen is loaded in tension until failure on a special testing machine that is able to plot the deformation of the specimen vs. the tension force. The failed specimen is shown in Figure 12.2b, and two types of force-deformation plots are given in Figure 12.3.

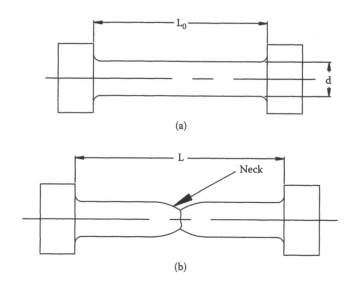

FIGURE 12.2 Typical tensile test specimen.

The tensile strength S_u and yield point S_y are obtained by dividing the respective forces by the nominal section areas of the specimen:

$$S_{y,1} = \frac{P_1}{A_0}; \quad S_{u,1} = \frac{P_{max,1}}{A_0}$$

$$S_{y,2} = \frac{P_2}{A_0}; \quad S_{u,2} = \frac{P_{max,2}}{A_0}$$

Here

P_1 (or P_2) = maximal force at which the plastic deformation of the material reaches some conventional value (mostly 0.2%)

$P_{max,1}$ (or $P_{max,2}$) = maximal tension force registered during the test

A_0 = area of the nominal cross section of the specimen

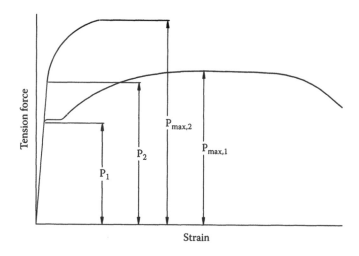

FIGURE 12.3 Stress–strain diagram.

In the course of the test, the specimen elongates (elastically and plastically), and its diameter decreases because the volume of metal remains nearly constant. But the decrease in the cross-sectional area is relatively small (at least until necking that takes place after the force P_{max} has been reached). Therefore, the S_y and S_u values are usually calculated as if the cross section remains unchanged.

The tensile strain δ is obtained from the formula

$$\delta = \frac{L - L_0}{L_0} 100 \%$$

where dimensions L_0 and L are the length of the specimen before and after the test (Figure 12.2). Because the elongation of the specimen is largely accounted for by the necking, the δ value depends on the length of the specimen. For steels, at $L_0 = 5d$ it is greater than at $L_0 = 10d$ by 22%.[3]

Materials have complicated structure, and the random orientation of crystals in grains and of grains relative to their neighbors cause certain scatter of mechanical properties, even if the specimens are cut from one blank. The strength can fluctuate depending on the following factors:

- Real chemical composition of the heat and its cleanliness
- Forging reduction ratio
- Location in the forging or casting the specimen is cut from
- Direction of the specimen axis: along the grain flow or across
- Peculiarities of the heat treatment procedure, location of the part in the furnace, and so on

For most machine elements, the stress must not exceed the yield point because plastic deformation leads to changes in their geometry. For this reason, the stress admissible at static loading is obtained by division of the yield stress by some safety factor that considers the possible uncertainties in the stress estimation (see chapter 13). But sometimes, when some small change in the geometry of the part is admissible, plastic deformation is acceptable and can be useful. Besides the cases described in Chapter 3, Section 3.3, screw connections can be mentioned. Though the plastic deformation in the root of the thread coil occurs at a relatively small load, it decreases neither the static strength, nor the fatigue strength of the bolt. This deformation increases the strength of material (strain hardening) and induces residual stresses directed opposite to the working stresses.

Tensile tests are very fast and relatively cheap. But they require the tensile-test machine and specimens made of the same material with the same heat treatment as the part to be calculated for strength. Much cheaper is the hardness test that in many cases can be performed directly on the part with no damage to its quality. Therefore, it is useful to apply the dependence between the Brinell or Rockwell hardness and the strength parameters of steels shown in Figure 12.4. These diagrams must be considered as approximate, though.

We have spoken so far about one-dimensional loading (tension–compression). The real state of stress can be more complicated. The material can be loaded simultaneously with normal and shear stresses or with normal stresses only but in two mutually perpendicular directions (biaxial stress). The yield stress in such cases can be determined from equations by Tresca or Hüber–Mises–Hencky. The latter theory is based on strain energy and conforms closely to the experimentally observed effects associated with the beginning and development of yield in metals. For a multiaxial stress, when there are all the possible components of stress, the yield criterion looks as follows:

$$\sigma_{eq} = \frac{1}{\sqrt{2}} \left[(\sigma_x - \sigma_y)^2 + (\sigma_x - \sigma_z)^2 + (\sigma_y - \sigma_z)^2 + 6\left(\tau_{xy}^2 + \tau_{xz}^2 + \tau_{yz}^2 \right) \right]^{1/2} = S_y \qquad (12.2)$$

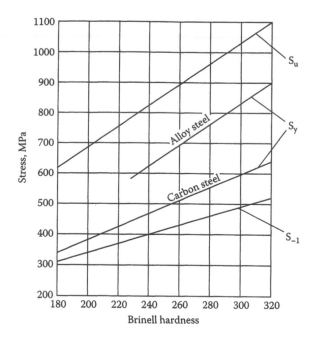

FIGURE 12.4 Approximate dependence on hardness of the steel mechanical properties.

where

σ_{eq} = equivalent stress

σ_x, σ_y, σ_z = main normal stresses

τ_{xy}, τ_{xz}, τ_{yz} = main shear stresses

If only one normal (σ) component and one shear (τ) stress component are applied to the material (for example, bending and torsion of a round bar), Equation 12.2 for equivalent stress takes the form

$$\sigma_{eq} = (\sigma^2 + 3\tau^2)^{1/2} \tag{12.3}$$

and the condition of the beginning of yield,

$$(\sigma^2 + 3\tau^2)^{1/2} = S_y \tag{12.4}$$

From this condition follow the relations for special cases:

- For pure tension ($\tau = 0$), the yield begins at

$$\sigma = S_y$$

- For pure torsion ($\sigma = 0$), the yield begins at

$$\sqrt{3}\,\tau = S_y; \quad \tau = S_{y,\tau} = 0.577\,S_y \tag{12.5}$$

where $S_{y,\tau}$ = yield point in shear.

Equation 12.5, which connects the yield points in tension and in shear, conforms to the experimental data.

From Equation 12.5 we have the following:

$$\frac{S_y^2}{S_{y,\tau}^2} = 3$$

Substituting this into Equation 12.4 instead of the factor 3, we have

$$\left(\sigma^2 + \frac{S_y^2}{S_{y,\tau}^2}\tau^2 \right) = S_y^2 \tag{12.6}$$

$$\frac{\sigma^2}{S_y^2} + \frac{\tau^2}{S_{y,\tau}^2} = 1$$

This is the condition for the onset of yield. If we want to prevent yielding, we have to insert some safety factor n. Let's introduce the idea of a partial safety factor (separately for the normal and shear stresses):

$$n_\sigma = \frac{S_y}{\sigma}; \quad n_\tau = \frac{S_{y,\tau}}{\tau}$$

In this case, Equation 12.6 becomes:

$$\frac{1}{n_\sigma^2} + \frac{1}{n_\tau^2} = 1$$

To have the overall safety factor n, the following condition should be fulfilled:

$$\frac{1}{n_\sigma^2} + \frac{1}{n_\tau^2} = \frac{1}{n^2}$$

or, similarly,

$$n = \frac{n_\sigma n_\tau}{\sqrt{n_\sigma^2 + n_\tau^2}} \tag{12.7}$$

12.1.2 Fatigue Strength (Stress Method)

The modes of failure at a repeated load and the number of cycles to failure depend on the stress magnitude. At a stress that is close to the tensile strength, the specimen deforms plastically and the number of cycles is very small. At lower stresses, when the cycles to failure number 10^2–10^4, there can be found both plastic deformation and an array of fatigue cracks. This mode of failure is called *low-cycle fatigue* (*LCF*), and in this case the life of the part should be calculated by the strain method (see Subsection 12.1.3). Decreasing the stress further decreases the plastic deformation. When the number of cycles to failure exceeds 10^5, the specimens fail in fatigue, with no visible plastic deformation, and the life prediction is usually made by the stress method (discussed in this section).

The LCF condition in machinery occurs rarely. For example, a shaft rotating with a speed of 1000 r/min makes 10^5 turns in less than 2 h. So in the vast majority of practical cases, the number of cycles exceeds N_0, and we have to check the parts for *high-cycle fatigue* (HCF) or for infinite service life.

The process of failure in HCF is much more complicated than that in static loading. It is characterized by the following features:

- Initiation and development of a crack happens without any visible deformation of the part. As a rule, there are no changes in the operation of the machine, until the crack extends over the majority of the cross section and brings about catastrophic failure.
- Onset of the initial crack, its propagation, and failure of the part occur at stresses much lower than the yield stress.
- The fatigue crack starts mostly from surface blemishes, such as surface asperities (roughness), nicks (caused by negligent storage and handling), corrosion, marking and stamping of all kinds (impact, chemical, or electrographic), marks of hardness tests (Rockwell, Brinell, or Vickers), fretting (see Chapter 2), etc.
- Another kind of common crack initiation place is the necessary fillets, keyways, grooves, splines, and other features that concentrate and focus the stresses just as the aforementioned accidental stress raisers do.

As was said, fatigue cracks mostly start from the surface of a part; therefore, the surface condition has a great effect on the results of the fatigue test. The surface of the specimen can be turned on a lathe, ground, polished (by hand, mechanically, or electrically). In all these cases, the condition of the surface layer and the residual stresses in it can be very different, depending not only on the kind, but also on the parameters of the machining process, and they can hardly be controlled. Because of this, the scatter of experimental points in fatigue tests is much greater than for tensile tests, and the determination of fatigue characteristics of a metal (such as fatigue limit) needs quite a lot of experiments to obtain reliable data.

The load fluctuation in time can be of different kinds. Figure 12.5 shows several typical cases:

1. Shear stress in the propeller shaft of a ship
2. Bending stress in the root of a gear tooth
3. Bending stress in the root of an idler pinion tooth
4. Bending stress in a rotating shaft bent by a moment of constant direction

Because the loading options are countless, it is common practice to test the specimens in fatigue under sinusoidal load. The admissible stresses obtained in these tests are then used for loading cycles different from the sinusoidal. So the sinusoidal diagrams shown in Figure 12.6a, Figure 12.6b, and Figure 12.6d can be substituted for the real loading diagrams represented in Figure 12.5a to Figure 12.5d.

It is convenient to represent the loading diagram as a sum of two components: mean stress σ_m (such as a static load) and harmonic component with the amplitude σ_a (see Figure 12.6). It is convenient because tests have shown that the damaging effect of mean stress is much less than that of the harmonic component (of course, only if the maximal stress σ_{max} doesn't reach the yield point). Therefore, it is conventional to determine the admissible stress amplitude of a certain loading cycle as admissible amplitude of a pure harmonic component (also called *symmetric cycle;* see Figure 12.6d) with the addition of some part of the mean stress (see the following text).

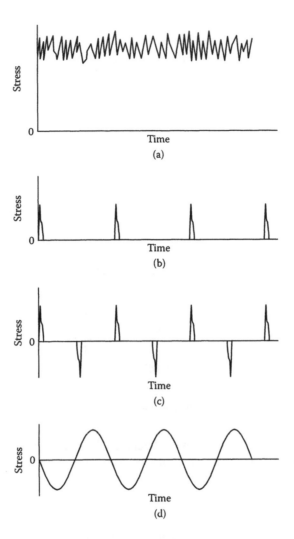

FIGURE 12.5 Examples of stress variation with time.

To label the loading diagrams with respect to proportion between σ_a and σ_m, parameter R was coined:

$$R = \frac{\sigma_{min}}{\sigma_{max}} = \frac{\sigma_m - \sigma_a}{\sigma_m + \sigma_a} = \frac{1 - \dfrac{\sigma_a}{\sigma_m}}{1 + \dfrac{\sigma_a}{\sigma_m}}$$

If $\sigma_a = 0$, $R = 1$ (static load). If $\sigma_a = \sigma_m$, $R = 0$ (pulsating cycle; see Figure 12.6b). If $\sigma_m = 0$, $R = -1$ (symmetric cycle; see Figure 12.6d). The limiting stresses S_{lim} for the cycles characterized by $R = 0$ and $R = -1$ are designated as S_0 and S_{-1}, respectively. The vast majority of fatigue tests are made with the symmetric loading cycle ($\sigma_m = 0$) realized in rotating bending. (See Figure 12.7; the bending deformation here is highly exaggerated.) So the S_{-1} value called the *fatigue limit* (or *endurance limit*) became the main parameter that rates the fatigue strength of material. In real

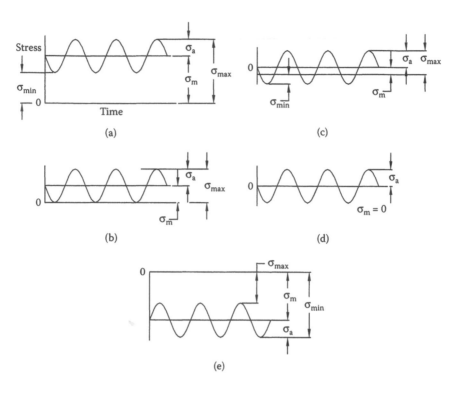

FIGURE 12.6 Mean stress, amplitude, and maximal and minimal stress of a cycle.

machines may happen load conditions with different R values, and the limiting stress amplitude of the harmonic component S_a is determined from the equation

$$S_a = S_{-1} - f(\sigma_m)$$

where $f(\sigma_m)$ is some function that accounts for the effect of mean stress. The function is still a matter of discussion at the time of printing this book.

The effect of the mean stress σ_m on the limiting value of the alternating stress has been studied experimentally. If the set of specimens is tested with a constant σ_m and stepwise-decreased σ_a, we finally receive (in coordinates $\sigma_a - \sigma_m$) a point that conforms to infinite life at this very σ_m value. A set of such tests with different magnitudes of σ_m gives several points, and we can draw a line

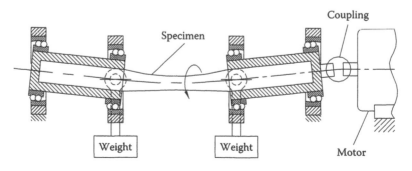

FIGURE 12.7 Rotating–bending endurance test machine.

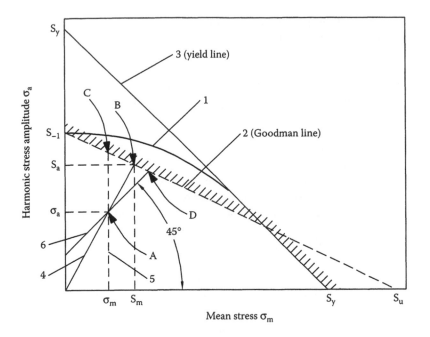

FIGURE 12.8 Goodman's diagram of the fatigue limit vs. the mean stress.

through these points. This is a boundary line (1 in Figure 12.8), and all the combinations of σ_a and σ_m above this line conform to the sloping portion of the $\sigma - N$ curve (1 in Figure 12.1), whereas all the points on line 1 conform to $S_{\lim}$ at different σ_m.

Although line 1 passes through most of the experimental points, it is not as certain as a line on paper. As was said, the results of fatigue tests have noticeable scattering, and the averaged line can be passed through the experimental points in several ways. Therefore, there are several simplifying suggestions about the boundary line equation, and the simplest is the Goodman diagram. It is a straight line (2 in Figure 12.8) that connects two well-known points: S_{-1} on the vertical axis (when $\sigma_m = 0$, i.e., symmetric cycle) and S_u on the horizontal axis (when $\sigma_a = 0$, i.e., static load).

Let's designate S_a and S_m, the vertical and horizontal coordinates of any point B, on line 2. In this case, the equation of line 2 is

$$S_a = S_{-1}\left(1 - \frac{S_m}{S_u}\right) \tag{12.8}$$

or, similarly,

$$\frac{S_a}{S_{-1}} + \frac{S_m}{S_u} = 1 \tag{12.9}$$

Because the maximal stress must not exceed the yield point S_y, additional constraint line 3 is plotted; its equation is

$$\sigma_m + \sigma_a = S_y$$

Thus, the sought-for limiting combinations of σ_a and σ_m are placed on the hatched sections of lines 2 and 3. But in the subsequent equations, line 3 doesn't play a part; it is just taken to mean that σ_{max} must not exceed S_y.

The admissible combination of σ_a and σ_m (point A in Figure 12.8) is less than the limiting combination of S_a and S_m because there should be a certain factor of safety. Let's calculate the safety factor for three different cases, reasoning from the Goodman diagram:

1. Where σ_a and σ_m are changing in the same proportion: For example, in a gear, when the torque transmitted grows, the tooth forces and bending moments applied to the shaft grow in the same proportion.
2. Where σ_m remains constant, but σ_a changes: The propeller shaft of a ship can be taken as an example. The mean shear stress at a given rotational speed is constant. The alternating shear stress is caused by torsional vibration, nonuniform fluid flow field, and other factors.
3. Where $\sigma_{min} = \sigma_m - \sigma_a = $ constant: This case belongs to bolted connections, where the minimal load equals the preload force and remains constant.

In the first case, the σ_a/σ_m ratio remains constant; therefore, the limiting point B is placed on line 4 (because all points of this line have the same ratio σ_a/σ_m). The equation of line 2:

$$y = -\frac{S_{-1}}{S_u}x + S_{-1} \qquad (12.10)$$

The equation of line 4:

$$y = \frac{\sigma_a}{\sigma_m}x$$

Point B of intersection of lines 2 and 4 has the vertical coordinate

$$S_a = \frac{S_{-1}}{1 + \frac{S_{-1}}{S_u} \cdot \frac{\sigma_m}{\sigma_a}}$$

Hence, the safety factor for the first case is

$$n_1 = \frac{S_a}{\sigma_a} = \frac{1}{\frac{\sigma_a}{S_{-1}} + \frac{\sigma_m}{S_u}} \qquad (12.11)$$

In the second case, the limiting point C lies on line 5 because all points on this line have the same σ_m value. The equation of line 5 is

$$x = \sigma_m$$

Substituting this in Equation 12.10, we obtain the vertical coordinate of point C:

$$y = S_a = S_{-1}\left(1 - \frac{\sigma_m}{S_u}\right)$$

From here, the safety factor for the second case is

$$n_2 = \frac{S_a}{\sigma_a} = \frac{S_{-1}}{\sigma_a}\left(1 - \frac{\sigma_m}{S_u}\right)$$

In the third case, line 6 drawn at 45° meets the condition $\sigma_m - \sigma_a = $ constant. The equation of this line is:

$$y = x + \sigma_a - \sigma_m$$

The vertical coordinate of limiting point D (the intersection of lines 2 and 6) is

$$y = S_a = S_{-1}\frac{S_u + \sigma_a - \sigma_m}{S_u + S_{-1}}$$

The safety factor for the third case is

$$n_3 = \frac{S_a}{\sigma_a} = \frac{S_{-1}}{\sigma_a}\frac{S_u + \sigma_a - \sigma_m}{S_u + S_{-1}}$$

It should be pointed out that the Goodman criterion, although prevalent, has a grave disadvantage: it overestimates the influence of mean stress on the fatigue strength of the metal. Specifically, experiments with bolts have shown[1] that the limiting stress amplitude S_a remains practically constant up to $S_m \leq 0.9S_y$. Among the other approximations of the experimental points, the following are mentioned.

Gerber's suggestion (1874) — a parabolic failure criterion:

$$\frac{S_a}{S_{-1}} + \left(\frac{S_m}{S_u}\right)^2 = 1$$

Or, in terms of S_a,

$$S_a = S_{-1}\left[1 - \left(\frac{S_m}{S_u}\right)^2\right] \tag{12.12}$$

The ASME suggestion — an elliptic failure criterion:

$$\left(\frac{S_a}{S_{-1}}\right)^2 + \left(\frac{S_m}{S_u}\right)^2 = 1$$

Or, again in terms of S_a,

$$S_a = S_{-1}\sqrt{1 - \left(\frac{S_m}{S_u}\right)^2} \tag{12.13}$$

Both these suggestions are used only rarely.

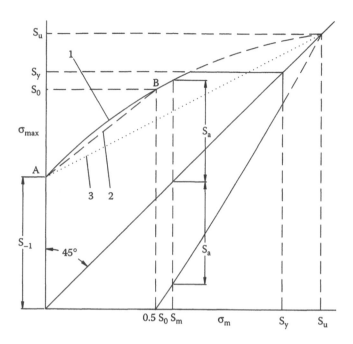

FIGURE 12.9 Serensen's diagram of maximal stress vs. mean stress.

In the engineering environment in Russia, the suggestion of S.V. Serensen[4] is in use everywhere. According to this, the limiting curve 1 (Figure 12.9) connects three control points: S_{-1}, S_0, and S_u. This diagram is plotted on the $\sigma_{max} - \sigma_m$ coordinates. Point A conforms to a pure harmonic cycle (S_{-1}), and point B, to a pulsating cycle (S_0). To simplify the equations, portion A–B of curve 1 is replaced by straight line 2, and then

$$S_{max} = S_{-1} + (1 - \psi_\sigma)S_m \tag{12.14}$$

where

$$\psi_\sigma = \frac{2S_{-1} - S_0}{S_0}$$

Substitution of $S_{max} = S_a + S_m$ into Equation 12.14 gives

$$S_a = S_{-1} - \psi_\sigma \cdot S_m \tag{12.15}$$

A similar equation is valid for the shear stresses:

$$\tau_a = \tau_{-1} - \psi_\tau \cdot \tau_m \tag{12.16}$$

For specimens without stress raisers the following values of ψ_σ and ψ_τ have been experimentally found: for carbon steels $\psi_\sigma = 0.1$–0.2, $\psi_\tau = 0$–0.1; and for alloy steels and light alloys $\psi_\sigma = 0.15$–0.30, $\psi_\tau = 0.05$–0.15.

In all cases, the greater values should be chosen for stronger materials. If $\sigma_m < 0$, it is recommended to take $\psi_\sigma = \psi_\tau = 0$. The maximal stress must not exceed S_y.

EXAMPLE 12.1

Alloy steel of HB 300, $S_u = 1050$ MPa, $S_{-1} = 500$ MPa. What is the limiting amplitude S_a, if the mean stress $S_m = 300$ MPa?

According to Goodman (Equation 12.8),

$$S_a = 500\left(1 - \frac{300}{1050}\right) = 357 \ MPa$$

According to Gerber (Equation 12.12),

$$S_a = 500\left[1 - \left(\frac{300}{1050}\right)^2\right] = 459 \ MPa$$

According to ASME (Equation 12.13),

$$S_a = 500\sqrt{1 - \left(\frac{300}{1050}\right)^2} = 479 \ MPa$$

According to Serensen (Equation 12.15),

$$S_a = 500 - (0.15 \cdots 0.30)300 = 455 \cdots 410 \ MPa$$

Who is right? We have already discussed the scattering of experimental data. Each of the mentioned authors may appear closer to the truth than the others in some specific cases, depending on the material properties and loading conditions. If you want to know more precisely the limiting stresses for a certain material under certain loading conditions, you have to undertake the suitable tests. Then you will know who is right. But if you don't have such a possibility, you are forced to take chances and gain experience. The authors got accustomed to Serensen's method in the course of several decades and take it as reliable.

Let's introduce the idea of equivalent stress $\sigma_{a,eq}$: This is the amplitude of a symmetric cycle, which has the same damaging effect on the material as the real loading cycle characterized by stresses σ_a and σ_m. It is clear that the safety factor n at the equivalent stress must be the same as at the real load cycle. Hence, the equivalent stress can be determined as follows:

$$\sigma_{a,eq} = \frac{S_{-1}}{n}$$

If we take Goodman's diagram and the first case, when the safety factor is determined by Equation 12.11, the equivalent stress

$$\sigma_{a,eq} = S_{-1}\left(\frac{\sigma_a}{S_{-1}} + \frac{\sigma_m}{S_u}\right) = \sigma_a + \frac{S_{-1}}{S_u}\sigma_m \tag{12.17}$$

The next, similar expression is valid for the shear stresses:

$$\tau_{a,eq} = \tau_a + \frac{S_{-1,\tau}}{\tau_u}\tau_m \tag{12.18}$$

In the same way can be determined the equivalent stresses for other cases, and also for other ways of plotting the limiting lines. Specifically for Serensen's approach,

$$\sigma_{a,eq} = \sigma_a + \psi_\sigma \sigma_m \tag{12.19}$$

$$\tau_{a,eq} = \tau_a + \psi_\tau \tau_m \tag{12.20}$$

Until now we have dealt with one-dimensional loads (tension–compression, or bending, or shear). Inasmuch as the fatigue phenomena are by nature associated with plasticity (see Section 3.1), it is assumed that Equation 12.2 to Equation 12.7 written for the static strength are valid for the fatigue strength as well. But the parameters σ_y and τ_y shall be replaced by S_{-1} and $S_{-1,\tau}$, respectively. In this case, the boundary strength equation (Equation 12.6) becomes

$$\frac{\sigma_a^2}{S_{-1}^2} + \frac{\tau_a^2}{S_{-1,\tau}^2} = 1 \tag{12.21}$$

This equation implies a symmetric cycle ($\sigma_m = \tau_m = 0$), in which σ_a and τ_a are the amplitudes of alternating stresses, which are synchronous and cophasal. Accordingly, Equation 12.4, when applied to fatigue, takes the form

$$\sqrt{\sigma_a^2 + 3\tau_a^2} = S_{-1} \tag{12.22}$$

Here we have

$$\frac{S_{-1}^2}{S_{-1,\tau}^2} = 3$$

This relation is verified experimentally, but if it is not exactly true, factor 3 in Equation 12.22 can be replaced by the stress ratio:

$$\sqrt{\sigma_a^2 + \left(\frac{S_{-1}}{S_{-1,\tau}}\right)^2 \tau_a^2} = S_{-1} \tag{12.23}$$

When the loading cycle is asymmetric ($\sigma_m \neq 0$ and $\tau_m \neq 0$), the strength condition can be written by analogy with Equation 12.6:

$$\left(\frac{\sigma_{a,eq}}{S_{-1}}\right)^2 + \left(\frac{\tau_{a,eq}}{S_{-1,\tau}}\right)^2 = 1 \tag{12.24}$$

Using, on the analogy of static strength, the partial safety factors

$$n_\sigma = \frac{S_{-1}}{\sigma_{a,eq}}; \quad n_\tau = \frac{S_{-1,\tau}}{\tau_{a,eq}} \tag{12.25}$$

we obtain the equation for the safety factor in the form

$$\frac{1}{n_\sigma^2} + \frac{1}{n_\tau^2} = \frac{1}{n^2}$$

or

$$n = \frac{n_\sigma n_\tau}{\sqrt{n_\sigma^2 + n_\tau^2}} \tag{12.26}$$

In the case of compound stress with all components acting, the equivalent amplitude $\sigma^*_{a,eq}$ and the equivalent mean stress $\sigma^*_{m,eq}$ should be determined first. By analogy with Equation 12.2,

$$\sigma^*_{a,eq} = \frac{1}{\sqrt{2}} \sqrt{(\sigma_{a1} - \sigma_{a2})^2 + (\sigma_{a1} - \sigma_{a3})^2 + (\sigma_{a2} - \sigma_{a3})^2} \tag{12.27}$$

$$\sigma^*_{m,eq} = \frac{1}{\sqrt{2}} \sqrt{(\sigma_{m1} - \sigma_{m2})^2 + (\sigma_{m1} - \sigma_{m3})^2 + (\sigma_{m2} - \sigma_{m3})^2} \tag{12.28}$$

The equivalent stress, which takes into account both the alternating and mean stresses, is determined by analogy with Equation 12.17 and Equation 12.19, written for uniaxial stress:

$$\sigma_{a,eq} = \sigma^*_{a,eq} + \frac{S_{-1}}{S_u} \sigma^*_{m,eq} \tag{12.29}$$

$$\sigma_{a,eq} = \sigma^*_{a,eq} + \psi_\sigma \sigma^*_{m,eq} \tag{12.30}$$

The safety factor is determined as before:

$$n = \frac{S_{-1}}{\sigma_{a,eq}}$$

The drawback of such a method is in the estimation of the influence of the compressive mean stress ($\sigma_m < 0$). From experiments, it is known that the compressive mean stress increases the admissible amplitude of the harmonic component. This effect is reflected in Equation 12.17 to Equation 12.20, respectively. In Equation 12.28 the mean stress $\sigma^*_{m,eq}$ is positive in principle, and this decreases the calculated safety factor (as compared with the real one) in the area of compressive mean stresses. The effort to consider more precisely the influence of the mean stresses[5] is made for the case of proportional loading, that is, when all kinds of stresses change simultaneously, and in the same proportion. Criteria for more complicated cases offered in the literature are based on Sines' criterion.

We guess that Equation 12.27 to Equation 12.30 can be used as they are; although they reduce the calculated safety factor, they are not too conservative. As stated earlier, in Equation 12.15 and Equation 12.16, it was recommended to take $\psi_\sigma = \psi_\tau = 0$ at $\sigma_m \leq 0$. This recommendation also decreases the calculated safety factor, though to a lesser extent. Another example provides the Goodman limiting line (dotted line 3 in Figure 12.9), which is drawn below the experimental points. Nevertheless, the Goodman approach has found wide application in strength calculations.

It should be taken into consideration that the inaccuracy of some limiting line or equation is only a small part of the many inaccuracies, uncertainties, and assumptions that accompany the calculations for strength and durability.

As we could see, the main characteristic of the material needed for fatigue calculation is the fatigue limit S_{-1}. This parameter can be determined only experimentally, but this research needs many specimens to make the results statistically reliable. In most cases, the engineer doesn't have the possibility to undertake this expensive and time-consuming work, and this basic parameter is obtained approximately from empiric formulas such as the following:[6]

For normalized and annealed steels:

$$S_{-1} = 0.454S_u + 8.4 \text{ MPa}$$

For carbon steels after quenching and tempering:

$$S_{-1} = 0.515S_u - 24 \text{ MPa}$$

For alloy steels after quenching and tempering:

$$S_{-1} = 0.383S_u + 94 \text{ MPa}$$

For high-alloy austenitic steels:

$$S_{-1} = 0.484S_u \text{ MPa}.$$

When these formulas are used, the error in the S_{-1} value doesn't usually exceed 15%.

12.1.3 LIMITED FATIGUE LIFE UNDER IRREGULAR LOADING (STRESS METHOD)

When the fluctuating stress exceeds the fatigue limit, the life span of the part is limited. If the stress amplitude is constant, the number of stress cycles to failure can be estimated from Equation 12.1, and the example of such a calculation has been provided earlier. But if the stress amplitude changes with time as shown in Figure 12.10, the less-damaging effect of smaller stresses should be considered.

Line 1 in Figure 12.10 presents the change of stress amplitude in time, and line 2 shows the change of the loading frequency (for example, this can be the rotational speed of a shaft). The load curve is sectioned so as to replace the curve by several stress levels. The greater the number of sections, the closer is the bar diagram to the load curve. Each stress level σ_i has its own loading frequency and time of application, and their product gives the number of cycles n_i at this stress magnitude. Here σ_1 is the maximal stress amplitude in the spectrum, and σ_i is any of the smaller amplitudes.

An approach to this problem was offered by A. Palmgren (1924) for ball bearings and then adapted by M.A. Miner (1945) for aircraft structures. The idea is based on the assumption that any cyclically loaded part has some resource of resistance that is not spent when the stress is less than the fatigue limit. But when the stress is higher, the resistance resource is used up in proportion to the number of cycles. For example, if $\sigma_1 = 600$ MPa and the number of cycles to failure $N_1 = 3 \cdot 10^5$ cycles, then

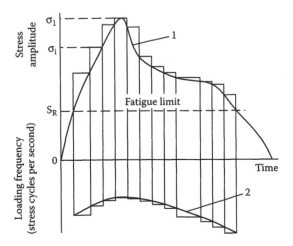

FIGURE 12.10 Spectrum of stresses and frequencies.

loading during $n_1 = 1 \cdot 10^5$ cycles takes up 33.3% of the resistance ability of the material. If afterward the same part is loaded, say, with $\sigma_2 = 550$ MPa, then $N_2 = 7.2 \cdot 10^5$ (provided that $m = 10$), and to use up the remaining 66.7% of the resistance ability, the duration of work under this load should be $n_1 = 7.2 \cdot 10^5 \cdot 0.667 = 4.8 \cdot 10^5$ cycles. In other words, if we symbolize as n_i the number of loading cycles at σ_i, and as N_i the number of cycles to failure at this stress, the criterion of exhaustion of the resistance resource should look as follows:

$$\sum \frac{n_i}{N_i} = 1 \qquad (12.31)$$

From the foregoing, it follows that the contribution of each loading step to the consumption of resistance resources depends on its "damaging ability" per cycle. For example, if at $\sigma_1 = 600$ MPa the number of cycles to failure $N_1 = 3 \cdot 10^5$ cycles, and at $\sigma_2 = 550$ MPa, $N_2 = 7.2 \cdot 10^5$ cycles, each cycle of σ_1 damages the part as much as 2.4 cycles of σ_2. Hence, the cycles of lesser load can be added to the cycles of maximal load according to their relative "damaging abilities." So the equivalent number of cycles n_e at maximal amplitude σ_1 is given by

$$n_e = n_1 + n_2\left(\frac{\sigma_2}{\sigma_1}\right)^m + n_3\left(\frac{\sigma_3}{\sigma_1}\right)^m + \cdots + n_i\left(\frac{\sigma_i}{\sigma_1}\right)^m + \cdots$$

Experiments have shown that Equation 12.31 is not exact, and

$$\sum \frac{n_i}{N_i} = C$$

where the C value may vary within the range from 0.25 to 4.[7] Momentary overload stress may reduce the C value to 0.05–0.1. For design purposes, $C = 1$ is usually taken, but it should be kept in mind that the real service life of the part may be ten times less than the calculated.

EXAMPLE 12.2

The fatigue limit of material, $S_{-1} = 500$ MPa, $N_0 = 3 \cdot 10^6$ cycles. The fatigue curve exponent $m = 10$. The part is cycled with the following parameters of load: $\sigma_a = 300$ MPa during 95% of the working time with the frequency of $f = 1000$ cycles per minute (cpm); $\sigma_a = 600$ MPa during 3%, $f = 800$ cpm; $\sigma_a = 650$ MPa during 1.5%, $f = 500$ cpm; and $\sigma_a = 750$ MPa during 0.5% of the working time, $f = 300$ cpm. The service life should be at least 100 h. Is the part strong enough?

First let's determine the number of cycles for the stresses that exceed the fatigue limit, beginning with the maximal stress:

$\sigma_{a1} = 750$ MPa; $n_1 = 100 \cdot 60 \cdot 300 \cdot 0.005 = 0.9 \cdot 10^4$ cycles
$\sigma_{a2} = 650$ MPa; $n_2 = 100 \cdot 60 \cdot 500 \cdot 0.015 = 4.5 \cdot 10^4$ cycles
$\sigma_{a3} = 600$ MPa; $n_3 = 100 \cdot 60 \cdot 800 \cdot 0.03 = 14.4 \cdot 10^4$ cycles

Now the equivalent number of cycles under the maximal load can be determined:

$$n_e = 0.9 \cdot 10^4 + 4.5 \cdot 10^4 \left(\frac{650}{750}\right)^{10} + 14.4 \cdot 10^4 \left(\frac{600}{750}\right)^{10} = 3.52 \cdot 10^4 \ cycles$$

The number of cycles to failure at maximal load

$$N = N_0 \left(\frac{S_{-1}}{\sigma_{a1}}\right)^m = 3 \cdot 10^6 \left(\frac{500}{750}\right)^{10} = 5.2 \cdot 10^4 \ cycles$$

According to this calculation, only 68% of the material strength resources are used up during 100 h of work. If the stresses given previously have been already multiplied by an appropriate safety factor (that means, the real stresses are not greater than the calculated ones), the result of this calculation can be considered as favorable.

12.1.4 FATIGUE LIFE (STRAIN METHOD)

This method was developed in the middle of the 20th century on the basis of low-cycle fatigue tests. As aforesaid, in this area the stress exceeds the yield point of the material, and the stress method, which doesn't account for this effect, may give an incorrect estimate of the load capability of the material. Later on, the strain method was extended to include HCF as well, but in that area the stress method, in our opinion, remains preferable. This subject is discussed in Section 12.3.

Before passing on to the basics of the strain method, we have to emphasize the differences in nomenclature vs. the stress method presented earlier. These differences are historically conditioned, similar to the differences between the metric and the inch systems. Here they are:

1. In the strain method, the cyclic strain $\Delta\varepsilon$ and the cyclic stress $\Delta\sigma$ are meant as double amplitudes (ranges). Respectively, the amplitudes are $\varepsilon_a = \Delta\varepsilon/2$ and $\sigma_a = \Delta\sigma/2$.
2. In the strain method, N means the number of load reversals, which is twice as many as the number of cycles. So the number of cycles equals $1/2\ N$.

Going to the strain method, we should recall that the phenomenon of fatigue is associated with plastic deformations in microvolumes. Among the multitude of grains forming the structure of the metal, there are grains with more imperfections in the crystal lattice (as compared to the neighbors). If this grain is so oriented that the slip plane and the shear stress are unidirectional, there occurs a plastic deformation in this grain. Thus, the macrodeformation of the metal may remain elastic.

At a cyclic loading, the "unlucky" grain is subjected to cyclic plastic deformation. The amplitude of this deformation is determined not by the load applied to the very grain, but by the deformation of the surrounding mass of the part. Because the mass works in the elastic range and obeys Hooke's law, the constant amplitude of load causes constant amplitude of deformation, too. Consequently, the considered grain is subjected to plastic deformation with constant amplitude. The behavior of this grain is of interest for the strength calculation, because the fatigue crack initiates in it.

Needless to say, experiments can't be performed with a grain, but with a specimen only. It is assumed that a smooth specimen, when tested at constant amplitude of strain (elastic + plastic deformation), is able to simulate the rupture of the metal grain under consideration. In other words, it is supposed that there is a similarity between the behaviors of the smooth specimen in the conditions of test stated previously and the microscopic volume of the real part under cyclic plastic deformation. This concept defines the design philosophy based on the linkage between the fatigue life of a part with an "unlucky" grain and that of the specimen cycled with the same level of stress. Accordingly, the expected fatigue life (until the time of formation of a microcrack) can be determined to sufficient accuracy when the history of deformation of the part and the strain-life fatigue properties of the specimen are known.

The nomenclature of data for cyclically deformable materials, construction of mathematical models, and test methods are widely represented in standards and special literature. The following definitions are used to form the basic relations.

In Section 12.1 (Figure 12.1 and Figure 12.2) it was stated that the yield point S_y and the ultimate strength S_u are obtained by dividing the respective tension forces by the cross-sectional area of the specimen A_0 before tension. Such stresses and σ–N (also known as S–N) diagrams (see Figure 3.9 [Chapter 3] and curve 1 in Figure 12.11) are used with the term "engineering" (engineering stress, engineering stress–strain curve and so on). Being hurt a bit by the shade of neglect with respect to engineers, we have to agree that the "engineering" (in other words, approximate) approach is not valid indeed for the strain method. Because the cross section of the specimen reduces with the increase of the tension force, the true stress σ will be obtained by dividing the tension force P at a given moment by the cross-sectional area A at the very same moment of time:

$$\sigma = \frac{P}{A}$$

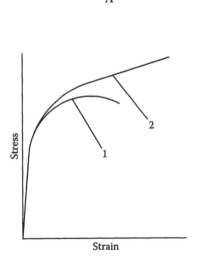

FIGURE 12.11 Engineering (1) and true (2) stress–strain diagrams.

Similarly, the idea of true deformation is introduced; it is based on the ratio of the "running" (i.e., instantaneous) elongation of the specimen to its running length:

$$\varepsilon = \int_{l_0}^{l} \frac{dl}{l} = \ln\frac{l}{l_0} \qquad (12.32)$$

From here, we derive the relation between the true deformation ε and the engineering deformation e:

$$\varepsilon = \ln(1+e) \qquad (12.33)$$

These relations are valid only before necking. After necking, the elongation becomes nonuniform over the length of the specimen. Fortunately, in practice no necking can occur, otherwise it would be one-cycle breakage and not low-cycle fatigue.

To link the true and engineering magnitudes of stress, let's assume that the volume of the specimen doesn't change during loading:

$$A_0\, l_0 = A\, l$$

Then, taking into account Equation 12.32 and Equation 12.33, we have

$$\varepsilon = \ln\frac{l}{l_0} = \ln\frac{A_0}{A} = \ln(1+e) \qquad (12.34)$$

The connection between the true stress σ and the engineering stress S is clear from the following definition:

$$\sigma = S\,\frac{A_0}{A} \qquad (12.35)$$

Because, from Equation 12.34, $A_0/A = 1 + e$, we obtain from Equation 12.35

$$\sigma = S\,(1+e)$$

The total true strain consists of two components, elastic and plastic:

$$\varepsilon_t = \varepsilon_e + \varepsilon_p$$

The elastic component is determined by Hooke's law and is given by

$$\varepsilon_e = \frac{\sigma}{E} \qquad (12.36)$$

where E = modulus of elasticity.

The plastic component is given in the form

$$\varepsilon_p = \left(\frac{\sigma}{K}\right)^{1/n} \tag{12.37}$$

where

K = strength coefficient
n = strain hardening exponent

So the total deformation of the specimen is given by the Ramberg–Osgood equation:

$$\varepsilon_t = \frac{\sigma}{E} + \left(\frac{\sigma}{K}\right)^{1/n} \tag{12.38}$$

All variables here have been determined. This relation for monotonic loading is presented graphically by curve 2 in Figure 12.11.

By monotonic is meant a loading with a gradually increasing force, as in a standard tensile test. The opposite of monotonic is cyclic loading.

To find the K value, in Equation 12.37 should be substituted the true fracture stress σ_f and true fracture strain ε_f, which are determined as follows:

$$\sigma_f = \frac{P_f}{A_f}$$

where

P_f = breakage force
A_f = cross-sectional area after breakage

$$\varepsilon_f = \ln\frac{A_0}{A_f} = \ln\frac{1}{1-RA}$$

where RA = reduction in the cross-sectional area, which is given by

$$RA = \frac{A_0 - A_f}{A_0}$$

From Equation 12.37,

$$K = \frac{\sigma}{\varepsilon_p^n} = \frac{\sigma_f}{\varepsilon_f^n}$$

Now let's have a short break from the tedious formulas and go to some speculations about the *cyclic* elastoplastic deformation. In this case, the structure of dependence between stress and strain is similar to that for monotonic loading. But the members are different, because the dependence $\sigma - \varepsilon$ at cyclic loading is different. This relation is obtained for each material by testing a set of specimens made of this material. Each specimen is cycled with its fixed amplitude of deformation, and the test lasts until the hysteresis loop stabilizes.

In the beginning of the cycling with a constant strain amplitude (called *transition period*), the stress in the metal may gradually increase (cyclic hardening, Figure 12.12a) or decrease (cyclic softening, Figure 12.12b), remain stable (Figure 12.12c), or even combine different modes. These effects are associated with the behavior of dislocations in the crystal lattice. Usually, soft materials with low dislocation density (such as aluminum alloys) show a tendency to cyclic hardening, and hard materials (for example, steels) are inclined to softening.

The transition period is short; then the hysteresis loop stabilizes, and, in this state, the effect seen is taken as the result of the experiment at a given amplitude of cyclic deformation.

A set of stabilized hysteresis loops obtained at different amplitudes of deformation is plotted on the same chart $\sigma - \varepsilon$ as shown in Figure 12.13. Line 1 connecting the extreme points of the loops represents the dependence $\sigma - \varepsilon$ at the cyclic deformation. This line is quite close to curve 2 in Figure 12.11, which represents the $\sigma - \varepsilon$ dependence at monotonic loading, and it is characterized with similar specific parameters that belong to cyclic loading. These parameters are provided with the sign (known as a *prime*):

σ'_y = cyclic yield strength
K' = cyclic strength coefficient
n' = cyclic strain hardening exponent

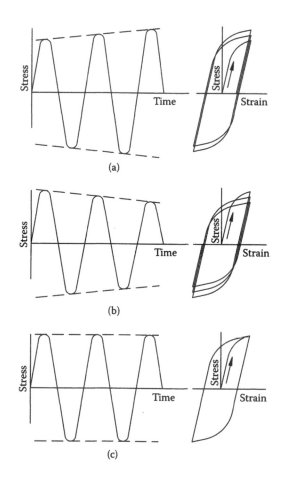

FIGURE 12.12 Hysteresis loops in the transition period of cyclic deformation.

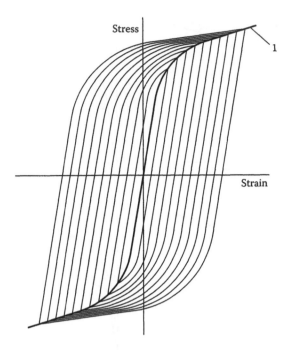

FIGURE 12.13 Construction of the cyclic stress–strain curve.

The cyclic yield strength is determined from the $\sigma - \varepsilon$ diagram as a stress at 0.2% of plastic deformation.

Figure 12.14 shows different variants of the correlation between the $\sigma - \varepsilon$ curves for monotonous and cyclic loading: cyclic softening (Figure 12.14a), cyclic hardening (Figure 12.14b), cyclically stable (Figure 12.14c), and mixed behavior (Figure 12.14d). Usually, metals with higher values of the monotonic strain-hardening exponent (n > 0.2) cyclically harden, and metals with n < 0.1 cyclically soften. One more rule of thumb: the material will harden if $S_u/S_y > 1.4$, and it will soften if $S_u/S_y < 1.2$. In the range from $S_u/S_y = 1.2$ to 1.4, the material may behave in any manner shown in Figure 12.14.

The total deformation, as in the case of monotonic loading, consists of elastic and plastic components:

$$\varepsilon = \varepsilon_e + \varepsilon_p = \frac{\sigma}{E} + \left(\frac{\sigma}{K'} \right)^{1/n'}$$ (12.39)

This is the equation of curve 1 in Figure 12.13. Equation 12.39 and Equation 12.38 are similar, because both the monotonic and the cyclic deformations are of the common nature. It is generally agreed that in cyclic loading the dependence between amplitudes of stress and deformation (σ_a and ε_a) is defined by the same curve 1 (Figure 12.13). Based on this assumption, this dependence looks similar to Equation 12.39:

$$\varepsilon_a = \varepsilon_{a,e} + \varepsilon_{a,p} = \frac{\sigma_a}{E} + \left(\frac{\sigma_a}{K'} \right)^{1/n'}$$ (12.40)

These dependences assume that the material behaves symmetrically with respect to tension and compression. This assumption is true for homogenous materials.

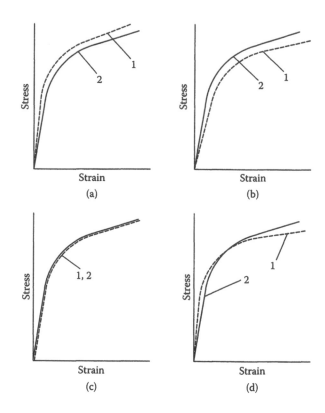

FIGURE 12.14 Differences between monotonic (1) and cyclic (2) stress–strain curves.

Because $\varepsilon_a = \Delta\varepsilon/2$ and $\sigma_a = \Delta\sigma/2$ (see previous text), Equation 12.40 can be written in the form

$$\frac{\Delta\varepsilon}{2} = \frac{\Delta\sigma}{2E} + \left(\frac{\Delta\sigma}{2K'}\right)^{1/n'} \tag{12.41}$$

This basic equation is widely used to describe the behavior of stresses and deformations at varying amplitudes of loading.

Coming to the technique of numerical estimate of the fatigue life, we should recall Basquin's formula (1910), which was chronologically the first:

$$\frac{\Delta\sigma}{2} = \sigma'_f (2N_f)^b \tag{12.42}$$

Here

$\Delta\sigma/2$ = true stress amplitude

σ'_f = fatigue strength coefficient

b = fatigue strength exponent

$2N_f$ = number of stress reversals to failure (1 reversal = 1/2 cycle)

Parameters σ'_f and b define the fatigue properties of the material. To sufficient accuracy, the fatigue strength coefficient σ'_f equals the true fracture stress σ_f. Fatigue strength exponent b is to be determined from the steady-state hysteresis loop; usually, it ranges from −0.05 to −0.12. Virtually,

Equation 12.42 is the same as Equation 12.1 which describes the sloping part of the Wöhler's curve, and it belongs to the elastic portion of the strain. In log-log coordinates it is a straight line.

The next decisive step, which is made to associate the elastic and plastic components of strain, was made by Coffin and Manson. They found that the plastic strain–life data (i.e., the $\varepsilon_p - N$ dependence) can also be presented in log-log coordinates as a straight line:

$$\frac{\Delta\varepsilon_p}{2} = \varepsilon'_f (2N_f)^c \tag{12.43}$$

where

$\Delta\varepsilon_p/2$ = plastic strain amplitude
$2N_f$ = number of reversals to failure
ε'_f = fatigue ductility coefficient
c = fatigue ductility exponent

Parameters ε'_f and c belong to the fracture properties of the material; ε'_f approximately equals the true fracture ductility ε_f and ranges from 0.5 to 0.7.

Equation 12.43 is very important because it enables bringing together all the assumptions made previously. Now, from Equation 12.42 and Equation 12.43, the relation between the total strain of a cycle and the number of cycles to failure can be written as follows:

$$\frac{\Delta\varepsilon}{2} = \frac{\Delta\varepsilon_e}{2} + \frac{\Delta\varepsilon_p}{2} = \frac{\sigma'_f}{E}(2N_f)^b + \varepsilon'_f(2N_f)^c \tag{12.44}$$

The two terms of the second member of Equation 12.43 can be represented in log-log coordinates as two straight lines (Figure 12.15). Line 1 corresponds to the first term (elastic component of deformation), and line 2, to the second term (plastic component). Line 3 conforms to Equation 12.44. It is seen that at large strain amplitudes (and small numbers of stress reversals), line 3 approaches asymptotically to the line of plastic deformations that make the main contribution to the failure. At small amplitudes, elastic line 1 only determines the number of cycles to failure.

Point A is called the transition point, and shows the number of stress reversals, $2N_t$, at which the influences of the elastic and plastic deformations are equal. The condition of their equality is obtained from Equation 12.44:

$$2N_t = \left(\frac{\varepsilon'_f E}{\sigma'_f}\right)^{1/(b-c)} \tag{12.45}$$

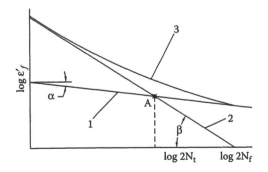

FIGURE 12.15 Strain–life curves: elastic (1), plastic (2), and summary (3).

In all, the fatigue properties of metals are determined by four parameters: ε'_f, σ'_f, b, and c. For certain materials and heat treatment conditions, these factors are given in handbooks and standards, for example, SAE J1099. But many materials are not covered by the database, therefore some empirical relations that express the strain–life properties of the material through its monotonic properties (such as Brinell hardness HB, tensile strength S_u, and modulus of elasticity E) have been developed. For practical use, the following relations are recommended:[8]

- Equation 12.44 can be approximated (suggested by Muralidharan and Manson) as follows:

$$\frac{\Delta\varepsilon}{2} = 0.623\left(\frac{S_u}{E}\right)^{0.832}(2N_f)^{-0.09} + 0.0196\cdot\varepsilon_f^{0.155}\left(\frac{S_u}{E}\right)^{-0.53}(2N_f)^{-0.56} \qquad (12.46)$$

- The fatigue strength coefficient:

$$\sigma'_f = 4.25\,HB + 225\ MPa; \qquad (12.47)$$

$$\sigma'_f = 1.04\,S_u + 345\ MPa \qquad (12.48)$$

- The fatigue ductility coefficient:

$$\varepsilon'_f = \frac{0.32(HB)^2 - 487HB + 191000}{E} \qquad (12.49)$$

- The fatigue strength exponent b varies from −0.057 to −0.140, and on average $b = -0.09$.
- The fatigue ductility exponent c ranges from −0.39 to −1.04, and on average $c = -0.60$. This value is usually taken for metals with pronounced yield ($\varepsilon_f \approx 1$). For high-strength steels ($\varepsilon_f \approx 0.5$), $c = -0.5$ is most acceptable.
- The transition fatigue life ($2N_t$, see Equation 12.45) can be obtained from the formula:

$$2N_t = 10^{5.755-0.0071HB} \qquad (12.50)$$

The last equation, together with Equation 12.45 and Equation 12.47 to Equation 12.49, may be useful for more exact estimation of factors b and c.

Factor b defines the slope angle of line 1 in Figure 12.15 ($b = \tan\alpha$), and it is clear that the calculated durability of the part is influenced very strongly by this factor in the area of high-cycle fatigue. Substituting in Equation 12.42, $b = -0.06$ and $b = -0.14$, we obtain:

$$2N_f = \left(\frac{2\sigma'_f}{\Delta\sigma}\right)^{16.7+7.1}$$

That means that the inaccurate estimation of the b value may lead to great change in the calculated durability.

In the area of low-cycle fatigue, line 2 is the determining element, and factor c characterizes the slope angle ($c = \tan\beta$). From Equation 12.43, if $c = -0.6$ or $c = -0.5$, the number of cycles to failure

$$2N_f = \left(\frac{2\varepsilon'_f}{\Delta\varepsilon_p}\right)^{1.67+2.0}$$

Hence, in this area the variation of the c value leads to a relatively small change in the calculated durability.

The relations obtained till now were valid for a symmetric cycle ($\sigma_m = 0$). For asymmetric cycles, the influence of the mean stress is taken into account on the basis of the following experimental information:

1. The influence of the mean stress is substantial mainly in high-cycle fatigue. Compressive mean stress increases the fatigue life, and tension mean stress decreases it.
2. In the area of low-cycle fatigue, in which the plastic deformation becomes significant, the mean stress relaxes virtually to zero.

It is appropriate to explain here that in the macrovolume, the mean stress can't disappear, because it is caused by the load diagram: if the load cycle has a mean load, there must be also the mean stress in the loaded part. But in this case, we are considering a microvolume (or a grain) of the part that is subjected to elastoplastic deformation with a *constant amplitude*. The plastic deformation in this microvolume relaxes the mean stress during the first cycles of loading, and only the harmonic part of the stress remains in the sequel.

In conformity with the aforesaid, only the elastic component (the first term of Equation 12.44) shall be adapted to the mean stress:

$$\frac{\Delta\varepsilon}{2} = \frac{\sigma'_f - \sigma_m}{E}(2N_f)^b + \varepsilon'_f(2N_f)^c \tag{12.51}$$

This relation (suggested by Morrow) is valid for $\sigma_m > 0$ only. For negative mean stress, which gives a certain positive effect, a more conservative approach is accepted, namely, $\sigma_m = 0$.

Now we go to multiaxial stress. Application of yield criteria (for example, Tresca or von Mises) is useful also in this case. By von Mises, the equivalent stress expressed in terms of main stresses

$$\sigma_e = \frac{1}{\sqrt{2}}[(\sigma_1 - \sigma_2)^2 + (\sigma_2 - \sigma_3)^2 + (\sigma_1 - \sigma_3)^2]^{1/2}$$

Let's explore how to go to the strain–life curve for cyclic torsion, when the data for cyclic tension–compression are known. By analogy with Equation 12.44, the equation for torsion looks as follows:

$$\frac{\Delta\gamma}{2} = \frac{\tau'_f}{G}(2N_f)^b + \gamma'_f(2N_f)^c \tag{12.52}$$

It is assumed that factors b and c remain the same. Then, taking into account the relation between the normal and shear stresses and deformations, Equation 12.52 is transformed to the following:

$$\frac{\Delta\gamma}{2} = \frac{\sigma'_f}{G\sqrt{3}}(2N_f)^b + \varepsilon'_f\sqrt{3}(2N_f)^c \tag{12.53}$$

The conversion of the stresses is made on the basis of Equation 12.5. The conversion of the deformations is based on the expression for the equivalent deformation:

$$\varepsilon_e = \frac{\sqrt{2}}{3}[(\varepsilon_1 - \varepsilon_2)^2 + (\varepsilon_2 - \varepsilon_3)^2 + (\varepsilon_1 - \varepsilon_3)^2]^{1/2} \tag{12.54}$$

In this expression is used the fact that for completely plastic deformation, Poisson's coefficient $v = 0.5$; that is, the material is assumed to be incompressible. Because in torsion $\varepsilon_1 = -\varepsilon_3 = \gamma/2$, and $\varepsilon_2 = 0$, the equivalent deformation is

$$\varepsilon_e = \frac{\gamma}{\sqrt{3}}$$

If, instead of von Mises, Tresca's criterion is used

$$\tau_e = \left|\frac{\sigma_1 - \sigma_3}{2}\right|$$

Then, for pure torsion

$$\tau_e = \frac{\sigma_1}{2}$$

Because at pure torsion $\tau_e = \tau$, we obtain (assuming that the same relation is valid for the fatigue characteristics):

$$\tau'_f = \frac{\sigma'_f}{2} \tag{12.55}$$

The deformation should be determined considering that besides the elongation ε_1, there is also transverse contraction $-v\varepsilon_1$. The equivalent shear stress is given by

$$\frac{\gamma_e}{2} = \frac{\varepsilon_1 - (-v\varepsilon_1)}{2} = \frac{(1+v)}{2}\varepsilon_1$$

But in the case of pure shear $\gamma_e = \gamma$; therefore, on the same assumption as mentioned previously,

$$\frac{\gamma'_f}{2} = \frac{1+v}{2}\varepsilon'_f \tag{12.56}$$

Insertion of Equation 12.55, Equation 12.56, and $v = 0.5$ into Equation 12.52 gives

$$\frac{\Delta\gamma}{2} = \frac{\sigma'_f}{2G}(2N_f)^b + 1.5\varepsilon'_f(2N_f)^c \tag{12.57}$$

These two examples with von Mises' criteria (Equation 12.53) and Tresca's criteria (Equation 12.57) show that the choice of criteria changes the obtained equation of the strain–life curve. But the differences in results remain within the usual accuracy of such calculations.

12.2 STRENGTH OF MACHINE ELEMENTS

The strength characteristics of materials are obtained from experiments with relatively small specimens. The strength of the real parts is sometimes also checked experimentally, but this undertaking is time consuming; besides, the results are valid only for these parts and can hardly be used for other parts. Therefore, in the overwhelming majority of engineering applications, fatigue strength is calculated starting with the material properties and taking into account the differences between the specimen and the part. The following differences are usually considered.

12.2.1 SURFACE FINISH

The smoother the surface, the higher is the fatigue limit. It is well known from experiments that surface asperities and chemical injuries facilitate the initiation of fatigue cracks. Because standard specimens are fine-polished or mirror-polished, the surface factor for them $K_S = 1$. The influence of surface finish increases with increased strength of the material. As the tensile strength of steel S_u rises from 400 MPa to 1400 MPa, the K_S value decreases almost linearly from 0.95 to 0.85 for grinding, from 0.92 to 0.76 for fine turning, from 0.9 to 0.6 for coarse turning, and from 0.8 to 0.3 for unmachined surfaces (as forged).[7]

These values of K_S are approximate, and in the literature can be found different recommendations. But it is clear that the strength of a part can be decreased by 50% or even more because of just a scratch or a rust stain in the highly loaded area. The areas critical for fatigue strength are in many cases mirror-polished and protected from harming influences, whatever their nature is.

12.2.2 DIMENSIONS OF THE PART

It has been noticed that the bigger the part, the lesser is the admissible stress amplitude. This effect is usually attributed to several factors:

- Lesser stress gradient in larger parts (Figure 12.16). As Figure 12.16 shows, at the same depth from the surface, the part with a larger diameter has a higher stress than does the smaller part. The more stressed subsurface layer in the larger part enhances crack propagation from the surface into the interior. (It should be mentioned that the fatigue limit in tension, where the stress gradient is zero, is less than in bending by about 15–20%.[6])
- Inferior quality of the metal achievable in larger parts due to metallurgical problems (nonuniformity of mechanical properties over the cross section caused by nonuniformity of forging and heat treatment). The standard specimens cut from smaller and larger blanks have different strength. As the blank diameter grows from 10 mm to 500 mm, the tensile strength of the specimen decreases in average by 10% for carbon steels and by 15–20% for alloy steels.[6]
- Increased volume of highly stressed material and, consequently, the increased quantity of "unlucky" grains exposed to high stress; thus increases the probability of origination of a fatigue crack.

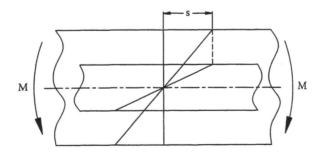

FIGURE 12.16 Stress distribution in bars subjected to bending loads.

The combined effect of these causes is expressed by factor K_d, and a recommended value can be only very approximate because of the aforementioned scatter of experimental points and the limited number of experiments with larger parts. Based on recommendations,[9] the size factor can be estimated from this formula:

$$K_d = \left(\frac{D}{d}\right)^{-0.034}$$ (12.58)

where

D = diameter of the part
d = diameter of the specimen that usually ranges from 5 to 10 mm

12.2.3 STRESS CONCENTRATION

The standard specimens that are made to determine the fatigue limit of a material are designed so as to diminish as far as possible the other influencing factors, such as surface finish, residual stresses, and stress raisers. But in the design of real parts the stress raisers mostly are unavoidable. The origin of local increase of stress called *stress concentration*, as well as the difference between the theoretical (K_t) and effective (K_e) values have been discussed in detail in Section 3.2 as applied to the *stress method*, in which the plastic portion of strain is negligibly small.

For the *strain method*, in which plastic deformation is considered, the influence of the stress raiser is expressed differently. Within the elastic range, the deformation is directly proportional to the stress. Hence, the increase of local stress by factor K_t leads to an increase in local deformation by the same factor:

$$\frac{\sigma_a}{S_a} = \frac{\varepsilon_a}{e_a} = K_t$$

where

σ_a and ε_a = local stress and strain amplitudes in the stress raiser area
S_a and e_a = nominal stress and strain amplitudes

In the yielding area, the deformation grows more than the stress; therefore, the individual rise factors for strain and stress should be considered separately:

$$\frac{\sigma_a}{S_a} = K_\sigma; \quad \frac{\varepsilon_a}{e_a} = K_\varepsilon;$$ (12.59)

Here $K_\sigma < K_t$, and $K_\varepsilon > K_t$. The relation between K_σ and K_ε has been suggested by H. Neuber (1962):

$$K_\sigma K_\varepsilon = K_t^2$$ (12.60)

This rule has been experimentally proved as applicable. From Equation 12.59 and Equation 12.60, we obtain

$$\sigma_a \varepsilon_a = S_a e_a K_t^2$$

or, similarly,

$$\frac{\Delta\sigma}{2}\frac{\Delta\varepsilon}{2} = \frac{\Delta S}{2}\frac{\Delta e}{2}K_t^2 \tag{12.61}$$

Because the nominal stress in the part doesn't exceed the yield point, the ΔS and Δe values are connected by Hooke's law:

$$\Delta e = \frac{\Delta S}{E}$$

Insertion of this relation into Equation 12.61 gives

$$\frac{\Delta\sigma}{2}\frac{\Delta\varepsilon}{2} = \frac{1}{E}\left(K_t\frac{\Delta S}{2}\right)^2 \tag{12.62}$$

Because $\Delta\sigma$ and $\Delta\varepsilon$ are connected by Equation 12.41, the joint solution of Equation 12.62 and Equation 12.41 gives the equation that connects the stress amplitude with the stress concentration factor:

$$\left(\frac{\Delta\sigma}{2}\right)^2 + E\left(\frac{\Delta\sigma}{2}\right)\left(\frac{\Delta\sigma}{2K'}\right)^{1/n'} = \left(\frac{\Delta S}{2}K_t\right)^2 \tag{12.63}$$

where
$\Delta\sigma/2$ = maximal stress amplitude in the stress raiser area
$\Delta S/2$ = nominal stress amplitude

This equation can be solved by iterations or by plotting the $\Delta\sigma - \Delta S$ curve for a certain K_t value. The effective stress concentration factor in this case is

$$K_e = \frac{\Delta\sigma}{\Delta S}$$

and it is different for each level of nominal stress.

12.2.4 USE OF FACTORS K_s, K_d, AND K_e

We have discussed here the three named factors that are usually taken into account when the admissible stresses are determined. The use of these factors is different in the stress and strain methods.

In the stress method, the calculation is based on the fatigue limit. If the fatigue limit of the metal obtained from experiments with specimens equals S_{-1}, the fatigue limit of a part made from this metal ($S_{-1,p}$) is

$$S_{-1,p} = S_{-1}\frac{K_s K_d}{K_e}$$

It's very easy, isn't it?

In the strain method, the situation is more complicated. It is agreed that the surface quality and dimensions are of no importance for plastically deformed microvolumes. Hence, factors K_s and K_d should influence only the elastic part of the deformation, that is, the first term of Equation 12.44. It is suggested[10] to change the fatigue strength exponent b to b' as follows:

$$b' = b + 0.159 \log(K_s K_d) \tag{12.64}$$

The stress concentration is included in the strength calculation as shown in Subsection 12.2.3.

12.3 COMPARATIVE CALCULATIONS FOR STRENGTH

EXAMPLE 12.3

A smooth round shaft of 100 mm in diameter, with no stress raisers, is exposed to cyclical bending. The shaft is made of steel SAE 4340 heat-treated to HB 275. The ultimate tensile strength $S_u = 1048$ MPa, and the 0.2% yield point $S_y = 834$ MPa. The surface finish is fine turning. How much is the admissible amplitude of bending stress σ_a, if the needed fatigue life amounts to 10^3, 10^4, 10^5, 10^6, 10^7, 10^8, and 10^9 cycles?

12.3A Calculation Using Stress Method

According to the empirical formulas given at the end of Subsection 12.1.2, the fatigue limit for alloy steel

$$S_{-1} = 0.383 S_u + 95 = 0.383 \cdot 1048 + 95 = 496 \ MPa$$

This result is valid for the specimens and should be multiplied by surface factor $K_s \approx 0.8$ (see recommendations in the preceding text) and size factor K_d, which is (from Equation 12.58)

$$K_d = \left(\frac{100}{5}\right)^{-0.034} = 0.90$$

All in all, the fatigue limit for the shaft

$$S_{-1,p} = S_{-1} K_s K_d = 496 \cdot 0.8 \cdot 0.9 = 357 \ MPa$$

To make calculations for $N \neq N_0$, we have to determine the number of cycles N_0 to the breakpoint of the $\sigma - N$ curve and the exponent m. The alloy steel SAE 4340 has the σ–N curve as shown in Figure 12.1c. Let's take $N_0 = 3 \cdot 10^6$ cycles, $m = 10$ (on portion 1, Figure 12.1), and $m_1 = 100$ (on portion 4). From Equation 12.1 we obtain for $N = 10^3$ cycles,

$$N_0 S_{-1,p}^m = N \sigma_a;$$

$$3 \cdot 10^6 \cdot 357^{10} = 1 \cdot 10^3 \cdot \sigma_a^{10};$$

$$\sigma_a = 357 \sqrt[10]{\frac{3 \cdot 10^6}{10^3}} = 795 \ MPa$$

In the same way, we obtain σ_a = 632, 502, and 398 MPa for N = 10^4, 10^5, and 10^6 cycles, respectively.

For $N = 10^7$ cycles

$$\sigma_a = 357 \sqrt[100]{\frac{3 \cdot 10^6}{10^7}} = 353 \ MPa$$

Similarly, for $N = 10^8$ and 10^9 cycles, σ_a = 345 and 337 MPa. Finally, the results of the calculation using the stress–life method are as follows:

Number of cycles to failure:	10^3	10^4	10^5	10^6	10^7	10^8	10^9
$\sigma_{a,s}$, MPa (stress method):	795	632	502	398	353	345	337

12.3B Calculation Using Strain Method

According to SAE J1099, the cyclic properties of this steel are as follows:

$$\sigma'_f = 1276 \ MPa; \ \varepsilon'_f = 1.224; \ b = -0.075; \ c = -0.714; \ K' = 1249 \ MPa, \ n' = 0.105$$

Modulus of elasticity $E = 1.9 \cdot 10^5$ MPa.
From Equation 12.64,

$$b' \doteq -0.075 + 0.159 \log(0.8 \cdot 0.9) = -0.075 - 0.023 = -0.098$$

Equation 12.44 with these data looks as follows:

$$\frac{\Delta \varepsilon}{2} = \frac{1276}{1.9 \cdot 10^5} (2N_f)^{-0.098} + 1.224 (2N_f)^{-0.714} = A + B$$

Substituting the required numbers of cycles to failure ($2N_f$ = 10^3, 10^4, 10^5, 10^6, 10^7, 10^8, and 10^9) in this equation, we obtain the following results:

$2N_f$	10^3	10^4	10^5	10^6	10^7	10^8	10^9
$10^3 \cdot A$	3.413	2.723	2.173	1.734	1.384	1.104	0.881
$10^3 \cdot B$	8.826	1.705	0.329	0.064	0	0	0
$10^3 \cdot \Delta \varepsilon / 2$	12.239	4.428	2.502	1.798	1.384	1.104	0.881

Term A represents the elastic portion of strain and term B the plastic portion. We can see that at low-cycle fatigue the deformation is mostly plastic, whereas at high-cycle fatigue the plastic component is negligible.

The admissible stress amplitude $\Delta \sigma / 2$ can be obtained from Equation 12.41. Substituting the K', n', and E values into this equation, we have the following relation between the strain and stress amplitudes for this kind of steel and heat treatment:

$$\frac{\Delta \varepsilon}{2} = \frac{\Delta \sigma}{2 \cdot 1.9 \cdot 10^5} + \left(\frac{\Delta \sigma}{2 \cdot 1249} \right)^{1/0.105} \tag{12.65}$$

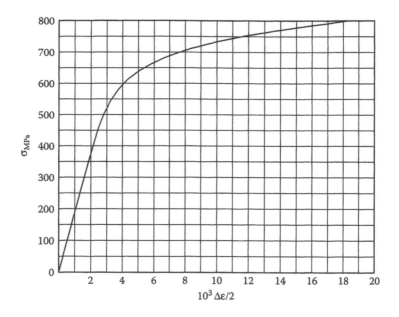

FIGURE 12.17 Plot of Equation 12.65.

Because this relation is a transcendental function, it can be solved by iterations or by plotting a diagram. The latter is shown in Figure 12.17. From this diagram, the following results have been obtained:

Number of cycles to failure:	10^3	10^4	10^5	10^6	10^7	10^8	10^9
$\sigma_{a,S}$, MPa (strain method):	755	616	462	341	263	210	168

12.3C Comparison of Methods

The results obtained using the two methods are united in Table 12.1.

TABLE 12.1
Admissible Stress Amplitudes for Shaft without Stress Raisers

Number of cycles to failure	10^3	10^4	10^5	10^6	10^7	10^8	10^9
$\sigma_{a,S}$, MPa (stress method)	795	632	502	398	353	345	337
$\sigma_{a,S}$, MPa (strain method)	755	616	462	341	263	210	168

EXAMPLE 12.4

Here is considered the same shaft as in Example 12.3 but with a filleted section transition as shown in Figure 12.18. The theoretical stress concentration factor $K_t = 2.0$.

12.4A Calculation Using Stress Method

The admissible stress amplitude before the stress raiser area is the same as has been obtained in Example 12.3A:

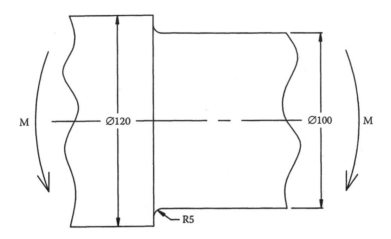

FIGURE 12.18 Shaft with a fillet.

Number of cycles to failure:	10^3	10^4	10^5	10^6	10^7	10^8	10^9
$\sigma_{a,S}$, MPa (stress method):	795	632	502	398	353	345	337

The effective stress concentration factor K_e is determined as stated in Section 3.2, assuming that the coefficient of sensitivity to stress concentration for this material $q = 0.9$:

$$K_e = 1 + q(K_t - 1) = 1 + 0.9(2 - 1) = 1.9$$

The nominal stress amplitude $\sigma_{a,SR}$ for the stress raiser area is given by

$$\sigma_{a,SR} = \frac{\sigma_{a,S}}{K_e}$$

The results for different numbers of cycles to failure are as follows:

Number of cycles to failure:	10^3	10^4	10^5	10^6	10^7	10^8	10^9
$\sigma_{a,SR}$, MPa (stress method):	418	333	264	209	186	182	177

12.4B Calculation Using Strain Method

The admissible strain amplitude depending on the number of cycles can be taken from Example 12.3B; it was calculated from Equation 12.44:

$2N_f$	10^3	10^4	10^5	10^6	10^7	10^8	10^9
$10^3 \cdot \Delta\varepsilon/2$	12.239	4.428	2.502	1.798	1.384	1.104	0.881

Now, when the $\Delta\varepsilon/2$ value is known, we have two equations (Equation 12.62 and Equation 12.63) with two unknowns: $\Delta\sigma/2$ and $K_t \cdot \Delta S/2$. Equation 12.63 in this case looks as follows:

$$\left(\frac{\Delta\sigma}{2}\right)^2 + 1.9 \cdot 10^5 \left(\frac{\Delta\sigma}{2}\right)\left(\frac{\Delta\sigma}{2 \cdot 1249}\right)^{1/0.105} = \left(\frac{\Delta S}{2} K_t\right)^2 \tag{12.66}$$

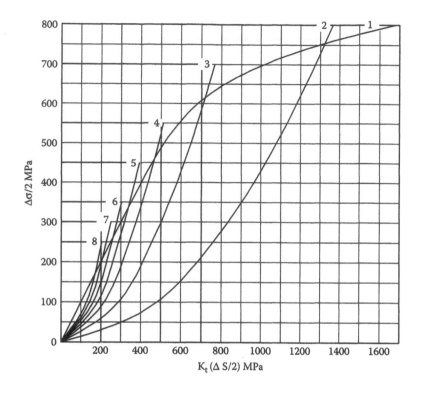

FIGURE 12.19 Plot of Equation 12.66 and Equation 12.67.

Equation 12.66 is plotted in Figure 12.19 (curve 1). From Equation 12.62 we have

$$K_t \frac{\Delta S}{2} = \sqrt{1.9 \cdot 10^5 \frac{\Delta \varepsilon}{2} \frac{\Delta \sigma}{2}} \tag{12.67}$$

Curves 2, 3, 4, 5, 6, 7, and 8 in Figure 12.19 represent Equation 12.67 at the $\Delta \varepsilon/2$ values taken for $2N_f = 10^3$, 10^4, 10^5, 10^6, 10^7, 10^8, and 10^9. The intersection points of these curves give us the $\Delta \sigma/2$ and $K_t \cdot \Delta S/2$ values for each number of cycles $2N_f$ indicated earlier. Then, the admissible nominal stress amplitude $\sigma_{a,S} = \Delta S/2$ is obtained by division of the $K_t \cdot \Delta S/2$ value by K_t. The results are as follows:

$2N_f$	10^3	10^4	10^5	10^6	10^7	10^8	10^9
$\Delta S/2$, MPa	662	360	235	169	131	105	86

12.4C Comparison of Methods

The results obtained using the two methods are united in Table 12.2.

TABLE 12.2
Admissible Stress Amplitudes for Shaft with Stress Raisers

Number of cycles to failure	10^3	10^4	10^5	10^6	10^7	10^8	10^9
$\sigma_{a,S}$, MPa (stress method)	418	333	264	209	186	182	177
$\sigma_{a,S}$, MPa (strain method)	662	360	235	169	131	105	86

From Table 12.1 and Table 12.2, we can conclude the following about the stress and strain methods of calculating fatigue strength:

1. It is known that at a high number of cycles (at $N_x > N_0 \approx 10^6$–$3 \cdot 10^6$) the materials have no (or very slight) decrease in the admissible stress (see Figure 12.1b and Figure 12.1c); the stress method is based on these experimental findings. The strain method, contrastingly, prescribes a noticeable decrease of the admissible stress throughout the N_x line, which doesn't agree with the experiments. So the strain method should not be used for HCF.

2. At low cycles, the stress method has not been developed enough to provide reliable results of life prediction. The parameters of the σ–N curve (N_0 and m) are represented very approximately. For instance, if in Subsection 12.3A of Example 12.3 we had taken $m = 6$ (instead of $m = 10$), the admissible stress amplitudes would have been 1356, 924, and 629 MPa (instead of 795, 632, and 502 MPa) at $N = 10^3$, 10^4, and 10^5 cycles, respectively. Therefore, the strain method that has been developed especially for the number of cycles less than N_0 is believed to provide a more accurate equation for the σ– N curve in this area and, consequently, a more exact life prediction than the stress method.

12.4 REAL STRENGTH OF MATERIALS

The real strength of the material the part is manufactured from may appear higher or lower than that taken in the calculations. The undesirable variation of the material strength (to a lower value) can be caused by unsatisfactory quality of the heat, imperfections of forged or rolled stock, low-quality heat treatment, or it can be just an incorrect material that got to production by mistake. Therefore, it is important to control the real quality and the specified strength parameters of the material.

The blank (stock) suppliers usually check all the parameters specified in the supply agreement: the quality of metal, chemical composition, and the strength characteristics. In important cases, the manufacturer of the machine parts arranges reinspection (on-receipt inspection) of the blanks to decrease the possibility of mistake.

One example of strict control of rolled stock taken from the aircraft industry consists of the following. Each bar is tested in the following manner. They cut a test piece from each end of a bar. The test pieces and the bar are marked identically. Then, each of the test pieces is checked for its chemical composition, and after the specified heat treatment is performed, the material is checked for its microstructure and mechanical properties. Provided that everything is okay, the bar is cut into pieces depending on the length of the parts to be produced, and the pieces are put in a sealed box along with the test documentation. The box is labeled with the part number and the material designation. Later on, the test documentation is kept in the file of the machine as long as it is in service.

Control of internal defects in the metal is mostly performed after rough machining by ultrasound or x-ray. But these methods can disclose only defects that have sufficient area in the direction of beaming. Therefore, after the machining is completed, the surfaces are checked additionally by magnetic or dye penetrant flaw detection for defects that measure up to 1 or 2 μm.

The hardness is usually measured on each part. Sometimes it is measured in several points. If the part is surface-hardened, the quality of the hardening (thickness of the layer, hardness and microstructure of the layer and core) is mostly controlled on a test-piece ("witness") made of the same material and treated together with the parts. All these methods of quality control are called *nondestructive testing* (NDT), because the parts remain unharmed.

Some of the tests can't be performed on the part by nondestructive methods. To obtain the needed data, the following processes are used:

- If the parts are relatively small and not too expensive, additional parts are manufactured using the same technology, from the same heat of metal and heat-treated together with the lot of parts. Then, one of these parts with the lowest hardness (or two, with the lowest and the highest hardness) can be picked out and cut for specimens for tests.
- If the parts are large scale and expensive, an additional forging can be made from the same heat of metal; it is desirable that the cross section of this forging be comparable with that of the part in the most stressed area. The test forging should be heat-treated together with the parts, then the specimens can be cut from it and tested.
- In the most important cases, each part is made with an over-measure or extended end that is cut off after the heat treatment to make specimens.

The parameters mostly checked under production conditions are the hardness, chemical composition, microstructure, tensile testing (with the estimation of tensile strength, yield point, and elongation), and test for impact toughness. The estimation of fatigue limit is performed rarely. All the needed kinds of control and criteria of acceptance or rejection shall be specified in the technical documentation.

REFERENCES

1. Birger, I.A., Shorr, B.F., and Iosilevich, G.B., *Strength Calculation of Machine Elements Handbook*, Mashinostroenie, Moscow, 1979.
2. Loeventhal, S.H., Factors that Affect the Fatigue Strength of Power Transmission Shafts and Their Impact on Design, NASA Technical Memorandum 83608, Lewis Research Center, Cleveland, OH, 1984.
3. Timoshenko, S., *Strength of Materials, Part II, Advanced Theory and Problems*, D. Van-Nostrand Company, Princeton, NJ.
4. Serensen, S.V., Kogaev, V.P., and Shneiderovich, R.M., *Loading Capacity and Calculation for Strength of Machine Elements*, Mashinostroenie, Moscow, 1975.
5. Sines, G., Failure of Materials under Combined Repeated Stresses Superimposed with Static Stresses, Tech. Note 3495, NACA, Washington, D.C., 1955.
6. Buch, A., *Fatigue Strength Calculation*, Trans Tech SA, Switzerland, 1988.
7. Kogaev, V.P., *Strength Calculation at Stresses Fluctuating with Time*, Mashinostroenie, Moscow, 1977.
8. Rössle, M.L. and Fatemi, A., Strain-controlled fatigue properties of steels and some simple approximations, The University of Toledo, OH, 2000.
9. *SAE Fatigue Design Handbook*, Vol. 4, Graham, J.A. (Ed.), Society of Automotive Engineers, 1968.
10. Stephens, R.I., Fatemi, A., Stephens, R.R., and Fuchs, H.O., *Metal Fatigue in Engineering*, 2nd ed., John Wiley and Sons, 2000.

13 Calculations for Strength

Here, we continue discussing the main elements of calculations for strength.

13.1 CHARACTERISTICS OF STRESSES IN THE PART

The problem of estimating the stress magnitude at the design stage is quite complicated. It consists of three problems solved sequentially:

1. Estimation of external loads applied to the mechanism
2. Determination of forces applied to the part
3. Estimation of stress field in the part and location of the most stressed areas

Let's consider these problems separately.

13.1.1 ESTIMATION OF EXTERNAL LOADS

A machine consists generally of a prime mover, an actuating mechanism, and a transmission between them. If the forces applied to the actuator are known, the load of all links of the machine can be calculated. It may be a static calculation or a complicated dynamic analysis using a computer program, but anyway it is possible. The problem is that the actuator's load in many cases can't be determined to sufficient accuracy because it faces all the variety and suddenness of the service conditions. The designer should not only know the routine loads, but also foresee the possible overloads that may harm the machine. Sometimes, the machine is protected against overload by additional devices, such as torque-limiting clutches, devices that shut off the electric motor when the current exceeds a preset limit, and others. But in many cases the safety devices are not exact enough (i.e., the minimal and the maximal load of their operation may differ too much), and this forces us to choose between two unpleasant options. The first one: if the minimal operating load of the limiter is adjusted too close to the working load of the machine, the latter becomes sensitive to overloads and often stops to rest, driving its operator mad. The second option: if the minimal operating load of the limiter is adjusted far from the working load, its limiting load is too high to provide any protection to the mechanism. In addition, the protecting device itself is an additional element of nonreliability that may disable the machine at a very inappropriate time. Therefore, the parts of the machine are mostly designed with some additional safety margin to make the machine more durable and foolproof. As a rough estimation, the static strength of the machine should be enough to withstand safely the maximal load developed by the motor. In other words, the external load may stop the machine but not break its mechanism.

When the weight of the machine is strictly limited, the load spectrum should be defined more precisely. The main method is the direct measurement of stresses in the machine parts during laboratory tests and field trials of the prototype. For similar machines with accumulated operating time, recommendations can be developed concerning the assumed load and the acceptable safety margin. If the machine is completely new, the designer is forced to spend more time and money on design, research, and development.

13.1.2 DETERMINATION OF FORCES APPLIED TO THE PART

Generally two independent calculations for strength are required, those for static strength and those for fatigue; the forces taken into consideration in these cases are different. For static strength calculation, the maximal forces applied to the part and the maximal stresses (they may not always be concurrent) should be determined. Static equilibrium equations are used for this purpose. Strictly speaking, it is incorrect, because in dynamic action the forces may be greater or smaller. But dynamic calculation is quite difficult and time consuming, so mostly static methods are used. The obtained forces are multiplied by some factor of dynamic amplification. Calculation for fatigue considers only cyclic load when the number of cycles is large enough (say, more than 100). If not, this load is considered as static.

In many cases, the load parameters can be defined only approximately, and the designer has to make some assumptions about it.

EXAMPLE 13.1

Figure 13.1 shows lifting drum 1 that is connected to shaft 2 and driven by gear 3 mounted on the shaft. The torque is transmitted through tooth-type coupling 4. The rotational speed of the shaft n = 100 r/min. The torque transmitted by the shaft is directly proportional to the lifted weight, and the maximal weight is known. Besides, it is known that once a year the lifting device is tested by lifting a weight of 110% of the maximal working value, and the maximal lifting time $t = 25$ sec. The mechanism is actuated 10–15 times a day, and the required service life is 20 years, with 300 working days a year. In 70% of the cases, the lifted weight amounts to 50% or less of the maximal value. How can one determine the load spectrum of the shaft?

The shaft is loaded with a torque and bending moments. The latter are caused by radial forces coming from gear 3 and drum bearing 5, and they can sum up or subtract depending on the design. For example, if Figure 13.1 shows the top view, the preferred side for the rope to hang down is the opposite of the gear mesh (the upper side of the drum in the drawing). In this case, the bending moments subtract from each other.

The calculation for static strength is based on the test load, when 110% of the maximal weight is lifted. In the beginning of lifting, the motor takes up the slack of the rope, and then it lifts the weight. The inertia force of the weight causes a dynamic increase in the rope force that depends on several factors: the lifting speed, the stiffness of the rope and the crane, and on the masses of the weight, and the crane. So the static strength calculation of the shaft should be based on the test weight lifting force multiplied by a dynamic amplification factor k_d. The dynamic load is applied during a relatively small time: say, it takes about a half-second (one turn of the shaft). So the number

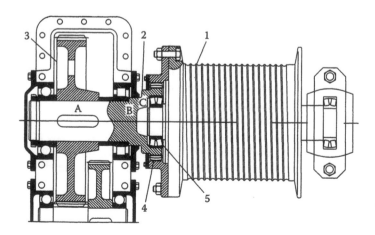

FIGURE 13.1 Lifting drum with a driving shaft.

of cycles under this load equals 20 (once a year). It should be noted that the torque and bending moments are applied to the shaft simultaneously, and both are directly proportional to the rope force.

For the fatigue calculations, weights of 50% of the maximal and less can be neglected with confidence. Assuming that in the remaining 30% of lifting cases the weight is maximal, and each lifting cycle takes 25 sec up and the same time down, we can calculate the number of cycles. When the shaft rotates, the torque remains nearly constant, and the shear stress is assumed to be constant as well. So the number of cycles for the shear stress is equal to the number of liftings. The shear cycle is pulsating ($R = 0$), and both the amplitude of shear stress τ_a and the mean shear stress τ_m equal to a half of the maximal shear stress.

The bending stress cycle lasts one turn of the shaft. This cycle is symmetric, and the bending stress amplitude is equal to the maximal bending stress. As we see, the bending stresses and the shear stresses are not in phase. To be on the safe side, we assume that they are in phase. Here are the calculations of the load spectrum:

Test load without dynamic amplification:

$$N_1 = 20(years) \cdot 1(time\ per\ year) \cdot 50(sec\ each\ time) \cdot 100(rpm)/60 = 1667\ cycles$$

Maximal working load with dynamic amplification:

$$N_2 = 20(years) \cdot 300(days) \cdot 15(times\ per\ day) \cdot 1(turn\ each\ time) \cdot 0.3(30\%) = 2.7 \cdot 10^4\ cycles$$

Maximal working load without dynamic amplification:

$$N_3 = 20(years) \cdot 300(days) \cdot 15(tpd) \cdot 0.3(30\%) \cdot 50(sec\ each\ time) \cdot 100(rpm)/60 = 2.25 \cdot 10^6\ cycles$$

The following simplifying assumptions have been made here:

In all cases, where the weight is greater than 50% of the maximal, it is assumed to be maximal.
Each lifting cycle takes the maximal time (25 sec in each direction).
The numbers of cycles have been calculated for the bending stresses, but we assume that the shear stresses are in phase with them.

All these assumptions are made so as to be on the safe side.

From these we understand that the designer, to obtain the loading spectrum, has to consider in detail the kinematics and operation procedure of the mechanism. Then the forces can be calculated or estimated by some means, including dedicated experiments.

There is not always a simple relationship between the load on the actuating mechanism and the forces applied to the parts. This relationship can be particularly hard to determine when the working process is fast or accompanied by impacts.

If the force and time of its application are known, the dynamic force can be estimated from a vibration analysis. Depending on the design of the mechanism and the accepted simplifications, the vibrating system can be of different number of degrees of freedom. For our purpose, it is enough to consider the simplest system with one degree of freedom shown in Figure 13.2. As is well known, the natural frequency of such a system with mass m on a spring of rate k is given by

$$f = \frac{1}{2\pi}\sqrt{\frac{k}{m}}\ Hz$$

(Here the spring rate k is in N/m and the mass m in kg.)

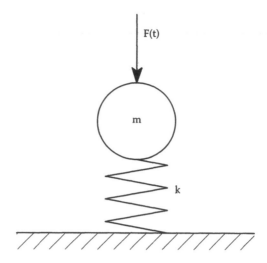

FIGURE 13.2 System with one degree of freedom.

The period of vibration is

$$T_0 = \frac{1}{f} = 2\pi\sqrt{\frac{m}{k}} \quad s \qquad (13.1)$$

The equations of motion with regard to force F are as follows:

$$m\ddot{x} + kx = F(t)$$

$$\ddot{x} + \omega_0^2 x = \frac{F(t)}{m} \qquad (13.2)$$

$$where \quad \omega_0^2 = \frac{k}{m} \quad (rad/s)^2$$

Let's consider the effect of constant force F_0 applied to the mass during time cell τ (Figure 13.3a):

$$F(t) = F_0 \quad @ \quad 0 \le t \le \tau$$

$$F(t) = 0 \quad @ \quad t \ge \tau$$

Because the load changes abruptly, the solution consists of two parts: for the region of $0 \le t \le \tau$ (when force F_0 is applied to the mass) and at $t > \tau$. For the first region the equations are

$$x(t) = \frac{F_0}{k}(1 - \cos\omega_0 t)$$

$$\dot{x}(t) = \frac{F_0\omega_0}{k}\sin(\omega_0 t) \qquad (13.3)$$

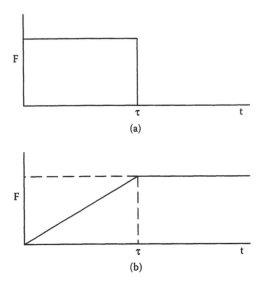

FIGURE 13.3 Kinds of loading.

The maximal displacement of the mass (and, consequently, the maximal values of spring deformation and spring force) takes place at the time point t_{max} when the speed equals zero and the mass begins its return motion:

$$\dot{x}(t_{max}) = 0$$

$$\frac{F_0 \omega_0}{k} \sin(\omega_0 t_{max}) = 0$$

Because the factors F_0, ω_0, and k are nonzero,

$$\sin(\omega_0 t_{max}) = 0$$

$$t_{max} = \frac{\pi}{\omega_0} \quad (\textit{the first root})$$

(13.4)

From Equation 13.1, Equation 13.2, and Equation 13.4 we obtain

$$t_{max} = \frac{T_0}{2}$$

If the time of load application $\tau \geq t_{max} = T_0/2$, the maximal displacement of the mass is achieved at $t_{max} = T_0/2$. From Equation 13.3 and Equation 13.4 we find

$$x_{max} = 2\frac{F_0}{k} = 2\delta_{st}$$

where $\delta_{st} = F_0/k$ = static displacement. In this way, we have obtained the well-known dynamic amplification factor $k_{dyn} = 2$.

If the time of load application $\tau < T_0/2$, the maximal displacement of the mass is less than $2\delta_{st}$. After the load ceases, the system oscillates freely with the starting conditions at $t = \tau$. As is known,[1] the equation of motion at this time is

$$x_2(t) = 2\delta_{st} \sin \frac{\tau \omega_0}{2} \sin \omega_0 \left(t - \frac{\tau}{2} \right)$$ (13.5)

Because the sine value can't be more than 1, the maximal displacement can't exceed the $2\delta_{st}$ value.

Let's estimate the effect of short impulses of load, say, at $\tau = 0.1 T_0$. Substituting in Equation 13.3 the value

$$t = 0.1 T_0 = 0.1 \cdot \frac{2\pi}{\omega_0} = \frac{0.2 \cdot \pi}{\omega_0}$$

we obtain the displacement at $t = \tau$:

$$x_1 = \frac{F_0}{k} (1 - \cos 0.2\pi) = 0.191 \delta_{st}$$

The mass continues moving, and the maximal displacement can be obtained from Equation 13.5 for the time when no force is applied to the mass. We substitute in this equation the following values:

$$\tau = 0.1 T_0; \quad \sin \omega_0 \left(t - \frac{\tau}{2} \right) = 1 \ (the \ \text{max} \ value)$$

Then,

$$x_2 = 2\delta_{st} \sin 0.1\pi = 0.618 \delta_{st}$$

This is the maximal deformation of the spring if the force is applied during $0.1 T_0$. Provided that the impulse of force is very small (say, $\tau = 0.01 T_0$), the maximal deformation of the spring equals only $0.063 \delta_{st}$. That is, the system (because of its inertia) doesn't respond to the short-time impulse.

Let's consider the case when the loading force grows linearly from zero to its finite amount F_0 as shown in Figure 13.3b. As is known, the displacement in the region of $0 \le t \le \tau$ is given by

$$x_1(t) = \delta_{st} \left(\frac{t}{\tau} - \frac{T_0}{2\pi\tau} \sin \omega_0 t \right)$$

In the region of $t > \tau$,

$$x_2(t) = \delta_{st} \left(1 - \frac{2 \cos \omega_0 (t - \frac{\tau}{2}) \sin \frac{\omega_0 \tau}{2}}{\omega_0 \tau} \right)$$

Because the cosine may change from -1 to 1, the maximal factor of dynamic amplification takes place when $\cos[\omega_0(t - \tau/2)] = -1$:

$$k_{dyn} = \frac{x_2(t)}{\delta_{st}} = 1 + 2\,\frac{\sin\frac{\omega_0\tau}{2}}{\omega_0\tau}$$

It is clear that the greater the time τ, the closer factor k_{dyn} is to 1. If we take

$$\sin\frac{\omega_0\tau}{2} = 1\,,$$

then factor k_{dyn}

$$k_{dyn} = 1 + \frac{2}{\omega_0\tau} = 1 + \frac{2T_0}{2\pi\tau} = 1 + 0.32\,\frac{T_0}{\tau}$$

At $\tau = 3T_0$, $k_{dyn} = 1.1$.

From the aforesaid it follows that:

1. The factor of dynamic amplification can't be greater than 2.
2. If the time the load takes to increase from zero to 100% is greater than the period of the lowest natural frequency of vibration by a factor of 3–4 or more, the dynamic amplification is negligible.

This good-looking solution is often hard to make use of. For example, a plough runs into a stone. You know the speed of the tractor, but you can't know the vibration characteristics of the multiple-blade plough, the magnitude of force F_0, and the time of its application τ. These parameters only can be calculated using a computer program for dynamic calculations (such as ADAMS). For this purpose, you have to make some assumptions about the weight of the stone and the resistance of the soil the stone is embedded in. The stress condition of the parts can be determined more exactly by measuring the stresses when the mechanism is manufactured and tested.

13.1.3 Estimation of Stresses in the Part

First, it should be made clear that usually the stresses are calculated not in the part, but in some scheme that reproduces to greater or lesser accuracy the shape of the part, its manner of fastening (i.e., the kind of attachment to the other parts), and the method of load application. The scarce capabilities of the "engineering" methods of stress calculation and the limitations of available computer programs impose corresponding limitations on the design of the analytical model. For example, bearing or gear forces are taken as point forces and placed approximately in the middle of the component's width (see Chapter 4, Section 4.4).

As applied to the finite element method (FEM), there are restrictions on the number of elements and, consequently, on the construction of intricately shaped parts and simulation of connections of all kinds: bolted, press fit (hub to shaft), slide fit (plain bearing or lug-to-pin connection), and the like. For instance, in Chapter 10, Subsection 10.3.1 have been analyzed several variants of bolted joints with only one bolt in each variant (see Figure 10.19 and Figure 10.20 in Chapter 10). Such a calculation of a full-scale bolted connection with tens of bolts may appear unrealistic. To overcome this difficulty, the designer can calculate first the load distribution between the bolts assuming that each bolt is just a point of rigid (or pliable) connection between the parts. Then the most loaded bolt can be calculated separately with minimum simplifications.

It is too difficult to build a model with no simplifications because there are factors that can hardly be determined. In the case of a bolted connection, such factors are, for example, the nonlinear compliance of the contact layers in all connections (between the parts, bolt head and part, nut and part, bolt and nut in the thread), the possible misalignment of the nut relative to the bolt thread under bending moment, and so on, not to mention the possible manufacturing tolerances that are in most cases neglected, and this is one of the simplifications. The possible influence of these factors can be analyzed, but if it appears substantial, the design should be changed to a better one.

When building the analytical model, the possible reductions of the number of elements should be made. For example, if the parts and their load are axial-symmetric, the FEM calculation can be performed on a sector that contains repeating elements of the design. If the parts and their load are symmetric about a line, the model can contain half of the real structure. The stress calculation is dedicated to reveal the most stressed area of the part. FEM calculation shows directly the place of maximal stress. The engineering method requires that checking calculations be made for all "suspicious" places where the load is large or there are stress raisers. For example, in shaft 2, Figure 13.1, sections A, B, and C are obviously such.

13.2 SAFETY FACTORS

It is clear from the preceding text that mathematical exactness is not one of the properties inherent in strength calculations. To begin with, the strength parameters of the materials are based not only on experimental data, which in themselves have statistical variance, but also on a number of hypotheses and assumptions, a good deal of which are logical or heuristic rather than experimentally proven. Many of the empirical coefficients that influence appreciably the calculated magnitude of admissible stress and fatigue life are defined for a limited number of materials only with sizable scatter of data. The strength parameters of other materials have to be estimated approximately. Factors such as sensitivity to stress concentration, size and surface factors, damage accumulation parameters (see Chapter 12, Subsection 12.1.3), as well as the load and stress estimation, are approximate.

In some respects, strength calculations are similar to polling before the presidential elections. The technology of polling presumes a certain statistical error, which depends on the percentage of voters questioned. But besides this error, in the space of time between the polling and the elections, some unplanned and unexpected events may occur, which may influence appreciably the voters' opinion. So the odds in favor of one candidate (his "safety factor") obtained in polling may appear insufficient, and another candidate will win. The more voters questioned in the polling and the less significant the events between polling and elections, the more exact will be the political forecast.

The same is true of strength calculations. The more characteristics and coefficients that are taken on the basis of direct experiments with the very part to be designed (instead of averaged empirical recommendations), the more exact will be the calculation and the smaller the necessary safety factor.

It has been already said, that the safety factor is the factor of lack of knowledge, a necessary precaution against the possible gap between the calculated and the real values of the material strength parameters and the magnitudes of operating stresses. Many factors influencing this gap have been mentioned, but many remain unnamed and even unknown.

Even in the most simple case, when a standard specimen is tested in a laboratory under uniaxial stress of constant amplitude, the load capacity and fatigue life can't be predicted exactly because it is probabilistic. Furthermore, the application of combined stress with varying amplitude to an intricately shaped part reduces the achievable accuracy of life prediction considerably. Therefore, the designer introduces a certain safety margin by increasing the calculated working load or decreasing the admissible stress by some factor called the *safety factor*.

In engineering practice the following values of safety factor are considered as satisfactory:[2]

- $k = 1.25–1.5$: For very reliable materials used under controllable conditions and exposed to loads and stresses that are definitely known. Such a safety factor is used when minimizing weight is of the highest importance.
- $k = 1.5–2.0$: For well-known materials used under stable conditions and subjected to loads that can be exactly determined.
- $k = 2.0–2.5$: For usual materials at usual loads that can be determined with usual accuracy.

It must be underlined that the use of a smaller safety factor presupposes better knowledge about all the components determining the life of the part. Without this, a decrease in the safety factor means an increase in the probability of a premature failure.

One more point gravely influencing the choice of a safety factor is a consideration of the possible consequences of failure. If it may immediately endanger the health of people or bring about heavy material losses, safety factors can be significantly more than what is stated in the preceding text. In most cases, we can hardly estimate abstractly the degree of lack of knowledge; therefore, the designer generally takes safety factors from experience in the appropriate branch of industry. This experience embraces ways of thinking and doing at all the stages of creation of a machine: research, design, calculations, manufacturing, trials, and field experience accumulation. The value of safety factor accepted in a certain machinery branch is that which provides with high probability the reliable operation of the equipment produced in this branch.

13.3 ERRORS DUE TO INAPPROPRIATE USE OF FEM

13.3.1 DESIGN PRINCIPLES AND PRECISION OF FEM

This method is based on division of the deformed body (or surface, or line) into small pieces called *elements* and solution of a set of equations that describe the joint behavior of the elements under load. As the basic unknowns in these equations are taken to be displacements, the strain and stress are determined from the displacement–strain equations and Hooke's law. To make the computational process easier, it is supposed that the displacement field (and, consequently, the stress field) within one element is described by simple polynomial functions. For instance, for a rod loaded in tension or compression the following linear function is used to describe the strain u of the element:

$$u(x) = u_1 \frac{x_2 - x}{h} + u_2 \frac{x - x_1}{h}$$

where

u_1 and u_2 = displacements of the end points of an element, called *nodes* (Figure 13.4)

$h = x_2 - x_1$ = length of the element

(Let's check this equation: at $x = x_1$, $u(x) = u_1$; at $x = x_2$, $u(x) = u_2$.)

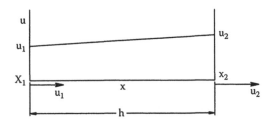

FIGURE 13.4 Distribution of displacements within a single element.

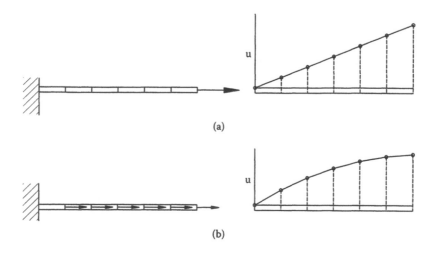

FIGURE 13.5 Distribution of load and strain in a tensioned bar.

The solution for the entire rod is a piecewise linear function. If the rod is loaded with a point force, this function is just a straight line formed by line segments (Figure 13.5a). If the tension force is evenly distributed along the rod, the elongation function is represented by parabola formed by straight line segments (Figure 13.5b). Thus, the unknown exact solution is represented by a linear approximation.

Let's go to strain and stresses. The stress (according to Hooke's law) is

$$\sigma = E\varepsilon$$

For the first element (from Figure 13.4),

$$\varepsilon_1 = \frac{u_2 - u_1}{h}; \quad \sigma_1 = E\frac{u_2 - u_1}{h}$$

For the second element,

$$\varepsilon_2 = \frac{u_3 - u_2}{h}; \quad \sigma_2 = E\frac{u_3 - u_2}{h}$$

For the third element,

$$\varepsilon_3 = \frac{u_4 - u_3}{h}; \quad \sigma_3 = E\frac{u_4 - u_3}{h}$$

and so on. As we see, within each single element the stress is constant, and between the elements it changes stepwise. In reality, the stress either remains constant (for example, when the load is applied as shown in Figure 13.5a) or changes gradually, without steps. So the accepted functions distort the stress field as compared with the exact solution.

Any of the FEM programs represent the results smoothed, and the user doesn't usually see the stress steps. But they can be seen when the suitable command is set into the computer. The magnitude of these steps is used to evaluate the quality of the solution. The less the magnitude of the stress steps, the closer is the result to the real stress field. Modern FEM programs enable the possibility of improvement of the calculation quality. There are two main ways of improvement:

1. Gradual raise of the degree of polynomials (use of polynomials of second, third, and higher degrees)
2. Size reduction of the elements in the areas of high stress gradient

In both cases the magnitudes of stress steps are used as a measure of the inaccuracy.

13.3.2 DESIGN OF MODEL FOR FEM COMPUTATION

The computational model is nothing more than a mathematical abstraction simplified as compared with the real part. The simplifications made when constructing the model should be chosen so that the results of the calculation would be close enough to reality. And yet the model should be adapted to the capabilities and peculiarities of the computer program.

All the calculated bodies are classified depending on the ratio of their dimensions: length, width, and height. If the length is much greater than width and height, this part is a well-known beam that is modeled as a line. Plates and shells have length and width much greater than height (thickness), and they are modeled as surfaces. If all dimensions of a body are of the same order, a three-dimensional model is usually constructed. Each successive type of model is more complicated than the previous one, and its calculation is more time consuming.

The tendency to simplify the model is practically justified. Specifically, the simplicity and beauty of the beam theory has been achieved because the three-dimensional problem was replaced by a one-dimensional problem. Depending on the ratio between the length and cross-sectional dimensions, the beam may have to be represented as a three-dimensional body, or as a line, or as a string (when the cross section is relatively small). Let's consider the limiting ratios of dimensions for these three possible presentations of a beam. Figure 13.6 shows a cantilever beam loaded with a transverse force F. The displacement (sag) under the force, with the correction by Timoshenko, is given by

$$w = \frac{Fl^3}{3EI} + \frac{Fl}{kGA}$$

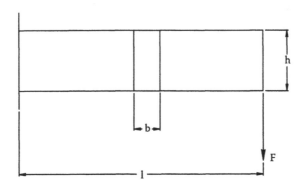

FIGURE 13.6 Cantilever beam.

where

  l = length of the beam
  E and G = modulus of elasticity in tension and in shear, respectively
  I = moment of inertia of the cross section
  A = area of the beam cross section
  k = correction factor that allows for the shape of the cross section

The first term is the displacement according to classical theory. The second term represents the shear deformations, and its magnitude determines the dividing line between body-modeled and line-modeled beams. For the sake of simplicity, let's take the cross section as a rectangle with height h and width b. Then,

$$w = \frac{Fl^3}{3EI}\left(1 + \frac{3}{k}\cdot\frac{E}{G}\cdot\frac{I}{Al^2}\right)$$

For this shape of cross section $k = 1.2$. Relation $E/G = 2(1 + v) = 2.6$. Thus, the second term of the equation is given by

$$\frac{3}{k}\cdot\frac{E}{G}\cdot\frac{I}{Al^2} = \frac{3}{1.2}\cdot 2.6\cdot\frac{bh^3}{12\cdot bh\cdot l^2} = 0.54\cdot\left(\frac{h}{l}\right)^2$$

Hence, the second term is determined by the ratio $(h/l)^2$. This ratio is the main criterion both in the theory of beams and in the theory of plates and shells, because it is the measure of their slenderness. Obviously, if $h/l = 0.1$, the second term can be neglected. If an error of 10% is acceptable, the classic theory of beams can be used up to $h/l = 0.43$.

Let's consider another extreme case, when the cross section of a beam is very small as compared to its length: the case of a string. If the string is not stretched, it sags under its own weight. To understand more precisely the difference between a beam and a string, let's consider two examples of beams with different grip conditions at their ends. In Figure 13.7a the beam is represented just as was done by the professor at the university. One of the supports has a roller, and this means that there is no axial force in the beam.

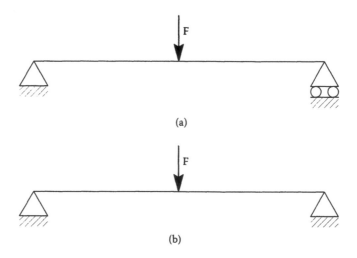

(a)

(b)

FIGURE 13.7 Beam supports.

If there is an axial force, the right support will just move with a minimal resisting force. In the real world, such rollers can be seen in steel bridges. The application of rollers here is governed mainly not by the wish to please the professor, but by the necessity to compensate for the difference in thermal expansion between the metal and the earth.

In shaft supports, the compensation of thermal deformations is achieved at the expense of unrestricted axial movement of the shaft relative to a bearing or of the bearing relative to the housing. In some cases, when the radial load on the bearing is very high, rollers can be placed between the bearing body and the housing to decrease the resistance to the axial displacement.

Figure 13.7b shows the same beam, but both supports are fixed. Seemingly, the stresses in these two beams should be different. This was suggested not only by our intuition but also by the efforts of the professor, not entirely in vain, at drawing these rollers. But linear static FEM calculation shows that the stresses are the same. What is the matter? The point is that this method has limitations because it is based on the assumption that the linear and angular displacements are recognized as negligibly small. Consequently, all the equations have been written for an absolutely rigid system, and when the transverse forces are considered, the real movability or rigidity of the support doesn't influence the calculated stress. Use of nonlinear static FEM calculation that allows for geometry changes while loading reveals the difference in deformation and stress pattern between these two kinds of support.

In fact, the displacements in real beams are relatively small indeed. The neglect of them greatly simplifies the structure and solution of the equations for displacements and stress in beams, whereas the error caused is very small. Recognition of the geometry changes as "small" (that can be neglected) or "large" (that should be taken into consideration) govern the choice of one of two schemes shown in Figure 13.7. It is generally agreed to recognize the transverse deformations as small if they don't exceed 20% of the beam height or 20% of the plate thickness (w < 0.2h, see literature cited[2]). FEM calculations show that even at $w = h$ the differences in results using both schemes are insignificant.

13.3.3 INTERPRETATION OF BOUNDARY CONDITIONS

Model construction includes also the choice of suitable application of loading forces and the adequate interpretation of the conditions at fixed points. The fixed points are also the points of application of forces to the part. These forces, together with the working load and inertia forces, form the system of forces and moments, the sum vectors of which equal zero (i.e., the system is in equilibrium).

The forces are applied to the part mostly through contact with other parts. (The exception is provided by forces of inertia and gravitation.) The strain and stresses in the part depend not only on the relative collocation of the force-transmitting areas, but also on their design. That is why it is important to model these areas as closely to the real design as possible.

In Chapter 4, Section 4.4 (Figure 4.11 to Figure 4.15), we have considered the choice of location of the points of load application for shafts modeled as lines. When the parts are modeled as two-dimensional or three-dimensional objects (in FEM calculation), the point application of a force results in too high stress and strain levels. Physically it is clear: a finite force applied to an infinitesimal area gives a theoretically infinite stress. But why does this not occur if the beam is modeled as a line?

The reason is that in the theory of beams the material is assumed to be rigid in the transverse direction, and the beam behaves in this direction as a rigid, nondeformable body. (The same assumption is used in the classic theory of plates: there is no deformation in the direction of their thickness.) This results in "nonsensitivity" of such models to concentration of the load; the force applied to a point or to a small area results in the same strain and stresses in the model. On the contrary, in two-dimensional and three-dimensional models, the bodies are deformable in all directions under consideration. As is known, if the force is applied at a point, the stress goes to infinity like function $1/r$ in a two-dimensional model and $1/r^2$ in a three-dimensional model (where r = distance to the point of force application). These functions are hyperbolas.

In FEM, as aforesaid, elements of the first order are mostly used, in which the displacement is approximated by bilinear functions (at best, by quadratic polynomials). Such elements try to represent the hyperbola by piecewise linear function and fail to do so satisfactorily in areas of high stress gradient, i.e., when the real stress changes considerably within one element. Size reduction of the elements improves the approximation but not radically. Better results are achieved by distributing the load between several nodes. Besides the improvement in calculation quality, such a presentation of load represents more closely the real force-transmitting contact of parts. Point and line are just mathematical abstractions, and no load can be transmitted through them in reality.

Problems with the selection of boundary conditions may sometimes occur even in seemingly easy cases. Figure 13.8 shows a cantilever beam 250 mm long, 50 mm high, and 1 mm thick, fixed in on the left end. The right end is loaded with a force $F = 100$ N. Modulus of elasticity $E = 2.1 \cdot 10^5$ MPa. Classical calculation gives the maximal tension-compression stress $\sigma = 60$ MPa and the sag $y = -0.238$ mm. FEM calculation gives us freedom of creation: we have to fix the left end of the beam against any kind of displacement, and we can do that in several ways. The following cases have been calculated:

Case 1: Each of the end nodes (1–9) are fixed against displacement in X and Y directions.
Case 2: Nodes 1, 5, and 9 are fixed against displacement in X and Y directions.
Case 3: Nodes 1, 3, 5, 7, and 9 are fixed against displacement in X and Y directions.
Case 4: Each of the end nodes (1–9) is fixed against displacement in X direction; in addition, node 9 is fixed in Y direction.
Case 5: Nodes 1,3,5,7, and 9 are fixed in X direction; in addition, node 9 is fixed in Y direction.

The results of the FEM calculations are tabulated in Figure 13.8. As is seen, the results obtained in cases 1 and 4 are close to the classical solution that is undoubtedly correct. The other cases show increased sag (up to 22%) and stress (greater by a factor of 1.63–2.35). Actually, calculation of this cantilever beam doesn't need the use of FEM. It is performed only to show the gentle reader that FEM calculation is not as simple as it appears, and some measures should always be taken to evaluate the validity of the results by other means — just as in any other method of calculation. (This is informally referred to as a "sanity check.")

There is no clear, universally applicable method to simplify boundary conditions. For example, the attachment with a bolt can be defined as an inhibition of movement in one (for instance, axial)

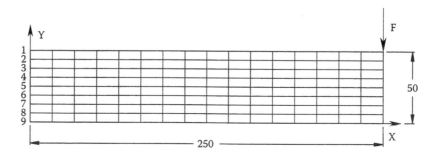

CASE	1	2	3	4	5
Max. von Mises stress, MPa	58.3	142	97.3	58.8	106
Max. X-X stress, MPa	61.2	141	99.9	58.1	98.0
Max. sag Y, mm	−0.244	−0.291	−0.264	−0.245	−0.266

FIGURE 13.8 Cantilever beam (FEM calculation).

direction, or a complete inhibition of movement. This inhibition can be replaced by a springy element that allows movement but provides resistance that is directly proportional to the displacement. A constant resistance to displacement can be imposed by application of friction forces. These movement limitations can be supplemented with an inhibition of turn of the connected parts at this point, and so on.

The complete inhibition of linear or angular displacement at the fixing points means that the adjoined parts are absolutely rigid. This assumption greatly simplifies the calculation, because there is no need to determine the compliance parameters of the fixing elements. In many cases this assumption is justified, but in each particular case it is worth checking whether it is valid.

It is clear that the introduction of pliable fixations in a statically determined scheme is meaningless because it doesn't change anything. In a statically indeterminate system, the calculation with rigid and pliable fasteners may give surprisingly different results. The influence of these boundary conditions on the stress pattern of the parts can be investigated by comparing FEM calculations with different boundary conditions.

EXAMPLE 13.2

A short beam is attached to a base at two points and is loaded by a force $F = 10,000$ N in the middle of its length, (see Figure 13.9). The section modulus W of this beam is given by

$$W = \frac{b h^2}{6} = \frac{10 \cdot 50^2}{6} = 4167 \ mm^3$$

The bending moment M

$$M = \frac{F L}{4} = \frac{10000 \cdot 100}{4} = 2.5 \cdot 10^5 \ N \cdot mm$$

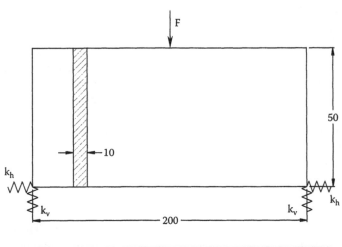

CASE	1	2	3	4	5	6
Max. tensile stress, MPa	25.2	31.3	46.0	59.4	46.0	59.4

FIGURE 13.9 Two-support beam (FEM calculation).

Thus, the maximal tension stress, according to classical beam theory (with the usual roller under one of the supports), is

$$\sigma = \frac{M}{W} = \frac{2.5 \cdot 10^5}{4167} = 60 \ MPa$$

This beam has also been calculated using FEM. The attachments have been assumed to be springy with different rates:

Case 1: $k_h = k_v = \infty$
Case 2: $k_h = k_v = 100,000$ N/mm
Case 3: $k_h = k_v = 10,000$ N/mm
Case 4: $k_h = k_v = 1,000$ N/mm
Case 5: $k_h = 10,000$ N/mm and $k_v = 1,000$ N/mm
Case 6: $k_h = 1,000$ N/mm and $k_v = 10,000$ N/mm

The results are tabulated in Figure 13.9. Comparing cases 3 with 5 and 4 with 6, we see that only the compliance in the horizontal direction plays a role in this case. This calculation shows that the concrete estimation of compliance of the attachments, even if it is rough, may be important.

13.3.4 IS THE COMPUTER PROGRAM CORRECT?

The computer programs are continually improving, and as long as the program is new, it may have some defects ("bugs") that have not been revealed. These defects may show up in very special cases. The usual user is much more likely to make his own errors because of misunderstanding of subtleties of the new program. Therefore, it is a usual practice to test the program by solving simpler problems that have been solved theoretically (known as a *closed-form* solution), and the formulas are available, so that the correctness of the result can be easily checked.

13.3.5 MORE ABOUT SIMPLIFIED ANALYTICAL MODELS

The simpler the calculation, the less the opportunities we have to make an error. Therefore, one of the ways to check the calculation is to make and calculate a simplified analytical model, even if your computer is able to calculate the model with no simplifications. The comparison of the results is useful. If the results are comparable, this adds to your confidence in your calculations. But if the results are dissimilar, there may be three options to think about:

1. One of the calculations (or both of them) may be erroneous; they should be checked for possible errors.
2. The simplifications may be incorrect or inappropriate to a certain case.
3. The computer program has bugs.

The simplifications are often unavoidable, because some problems (for instance, contact problems) may be too difficult to solve using the available computer program. Therefore, the computing engineer is forced to substitute, for example, a rigid fixing point for a real bolt joint. The possible influence of such forced simplifications on the stress condition of the parts should be considered.

Making simplifications, we also can transform the real, complicated problem into a simpler one that has an exact analytical solution (again, a closed-form solution), which can be calculated manually to obtain an approximate result. In Chapter 4, Example 4.5 the slope angle of the shaft was calculated approximately and using FEM, the difference was found to be as great as 19%.

Example 5.3 and Figure 5.9 (Chapter 5) demonstrate the strength calculation of an end plate. It was made twice: first under the simplifying assumption that the bolt force is distributed evenly over the bolt centers circle, and then using FEM. The difference in the maximal stress magnitude was of 25%.

In Chapter 7, Example 7.8 (in the end), the simplified calculation of oil heating has been used to check the result obtained from an exact formula: the results were identical. Here is one more example of the development and use of a simplified analytical model.

EXAMPLE 13.3

A ribbed round plate (Figure 13.10) is supported over its contour and loaded in the center with a force P = 10,000 N. According to the FEM calculation, the maximal tension stress (on the ribs) equals 140 MPa, and the maximal sag (in the center of the plate) equals 0.182 mm. Are these results correct?

Let's try to calculate this plate using the available formulas for round plates without ribs.[3] To make such a calculation, we have to replace the ribbed plate by an "equivalent" plate of constant thickness. One sixth of the cylindrical section of the real plate looks as shown in Figure 13.11a. The less the radius of this section, the less is the length L between points A and B and the stiffer the section per unit of length. Let's calculate the parameters of the equivalent section at the mean diameter $d_1 = 150$ mm. Now, we are going to determine the moment of inertia I and the section modulus W of the real section (about the X–X axis) using one sixth of it presented in Figure 13.11a. The length of arc A–B is

$$L = \frac{\pi d_1}{6} = \frac{\pi \cdot 150}{6} = 78.5 \, mm$$

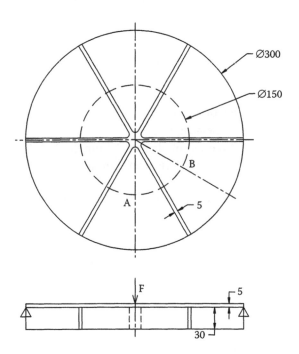

FIGURE 13.10 Ribbed plate.

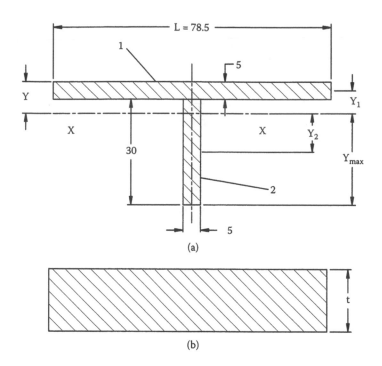

FIGURE 13.11 Real section (a) and equivalent section (b) of a ribbed plate.

As is known, the neutral line of a section passes through its center of gravity (CG). The coordinate Y of the CG is obtained from the equilibrium equation:

$$A_1 Y_1 = A_2 Y_2$$

where
$A_1 = 78.5 \cdot 5 = 392.5$ mm^2 = area of element 1
$A_2 = 30 \cdot 5 = 150$ mm^2 = area of element 2
$Y_1 = Y - 2.5$ mm = distance of the CG of element 1 from the $X–X$ line
$Y_2 = 20 - Y$ mm = distance of the CG of element 2 from the $X–X$ line

From here, $Y = 7.34$ mm. And now the moment of inertia of this cross section can be obtained from the following formula:

$$I = I_1 + I_2 + A_1 Y_1^2 + A_2 Y_2^2$$

where

$$I_1 = \frac{78.5 \cdot 5^3}{12} = 818 \; mm^4; \quad I_2 = \frac{5 \cdot 30^3}{12} = 11250 \; mm^4$$

$$I = 818 + 11250 + 392(7.34 - 2.5)^2 + 150(20 - 7.34)^2 = 45292 \; mm^4$$

The section modulus is given by

$$W = \frac{I}{Y_{max}} = \frac{45292}{27.66} = 1637 \ mm^3$$

The equivalent thickness of the plate t shown in Figure 13.11b should be calculated twice: with respect to rigidity (t_r) and to strength (t_s). In the first case, the real section and the equivalent one should have the same moment of inertia I. For one sixth of the cylindrical section, it is

$$\frac{78.5 \cdot t_r^3}{12} = 45292 \ mm^4$$

and from here thickness t_r of the equivalent plate is

$$t_r = \sqrt[3]{\frac{45929 \cdot 12}{78.5}} = 19.1 \ mm$$

The maximal sag of the equivalent plate is

$$w_{max} = P \frac{(3+v)a^2}{16\pi(1+v)D}$$

where
  $P = 10,000 \ N$ = force applied to the plate
  $v = 0.3$ = Poisson's ratio
  $a = 150 \ mm$ = radius of the plate

$$D = \frac{Et^3}{12(1-v^2)} = \frac{2.06 \cdot 10^5 \cdot 19.1^3}{12(1-0.3^2)} = 1.31 \cdot 10^8 \ N \cdot mm$$

(Here, $E = 2.06 \cdot 10^5$ MPa = modulus of elasticity of the material.)
  The maximal sag in the center of the plate is

$$w_{max} = 10000 \frac{(3+0.3)150^2}{16\pi(1+0.3)1.31 \cdot 10^8} = 0.087 \ mm$$

This result amounts to one half of what has been obtained using FEM (0.182 mm).
  For the stress calculation, the real section of the plate and the equivalent one should have the same section modulus W. For one sixth of the cylindrical section, it is given by

$$\frac{78.5 \cdot t_s^2}{6} = 1637 \ mm^3$$

From here,

$$t_s = \sqrt{\frac{1637 \cdot 6}{78.5}} = 11.2 \ mm$$

The maximal tension stress in the center of the equivalent plate is

$$\sigma_{max} = \frac{P}{t_s^2}(1+\nu)\left(0.485\ln\frac{a}{t_s}+0.52\right) = \frac{10000}{11.2^2}(1+0.3)\left(0.485\ln\frac{150}{11.2}+0.52\right) = 184 \; MPa$$

This stress is greater by about 30% than that determined using FEM (140 MPa). So in this case the simplified model was a rough approximation. Nevertheless, it provides us with the order of stress and strain magnitudes and thus gives us the desired check of the FEM calculation.

13.3.6 CONSIDERATION OF DEFORMATIONS

The computerized calculation for strength offers the unique ability to visualize the deformation pattern. Each designer imagines these deformations, and, looking at the computer simulation, can find some discrepancies between the imagined and observed pictures. By this means, gross mistakes can be easily found. For example, if you see some displacement in a point that simulates a bearing or a rigid connection and has been defined as an immovable attachment point, that means there was a mistake in the boundary conditions data.

The computer picture is not always incorrect; the designer's imagination can be false, too. This case is nevertheless interesting because it improves our intuition, and supports the notion that visualizing the deformation pattern is not only a useful tool for fault tracing but also an enthralling occupation.

The deformation pattern is associated closely with the boundary conditions. The latter are simplified as compared with reality, and this simplification may be irrelevant. Observation of the deformations may provide the designer with "food for thought" on the validity of these simplifications.

It was shown previously that the boundary conditions may influence considerably the results of the stress and strain calculation. The acceptable choice of boundary conditions can be made relying on the FEM analysis of different options. In some cases, the design of the connection should be changed so as to decrease the influence of boundary condition uncertainties on the stressed state of the part. Generally, to make the configuration more easy and certain for strength calculation increases its reliability.

13.4 HUMAN ERROR

You have done a calculation. No matter what kind of calculation it is, the result is the same: you see formulas and figures, and you don't know whether it is correct or false. If you have an experienced supervisor, you go to him. You give him your opus and say as confidently as you can, "Here you are! That'll do?" But the supervisor looks frostily; to check your work, he has to do it himself anew, but he has his own work to do. He asks you, "Did you check your calculation?" A good question, isn't it? If you had asked for advice, we would have advised you to check your calculation thoroughly before being asked this question. But how to do that? When the calculation is complicated (especially when it is hidden inside a computer that makes 1 million operations per second), the computing engineer often feels himself unable to evaluate its correctness. And then he says, "The calculation shows that ..." and other words. The problem is that, whatever the calculation shows, you, and only you, are the one who is responsible for the results of the calculation you had been charged with. Conclusion: you must somehow check your calculations.

The error you are looking for may appear in any place. It can be just an arithmetic error, false dimensioning, incorrect definition of boundary conditions, and so on. All this should not drive you to despair; you are able to find any possible errors and to feel completely confident about the correctness of your calculations.

13.4.1 ARITHMETIC

Arithmetic errors are among the most frequently occurring. One of the simplest ways to find them is to make the calculation twice or thrice. If each time you have different results, you have to take a rest and repeat the calculation in a day or two. Another way is to simplify the calculation to avoid outrageous mistakes. For example,

$$\frac{\pi \cdot 203750 \cdot 18,82 \cdot 0.0093}{0.294 \cdot 15800 \cdot 25.4} = 0.95$$

Is this correct? We can write the same figures approximately, and in a simplified form, so as to enable mental calculation:

$$\frac{3 \cdot 2 \cdot 10^5 \cdot 2 \cdot 10 \cdot 10^{-2}}{3 \cdot 10^{-1} \cdot 2 \cdot 10^4 \cdot 2 \cdot 10} \approx 1$$

The next day, you can make the two steps of calculation again to be completely confident of its correctness.

13.4.2 Units (Dimensions)

Before you start making arithmetic errors, you have to decide about the units (dimensions) you use for the equation members. *You should never write any figure without its dimension!* For example, you must not write "the speed is 10," or "the time needed for this operation is less than 3," or "the force equals 5000." Possibly, you remember that the speed is measured in meters per second (m/sec), time in hours (h), and force in Newtons (N), but it should be written each time you write the figure. In this way, you keep feeling the physical essence of the problem and don't let your mind stray from it. For example, if you obtain that the speed equals 5000, it may not attract your attention. But if you see 5000 m/sec, you suddenly remember that it is quite close to the orbital velocity, and it looks a bit unreal for the railway train you are working on.

And this is not enough. When you use a formula, the result depends on the dimensions of the formula members. Therefore, each time you substitute the figures into formulas, it is worth writing separately their dimensions and checking the dimension of the result. Let's take, for example, Equation 7.4 and put the torque T_1 in N·mm, W in mm, and d_1 in mm, the gear ratio i being a nondimensional quantity:

$$K_H = \frac{2T_1(i \pm 1)}{W\,d_1^2\,i} = \frac{N \cdot mm}{mm \cdot mm^2} = \frac{N}{mm^2} = MPa$$

If the torque was taken, say, in N·m or lb·in, you can immediately see that the resultant dimension looks unfamiliar, and you will change the dimensions (and, respectively, the figures) to obtain a more reasonable result.

A good way to check the dimensions and the arithmetic at once is to repeat the calculation in different units.

Example 13.4

Spur gear $d_1 = 100$ mm, $d_2 = 200$ mm, $W = 50$ mm transmits load of $P = 150$ kW at $n_1 = 1500$ r/min. Gear ratio $i = 2$. Torque applied to the pinion is

$$T_1 = 9555\frac{P(kW)}{n_1(rpm)} = 9555\frac{150}{1500}\,N \cdot m = 955.5\,N \cdot m = 9.56 \cdot 10^5\,N \cdot mm$$

$$K_H = \frac{2 \cdot 9.56 \cdot 10^5(2+1)}{50 \cdot 100^2 \cdot 2} = 5.74\,MPa$$

Now, let's go to another system of dimensions: horsepower (hp), pounds (lb), and inches (in.).

$$P = 150 \ kW = 150 \cdot 1.36 \ HP = 204 \ HP$$

$$W = \frac{50}{25.4} = 1.968 \ in; \quad d_1 = \frac{100}{25.4} = 3.937 \ in$$

$$T_1 = 6.22 \cdot 10^4 \ \frac{P(Hp)}{n_1(rpm)} = 6.22 \cdot 10^4 \ \frac{204}{1500} = 8460 \ lb \cdot in$$

$$K_H = \frac{2 \cdot 8460 \cdot 3}{1.968 \cdot 3.937^2 \cdot 2} \ \frac{lb \cdot in}{in \cdot in^2} = 832 \ psi$$

$$1 \ psi = 7.02 \cdot 10^{-4} \ kg \ / \ mm^2 = 6.89 \cdot 10^{-3} \ N \ / \ mm^2$$

$$K_H = 832 \cdot 6.89 \cdot 10^{-3} = 5.73 \ MPa$$

Such a method provides a good way of control, and it also gives us practice in dealing with different dimensions.

13.4.3 Is This Formula Correct?

The calculation can be erroneous not only because of your own errors, but also because the formulas in the books may have misprints. Every time you use a formula, you should be suspicious of and check it. You can try and develop it by yourself, or find it in other books (if there are such), but the first thing is to check the dimensions. For example, if you see the formula

$$F(\sigma) = \frac{1}{6}[(\sigma_1 - \sigma_2) + (\sigma_2 - \sigma_3) + (\sigma_1 - \sigma_3)] - \frac{1}{3}\sigma_e^2$$

you can immediately notice that in the first term of the equation the power of the stresses is 1 (say, MPa), and in the second term the stress is squared (MPa²). In a perfect formula, terms with different dimensions can't be added or subtracted, so this equation is surely false.

When the formula is empirical, the check on dimensions may be useless. For example, Equation 6.8 for elastic deformation of ball bearings

$$\delta_1 = (0.153 - 0.00044d) \ F_r^{2/3} \ \mu m$$

has no physically reasonable dimensions. To obtain the true result, we must know exactly the dimensions of all the equation terms; in this very case, diameter d must be in mm and force F_r in N, then deformation δ_1 is obtained in μm.

Empirical formulas can't be checked by developing; even comparison with other books can be useless because the false formula can be reprinted in several books with the same error. In such cases, it is important to find somebody who is experienced in using this formula or to consult the author of this formula. If not, some experiments should be made to check the validity of the formula. If you have checked the formula and used it successfully, you can keep this formula (and also the calculation as an example) for your following professional life.

Sometimes an empirical formula can be checked by analysis of a simplified analytical model. For example, the elastic deformation in the rolling contacts of straight roller bearing NU310E under radial load $F_r = 10,000$ N, when calculated using the empirical formula in Equation 6.9 (Chapter 6), is given by

$$\delta_1 = \frac{0.065}{d} F_r = \frac{0.065}{50} 10000 = 13\,\mu m$$

In Chapter 6, Example 6.2 (Subsection 6.2.2) is shown how this deformation can be calculated. The summary deformation in rolling contacts

$$\Delta_{C,i} + \Delta_{C,o} = 0.0085 + 0.0085 = 0.017\,mm = 17\,\mu m$$

Thus, this empirical formula is okay.

You may ask, "Okay, but what shall I take for my consequent calculations: 13 μm or 17 μm?" The point is that both the empirical formula and the checking calculation are approximate, and it is worth taking the worst result. Virtually, the difference of 4 μm is not significant in general applications. When it is important, indeed, both the empirical formulas and approximate calculations may be unacceptable, and experimental research of the deformation should be undertaken.

REFERENCES

1. Harris, C.M., *Shock and Vibration Handbook,* 2nd ed., Harris, C.M. and Crede, C.E., Eds., McGraw-Hill, New York, 1976.
2. Juvinal, R.C., *Engineering Considerations of Stress, Strain and Strength,* McGraw-Hill, New York, 1967.
3. Timoshenko, S. and Woinowsky-Krieger, S., *Theory of Plates and Shells,* McGraw-Hill, New York, 1959.

14 Finale

> Pussy cat, pussy cat, where have you been?
> I've been in London to look at the queen.
> Pussy cat, pussy cat, what did you there?
> I frightened a little mouse under her chair!

Our book is finished, and it is a pity, as we have been able to show you only as much as a little mouse from the "Great Kingdom of Machinery." We sought to tell you some basic facts about typical machine elements: their life, design considerations, and fundamentals of calculations. Evidently, we were unable to do more (in the context of the declared subject matter) because our time resources, knowledge, and experience are limited. But, the abilities and experience of the fellowship of mechanical engineers are immensely greater, and each of us has had the happy chance of putting a brick in the Great Chinese Wall of collective knowledge.

Our goal was to give our readers (especially those whose experience is limited) some ideas to form the basis for their own research effort. We would highly appreciate any feedback on this book from you: remarks, interesting cases, and examples backing or contradicting what is written in this book. To learn from your experience would be a great reward for our work.

"The worst beginning is better than the best end," said Shalom Aleichem. This book has been finished, with our regrets. But your creative accomplishments and ours as well begin every morning, and the end is far away. Let's go forward!

Index

Milton Keynes UK
Ingram Content Group UK Ltd.
UKHW051942071024
449327UK00026B/2131